THE NEW NATURAL

A SURVEY OF BRITISH NATURAL HISTORY

VEGETATION OF BRITAIN

AND IRELAND

THE NEW NATURALIST LIBRARY

VEGETATION OF BRITAIN AND IRELAND

MICHAEL PROCTOR

Collins

This edition published in 2013 by Collins,
an imprint of HarperCollins Publishers

HarperCollins Publishers
77–85 Fulham Palace Road
London W6 8JB
www.collins.co.uk

First published 2013

© Michael Proctor, 2013

A CIP catalogue record for this book is available
from the British Library.

Set in FF Nexus

Edited and designed by
D & N Publishing
Baydon, Wiltshire

Printed in Hong Kong by Printing Express

Hardback
ISBN 978-0-00-220148-3

Paperback
ISBN 978-0-00-2201490

Contents

Editors' Preface

A HUNDRED YEARS AGO Arthur Tansley edited a book entitled *Types of British Vegetation*, written by the botanists of the Central Committee for the Survey and Study of British Vegetation. This book of 1911 was the first account of its kind, written for the benefit of a small and select international group of botanists invited for a countrywide excursion, and describing associations of plants that define our vegetation. He wrote a greatly enlarged account of the vegetation of Britain and Ireland in 1939, a landmark in the study of our vegetation. It was made possible by the strong tradition of the study and recording of our flora, which goes back long before this synthesis, maintained by a multitude of field botanists, very often county-based. It is their local knowledge that has been essential for understanding our vegetation and its composition. Since Tansley's time, the vegetation of Britain and Ireland has been studied in much more depth, with results often presented in the many New Naturalist books written about the natural history of the regions or more generally about plants or particular groups of plants, such as the early volume on *Wild Flowers* (1954), which ran to many editions. The present book is indeed timely in the sense that so much has been learned in recent years about our changing flora and its distribution, stimulated by the mapping projects of the Botanical Society of the British Isles and county botanical recorders. The more recent recognition of the significance of biodiversity has also promoted our interest in the flora and its distribution, especially in respect of the importance of the nature of environments in determining distribution, perhaps less well-understood and understudied in earlier years. The wealth of information now gained needs to be seen against the wider background of the plant communities that form our vegetation. Such needs have resulted in classification of our vegetation in more formal ways, remembering that any ordering presents just a necessary snapshot in a changing world. Michael Proctor has an unrivalled and practical life-time knowledge of our flora and vegetation, with expertise about both higher and lower plants, also reflected in his earlier New Naturalist book with his colleague Peter Yeo on *The Pollination of Flowers* (1973). He is also a very experienced photographer – his photography appeared in the original New Naturalist *Wild Flowers* – as will be seen in the illustrations. The treatment of the major classes of vegetation, following discussion of the essential historical aspects of the subject, places local floras in the wider context of the vegetation of Britain and Ireland, an original aim of Tansley, now realised so well by Michael Proctor.

Author's Foreword and Acknowledgements

THIS IS PRIMARILY A BOOK about the common plants that give the colours to our countryside. What makes some plants common is as interesting as what makes others rare. The most interesting plants of all are often those that are neither very common nor very rare, but are characteristic of particular soils or particular situations in the landscape. They prompt that classic *New Naturalist* question, 'why?'

I have always been fascinated by vegetation, and by the subtle, and sometimes striking, colours with which it paints our landscape. My mother had a cherished copy of Bevis & Jeffery's 1911 book *British Plants: their Biology and Ecology* from her schooldays, which I read avidly, and soaked up the findings of early British plant ecology. 'Oak–birch heath' was recognisable on Harrow Weald Common within easy reach of our home in Harrow, and I knew (or could visualise) many of the other vegetation types described. The Chilterns lay to the north of us, and my aunt lived in Woking close to Surrey heathland, which was much more extensive then than now. In 1946 we moved to Hampshire, so in my mid-teens I found myself within easy cycling distance of the New Forest on one side and Purbeck on the other. Undergraduate years in Cambridge brought me into contact with inspirational teachers (of whom I owe a particular debt to Harry Godwin and Max Walters) and fellow students, and gave me a new landscape to explore. Subsequent postgraduate work provided the incentive and opportunity to get to know better the British chalk and limestone, and to make the acquaintance of Scotland, Connemara and the Burren. A first job in the Nature Conservancy in Bangor, and hill walking in Snowdonia in every month of the year, brought home the varied colours of the hill grasslands, at their best and most distinctive in autumn. In the high summer months they all too often merged into a near-uniform green, a dead loss for landscape (and ecological) photography!

The present book has had a long gestation. In 1968, I revised the late Sir Arthur Tansley's *Britain's Green Mantle* for a second edition. About 1975, Allen & Unwin, the publishers, wrote to me asking whether I would consider revising it again for a third edition. I demurred, for two reasons. First, so much new material was appearing that I feared that so much new wine would burst an excellent old bottle. Second, in 1975, we were just embarking on fieldwork for a National Vegetation Classification (NVC) and it seemed inappropriate to embark on a new popular book until the NVC was finished. In the event the first of the five volumes of *British Plant Communities* appeared in 1991, and the last in 2000, rounding off a no-less-astonishing achievement on the part of John Rodwell, who saw the project through and wrote almost all the text, than Tansley's *British Islands and their Vegetation* half a century earlier. All thought of a new semi-popular book on British and Irish vegetation had been put on hold during the NVC project. When the New Naturalist on the *Natural History of Pollination*, which I had written jointly with Peter Yeo and Andrew Lack, was published in 1996, Collins asked me if I had another New Naturalist I would like to write. Fifteen years have passed since then, largely because I had greatly underestimated how long it would take to work up accumulated research data (some of which finds its place here) on chemical analysis of mire waters, and bryophyte physiology. A positive spin-off of the delay is that, with a new millennium, it is easier to put the long shadow of Tansley behind us, and take a fresh look at British and Irish vegetation in the light of twenty-first-century realities.

This book is intended to be readable, informative and sometimes thought-provoking. It is in no way a textbook. It is aimed at anyone interested in natural history or in our countryside. I am aware of writing for two constituencies, those who habitually use Latin names of plants, and those who are happier with names in the vernacular. In my experience few people are fully bilingual in this matter! Continental readers (of whom I hope there will be many) are likely to find Latin names more accessible than English names. Both Latin and English names of vascular plants follow the second edition of Clive Stace's *New Flora of the British Isles* (1997), the same author's *Field Flora of the British Isles* (1999) and the *New Atlas of the British and Irish Flora* (Preston *et al.* 2002). I have not adopted the names in the third edition of Stace's *Flora* (2008), because it seemed more important to preserve the consistency between this book, the *New Atlas* and the *Field Flora*. Latin names of bryophytes follow the second edition of Tony Smith's *Moss Flora of Britain and Ireland* (2004), Jean Paton's *The Liverwort Flora of the British Isles* (1999) and the British Bryological Society's *Mosses and Liverworts of Britain and Ireland* (Atherton *et al.* 2010); I have deliberately not added to the Tower of Babel by giving English names of mosses and liverworts, which very few people know or use. Lichen names follow Frank Dobson's *Lichens* (2011).

I have made use of 'boxes' in some chapters for technical matter that some readers will feel short-changed without, but others may prefer to pass by. Much the same might be said of the lists of common or characteristic species of many of the communities. Read past them if you feel they interrupt the flow of the narrative; there will be other readers who value them – or maybe you yourself will turn back to them on another occasion. I have tried not to burden the text with too many references, and have not attempted to be comprehensive. Readers needing more detail will find many references in Rodwell's five volumes, or in the books listed here on individual vegetation types or topics. Vegetation is a compellingly visual subject, whether at the level of the individual plant or of the landscape, so this book contains a lot of pictures, accumulated over the years from the 1950s onwards. Vegetation changes, and consequently some of the pictures have archival interest, so I have generally dated them.

Any author of a book of this kind must owe a debt to his predecessors. Throughout, I have leaned heavily on John Rodwell's *British Plant Communities* (1991–2000). The present book takes a broad-brush approach to British and Irish vegetation, which means that I have ignored many of his sub-community distinctions. Where it seemed appropriate I have added NVC numbers as superscripts to the community description. Obviously, NVC numbers strictly apply to Britain alone. In many cases, Irish plant communities are essentially 'the same' as parallel communities on the sister island, although there are Irish communities that have no close parallel in Britain. My debt to Tansley is less immediately obvious but nonetheless great. In historical matters I have made extensive use of Oliver Rackham's books, notably *The History of the Countryside* (1986). Other books that have been a source of inspiration in various ways are those by Praeger (1934), Westhoff and Den Held (1975), Oberdorfer (1949, 1973), Wilmanns (1978), Ellenberg (1988) and not least, that splendid cooperative project the *New Atlas of the British and Irish Flora*, a wonderful source of ecological and plant-geographical insights.

Many individuals have helped over the years, in one way or another, directly or indirectly, in the preparation of this book, and I cannot hope to name them all. John Birks, Margaret Bradshaw, John Cross, Gerry Doyle, Wanda Fojt, Philip Grime, Daniel Kelly, Colin Legg, Richard Lindsey, Roger Meade, Martin Page, Adrian Pickles, Donald Pigott, Chris Preston, Tim Rich, John Rodwell, Jim Ryan, Micheline Sheehy Skeffington, David Streeter, Peter Tyler, Richard West and Bryan Wheeler have all contributed, by reading and commenting on chapters, and by answering my questions, or have helped materially in various other ways. I thank Colin Harrower of the Biological Records Centre for printing distribution maps, the British Schools Exploring Society on whose 1971 expedition to Iceland Figures 20 and 23 were taken, Patricia Macdonald, and Lorne Gill and Betty Common of SNH for their ready help in locating aerial photographs, Iain Thornber for the air photographs of Claish Moss, the University of Exeter for my continuing University Fellowship and for library and computer facilities, and Cambridge University Press, Wiley-Blackwell and other publishers acknowledged individually for permission to use copyright material. Julia Koppitz of HarperCollins and the production team, especially David Price-Goodfellow and Hugh Brazier of D & N Publishing, were a pleasure to work with, and deserve due credit for their contribution to the finished product. Last but not least, I am grateful to Janet Betts for encouraging me to finish this book, and for being 'the wind beneath my wings' while I did so. Responsibility for opinions expressed in the book, and for any errors or infelicities the reader may encounter, is my own. Writing this book has been a long haul, but an enjoyable one; it has taught me a lot, and reminded me of some things I had forgotten I ever knew. I hope my readers may find no less pleasure in reading it.

As this book goes to press, ash dieback disease (*Chalara fraxinea*) has just been confirmed in numerous counties. This fungal disease first appeared in Lithuania and east Poland in the 1990s, and has since devastated ashwoods over most of eastern, central and northern Europe. It could have as great an impact on our landscape as myxomatosis or Dutch elm disease, perhaps more so. We shall see its effects over the next few years.

CHAPTER 1

Britain and Ireland: Two Western Islands

I N NOVEMBER 1996 I looked down from a small seaplane at a western coast, with mingled feelings of familiarity and strangeness. Forest covered the landscape, broken here and there by windswept crags and by stretches of wet tussock-sedge moorland. Behind a long sandy shore, there were expanses of patchily scrub-covered dunes swept by the winds from the ocean. Along the lower course of the river and its estuary below us, forest came down to the water's edge, and the water of the river itself was dark from the peaty soils of the country inland. Irresistibly, landscape and atmosphere reminded me by turns of western Scotland, Connemara and Kerry, the dunes of our west coasts, and the wooded estuaries of south Devon and Cornwall. Yet this landscape – southwestern Tasmania – was different. It bore few of the signs of the millennia of intense human occupation that are such an inseparable (but often unrecognised) part of our own familiar European scene. And so it gives something of a picture of what western Britain and Ireland might have been like before our ancestors began to shape the tamed and settled countryside we live in now.

To a visitor from central Europe, Britain and Ireland also combine the familiar and the strange. Eastern England is in effect a continuation of the north-European plain; indeed it *was* continuous with the north-European plain until rising sea level filled the North Sea some 8000 years ago. A visitor from Germany, the Netherlands or northern France will find relatively few surprises in the flora and vegetation of east and southeast England or the Midlands, and conversely anyone who knows the British flora will find much that is familiar in neighbouring parts of the Continent, or even as far east as Poland or Hungary. But towards the Atlantic seaboard the contrasts multiply, and Devon and Cornwall, west Wales, western

FIG 1. Bluebells (*Hyacinthoides non-scripta*) in flower in an oakwood. Pen Hill, Sidbury, Devon, May 1978.

FIG 2. Dry heath with bell heather (*Erica cinerea*) and western gorse (*Ulex gallii*), Mynydd Bodafon, Anglesey, August 1975.

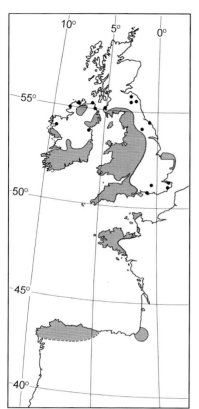

FIG 3. The world distribution of western gorse (*Ulex gallii*), a 'flagship' Atlantic species with its headquarters in Britain and Ireland. Other common Atlantic species, including bell heather, bluebell and common gorse (*Ulex europaeus*) have wider distributions but do not reach central Europe. Marsh St John's-wort (*Hypericum elodes*) has a very similar distribution to western gorse with us. Other species are more restricted to the Atlantic seaboard, for instance western butterwort (*Pinguicula lusitanica*), Cornish moneywort (*Sibthorpia europaea*), and several species confined to the Iberian Peninsula and Ireland, including St Patrick's cabbage (*Saxifraga spathularis*) and the rare Mackay's heath (*Erica mackaiana*, confined to a small area in north Spain and scattered sites in western Ireland).

Scotland, and above all western Ireland, can be as great a contrast with home for a central European botanist as the Alps or southern Europe are for us. We get excited about gentians, edelweiss, continental 'steppe' plants and Mediterranean orchids. To many continental visitors our bluebells, gorse and bell heather (Figs 1–3) are more exciting than the rarities we ourselves most cherish.

Of course there is no sharp divide between 'continental' and 'oceanic' vegetation or plant distributions, and broad-scale differences in climate often interact with small-scale variations in topography and geology in shaping the detailed distribution of plants and vegetation. All of Britain is to an extent 'oceanic' relative to central Europe, and Ireland as a whole is more 'oceanic' than Britain. But both islands embrace big contrasts within themselves. The ferry crossing over the Irish Sea takes the traveller to a landscape that he may feel has more in common with the coast of southern England than with the western shore he has just left. At Rosslare or Dun Laoghaire oceanic western

Ireland is still a long way away. A mountain barrier often sharpens contrasts. There is a dramatic difference between the soft browns and greens of the mild, moist western slopes of the Scottish Highlands and the heather-clad hills of the eastern side. Similar, if less striking contrasts can be seen across Wales, or even between the east and west sides of Dartmoor.

CLIMATE

Temperature: means and extremes

Britain and Ireland have a relatively 'oceanic' climate compared with continental Europe. The average range of temperature between summer and winter is generally around 10–12 °C – a degree or two less in such places as Scilly or southwest Ireland, and locally slightly more in parts of the Midlands and southeast England. This compares with an annual range of about 15 °C in Brussels, 19 °C in Berlin, 22 °C in Warsaw and 28 °C in Moscow. Lowland (sea-level) mean July temperatures in Britain and Ireland range from about 17.5 °C round the Thames estuary to about 12 °C in the Shetlands, and mean January temperatures from 3 °C or less in northeast England and much of Scotland to over 7 °C in small areas of west Cornwall and southwest Ireland.

Given reasonably adequate rainfall, temperature is the factor that has the biggest effect on plant distribution. Temperature may act on plants in various ways. In general, the growth rate of plants increases with temperature, and many plants have distributions in Britain that suggest they may be limited

FIG 4. Field rose (*Rosa arvensis*), a common species in hedges in southern Britain and Ireland, with characteristic long styles and sparsely lobed sepals. Morchard Bishop, Devon, June 1988.

northwards by average summer temperature. Examples are field rose (*Rosa arvensis*) (Figs 4, 5a) and yellow archangel (*Lamiastrum galeobdolon*) (Fig. 5b). A plant's northern limit may be determined not so much by vegetative growth as by the need for warmth to set viable seed, as in dwarf thistle (*Cirsium acaule*) and small-leaved lime (*Tilia cordata*). For at least some species the limiting factor is not so much the average temperature of any particular month, as a combination of temperature and the length of the growing season. Measures expressing this are often used to predict the performance of farm crops, usually based on the assumption that growth is proportional to the product of time and temperature above a threshold value, usually 6 or 10 °C; thus, sweet corn (maize) requires about 750 degree-days above 10 °C between sowing and picking.

Another group of plants has distributions that appear to be determined by winter temperature. A classic example is wild madder, whose northern limit across much of southern and western Europe lies close to the 5 °C mean January isotherm (Fig. 6). It is a common Mediterranean species, and in Britain and

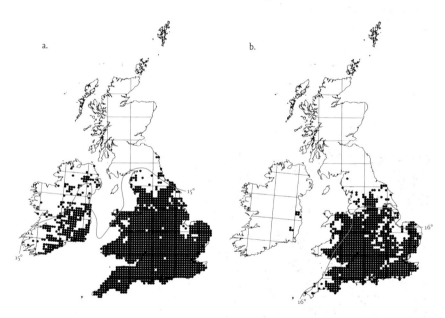

FIG 5. Distribution of (a) field rose and (b) yellow archangel in Britain and Ireland. Field rose is common south of the mean July isotherm for 15 °C, rare to the north of it. Yellow archangel has a similar but more restricted distribution; its northern limit roughly matches the 16 °C mean July isotherm. Summer temperature might account for the rarity of this species in Ireland.

FIG 6. Wild madder (*Rubia peregrina*). The northern limit of wild madder in Britain and Ireland roughly coincides with the 6 °C mean January isotherm.

FIG 7. Wall pennywort (*Umbilicus rupestris*). The distribution of wall pennywort is mostly to the south and west of the 5 °C mean January isotherm.

Ireland it is a plant of the south and west. A good many other species have northern limits similarly parallel to the mean winter isotherms in our part of Europe, such as wall pennywort (Fig. 7), which has a rather wider range, and Cornish moneywort (*Sibthorpia europaea*), which is more confined to the southwest than wild madder. Probably the main need for all of them is a long winter growing season. Species that are frost-sensitive show a different pattern, with northern limits more closely related to the trend of average *minimum* temperatures.

Holly (*Ilex aquifolium*) and ivy (*Hedera helix*) are both plants that need a mild winter. They have been the traditional evergreens at midwinter festivities in our islands since pre-Christian times. Neither grows in central Europe, with its harder winters, and their place is taken by the evergreen spruce and fir of the forest, the *Tannenbaum*, which Prince Albert brought from his native Germany to become our familiar (non-native!) Christmas tree.

Mountain plants (Chapter 17) appear to shun high summer temperatures, and perhaps particularly high summer maxima (Figs 241, 242). This is probably seldom if ever due to simple physiological intolerance, and is most likely to reflect either exclusion from low altitudes by the competition of better-adapted lowland species, or intolerance of the drying that higher temperature promotes.

Effects of latitude, altitude, slope and aspect

Latitude is a major factor determining temperature. Orkney or Iceland may be fascinating holiday destinations, but we go there in full knowledge of their cool climate, and if we want warmth and sunshine we make for the Mediterranean, the Canary Islands or the Caribbean. But other factors are important too. Temperature typically falls by about 0.6 °C for every 100 m of altitude above sea level; the high mountains of central Africa and the Andes carry permanent snow even though they lie close to the equator. Slope and aspect (Fig. 8) also have big effects for plants growing close to the ground – and for other organisms too, including ourselves. Southern species that reach their northern limit with us, like the horseshoe vetch (Fig. 9) of the chalk downs, or the rare purple gromwell (*Lithospermum purpureocaeruleum*) typically grow here on warm south-facing slopes. Vineyards in southern England and the nearby Continent are sited on warm south- to west-facing slopes. Conversely, arctic and alpine plants are at their best and most abundant on the cliffs of the cool north- and east-facing slopes of our mountains. Other factors have generally smaller but still significant effects on temperature. In particular, water has a higher heat capacity than rocks and soils, so the sea has a moderating effect on the temperature of neighbouring land. On a broad geographical scale we see this in the cooler summers and

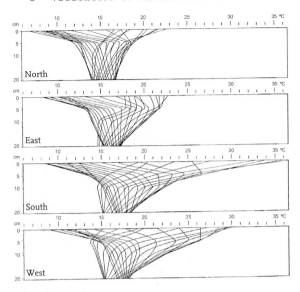

FIG 8. Hourly temperature at the ground surface and at depths of 5, 10 and 20 cm during a cloudless day and night from 0900 h on 18 June to 0800 h on 19 June 1955 at sites facing the four cardinal points of the compass at Pen Dinas, Aberystwyth. All four measuring sites were bare ground, altitude *c.* 60 m, slope 21–23.5°. North and east slopes begin to warm up first in the early morning, the highest temperature is reached on the south slope, and the west slope is warmer than the east because it gets sunshine in the afternoon when the air is already warm. There is much less difference between the night minimum temperatures than between the daytime maxima. Notice how the temperature of the soil lags behind that of the surface, particularly on the south and west slope. From Taylor 1967.

milder winters of Britain and Ireland compared with central Europe. It accounts for the (S)W–E trend of the isotherms for mean July air temperature in Figure 5, and the NW–SE trend of mean January air temperature in Figures 6 and 7. But it is apparent on a smaller scale too, showing up very clearly in maps of daily minimum temperature, with a particularly steep gradient inland along the Channel coast from Devon to Kent; often on a winter morning the ground is white from a light frost in Exeter, while Exmouth, only some 15 km away, has had a frost-free night. The moderating effect of the sea on summer temperature probably lies behind the occurrence of mountain plants such as roseroot (*Sedum rosea*) and mountain avens (*Dryas octopetala*) close to sea level on the west coasts of Scotland and Ireland.

The altitudinal tree limit
Britain and Ireland have less forest than almost any country in continental Europe, though we know that they were largely forested in the past. The open character of our uplands is mostly due to long traditional use as grazing for farm stock, especially sheep. The upper limit of forest in Britain would probably

FIG 9. Horseshoe vetch (*Hippocrepis comosa*) and white rock-rose (*Helianthemum apenninum*), in a turf rich in dwarf sedge (*Carex humilis*) and the rare grass *Koeleria vallesiana*, on the south-facing slope of Brean Down, Somerset.

be at about 600–650 m above sea level in our present climate (Chapter 17). In temperate regions of the world, the tree-line corresponds roughly with the 10 °C mean July isotherm, so our low tree-line is a natural consequence of the cool summers in our oceanic climate.

Wind

Continental visitors often remark on the extraordinary wind-cut shapes of trees in exposed western districts of our islands (Fig. 10), while we wonder at the elegant straightness of trees in the forests of central or northern Europe, or in many parts of North America. Wind has a number of effects on plant growth, of which the dramatic mechanical effects of exceptional storms like those of 16 October 1987 in southeast England and of 25 January 1990 over much of the rest of Britain, tearing off leaves and branches and uprooting whole trees, are probably in the long run the least important. Wind has two familiar and all-pervasive effects. During daytime, especially on sunny days, plants in the open are generally warmer than air temperature, and they are chilled (as we are) by wind. Also, wind accentuates drying; what is welcome on the clothesline increases water stress for plants under dry conditions. These two factors are responsible for a large part of the ecological effects of wind, including the wind-cut forms of shrubs and trees.

Rainfall

Water is essential for all life. Plants generally depend on the water in the soil, and on the rainfall that is its ultimate source. Average annual rainfall ranges very widely in Britain and Ireland, from around 500 mm (20 inches) in the driest

FIG 10. Wind-cut beech on hedgebanks, Bursdon Moor, Devon.

parts of the east Midlands to 2500 mm (100 inches) or more in mountainous areas such as Snowdonia, the English Lake District, western Scotland, and Kerry and Connemara. However, averages hide a great deal of variability. The seasonal distribution of rain varies between different parts of the country. Lowland areas of both Britain and Ireland show a rather even distribution of rainfall from month to month. Western districts of both islands show on average a strongly marked rainfall minimum in spring and early summer (April to June), and a maximum in autumn and winter (October to January); the average rainfall in the wet autumn months in western Scotland, western Ireland or southwest England is often twice the average in the dry early-summer months. Measurable rain (> 0.2 mm) falls on more than 250 days of the year in Kerry, Connemara, Donegal, the western Highlands of Scotland, Skye and the Outer Hebrides, but the Channel coast and much of eastern England have less than 175 'rain days' defined in this way (Fig. 203). Rain is actually falling for around 500 hours a year in the Midlands and East Anglia (only some 6% of the time), but for over 1500 hours (17% of the time) in much of the western Highlands.

Microclimate: 'the climate near the ground'
It is easy to talk about the climate of a place, forgetting that 'climate' is an abstraction – a summing up of the weather in all its never fully predictable and sometimes exasperating variety over a period of years. Climatic data are conventionally calculated over a 30-year period of observations, but even this sidesteps the fact that we know climate is constantly changing. It has changed dramatically in the past and there is no reason to suppose it will not continue to change in the future.

It is also easy to forget that the 'climate' experienced by a plant (or any other organism) is never the same in two different places, however close together they are. Every surface – rock, soil, plant or animal – influences and is constantly exchanging heat energy and often water vapour and other gases with its surroundings. Close to plant, soil or rock surfaces (within a millimetre or so), movement of heat and vapour takes place by slow molecular diffusion in the 'laminar boundary layer' of air next the surface; farther away, it takes place in varying degrees faster by turbulent mixing. The result is that at every scale there are gradients of light, temperature, humidity, and often of carbon dioxide as well. An individual plant or animal experiences not the solar radiation, temperature or humidity that a meteorological observer reads off his instruments, but its own microclimate. The site for a meteorological station has to be carefully chosen to avoid local microclimatic peculiarities, and temperature and humidity measurements are made inside a standard white-painted louvred box at a standard, human-scale, height above the ground. It could be said that normal climatic data record the microclimate of Stevenson screens – and that few organisms (apart from the occasional spider) live in Stevenson screens to experience it!

In sunny weather, foliage is in most cases above air ('shade') temperature, sometimes 10 °C or more higher than air temperature in short vegetation near the ground, though evaporation from healthy leaves generally keeps their temperature safely below 35–40 °C. Dry rock and soil surfaces often reach 50 °C or more on clear sunny days. In the northern hemisphere, solar radiation is greatest on south-facing slopes, and in our latitudes this can lead to some dramatic microclimatic contrasts, greatest in short grassland or herbaceous vegetation. The energy input from the sun on a clear summer day is in the region of a kilowatt per square metre. A third or half of this may be dissipated by re-radiation to the environment, leaving the rest to be accounted for by a combination of the latent heat of evaporation as the plant loses water, and convective cooling by the air, which is limited by relatively high diffusion resistance in a short turf. Plants in these situations (such as dwarf thistle and common rock-rose, *Helianthemum nummularium*) typically have deep and extensive root systems and lose water rapidly even under apparently dry conditions. When they are water-stressed, the older leaves wilt and die; the plant (willy nilly) continues to use water but trims its outgoings to match its income.

Tall herbaceous plants, shrubs and trees can more easily lose heat direct to the air, so they remain much closer to air temperature regardless of transpiration. In overcast weather or in shade foliage tends to be cooler than the air, and the air under a woodland canopy or in shady rock crevices is generally both cooler and more humid than the air outside. On clear nights, foliage and the ground surface lose heat by radiation, and become colder than the air. If leaves cannot draw heat from warmer surrounding air the surface temperature may fall below freezing, resulting in a 'ground frost'. Ground frosts particularly affect sites open to the sky but sheltered from the wind, and on clear 'radiation nights' cold air can drain down and accumulate in 'frost hollows', so sheltered places are often not the most favourable for cold-sensitive plants. These are not always the plants one might first think of as 'tender'. Young growth of bracken (*Pteridium aquilinum*) is not uncommonly damaged by late ground frosts in May and June.

Microclimate is a complex and fascinating subject, but enough has been said to caution against too-simple interpretation of plant distribution in terms of standard climatic data, and to show something of the intricate ways in which climatic factors can interact in their effects on plants and vegetation.

THE GEOLOGY OF BRITAIN AND IRELAND

For so small an area, Britain and Ireland preserve a remarkably complete
record of geological history. The rocks range in age from the ancient Lewisian
granulites and gneisses of northwest Scotland (Fig. 11), the oldest dating from
some 2600 million years ago, to the geologically recent gravels and clays of
the Hampshire and London basins and East Anglia, a few tens of millions of
years old. Few major geological periods have failed to leave some mark on the
geology of our islands. Taking a very broad-brush view, the oldest (Precambrian)
rocks and the highly folded and altered rocks of the Scottish Highlands and
the mountains of Donegal and Connemara are generally hard, often underlie
mountainous topography, and give rise to thin acid soils (though they include
some calcareous schists and some hard limestones).

Much the same is true of the Lower Palaeozoic sedimentary shales, slates and
grits (with some volcanic rocks) that make up most of the Southern Uplands of

FIG 11. Gruinard Bay, Wester Ross, June 1981. The raised beach in the foreground owes its
origin to the continuing isostatic rise of the land, relieved of the burden of ice of the last
glaciation. The hills in the middle distance are made up of Lewisian gneiss, among the oldest
known rocks on earth; the cloud-covered mountains beyond are carved from more recent (but
still ancient) Torridonian and Cambrian rocks, and the Moine schists.

Scotland, the Lake District and northern and central Wales (though again there are some significant limestones, notably the extensive Cambrian limestone in northwest Scotland). The Old Red Sandstone, formed under desert conditions in the Devonian period, is also generally hard and acid, and often underlies upland country, as in the Brecon Beacons, on Exmoor and in Cork and Kerry; there is marine limestone of similar age round Torbay and Plymouth. Rocks laid down in the ensuing Carboniferous period are a major element in our landscape. Carboniferous Limestone is the rock of the Derbyshire and Yorkshire Dales, much of the North Wales coast, the Mendip Hills, and the limestone crags that rim the South Wales coalfield. In Ireland, Carboniferous Limestone underlies almost all of the midland plain, and is spectacularly developed in the Burren district of Co. Clare (Fig. 12). Above the Carboniferous Limestone lies the hard sandstone known as the Millstone Grit, which caps the summits of the 'Three Peaks' around the head of Ribblesdale, and broad tracts of the Pennines, where it is usually covered by peaty cottongrass moor. The shales, grits and coal seams of the Coal Measures that follow the Grit were eroded off the summits and make little impact on the landscape in their own right, but they provided the major resources that powered the industrial revolution in nineteenth-century Britain.

FIG 12. Carboniferous limestone country; looking north from Mullaghmore, the Burren, Co. Clare. Fragmentary limestone grassland, hazel scrub, and small lakes, fens and turloughs in wet hollows in the flat limestone stretching away to the northeast.

FIG 13. The chalk escarpment of the Chilterns looking eastwards from Ivinghoe Beacon, September 1981. The white figure of the Whipsnade Lion can be seen at the right of the picture. Flat arable country on soft Lower Cretaceous and Jurassic strata underlying the chalk.

The Permian and Triassic 'New Red' rocks are softer than their Devonian precursors; they include sandstones, marls, beds of water-worn pebbles, and locally beds of salts that accumulated by evaporation (providing a starting point for the chemical industries of Teesside and Merseyside); they embrace all the ingredients of a fossilised desert landscape. New Red rocks occur from the Scottish lowlands to Devon, occupying broad belts of country on both sides of the Pennines and covering much of the north and west of the English Midlands. East of the Pennines, Permian rocks also include the marine Magnesian Limestone, stretching from near Sheffield to the coast of County Durham. The uppermost (Rhaetic) rocks of the Trias were also laid down in the sea, marking the transition to the long period of marine deposition of the Jurassic.

Sedimentary rocks from the Jurassic onwards appear in only tantalising glimpses in Scotland and Ireland, but dominate the geology and landscape of southern and eastern England. There they form an almost unbroken succession, dipping gently southeastwards from a somewhat sinuous line stretching from Lyme Bay to the North Yorkshire coast, and very little deformed by earth movements except in the extreme south from Dorset to Kent. The harder rocks, such as the Jurassic oolitic limestone of the Cotswolds and North York Moors, and the chalk, form escarpments generally facing northwest (Fig. 13), the softer rocks forming tracts of lower country in between. The youngest rocks are the sands, gravels and clays of the Hampshire and London basins (in our context chiefly notable as the beds that underlie the sandy and gravelly heaths of Surrey

and Hampshire), and the still later shelly sands and gravels of the 'Crag' in East
Anglia.

Mountain-building

The complexity of the geology as a template for our present-day landscape is all
the greater because Britain and Ireland have been close to the axis of two major
mountain-building episodes, and close enough to be influenced by another.
The Caledonian episode, which took place some 400–500 million years ago,
created a vast mountain chain of which the shattered remnants, torn apart by
the opening of the Atlantic Ocean, form the mountains of Scotland and western
Ireland, Scandinavia, East Greenland, Newfoundland and the Appalachians.
The NE–SW orientation of the Caledonian folding is still obvious on the map
of Scotland. The Hercynian (or Variscan) episode, some 300 million years ago,
created an east–west mountain range extending from central Europe to eastern
North America (which then adjoined Europe), whose remains determine much
of the underlying structure of France, northern Spain and southern Britain and
Ireland. Its most conspicuous legacies in our present-day landscape are the line
of granite intrusions in the southwest peninsula of England, from Dartmoor
(Fig. 14) to the Isles of Scilly, and the east–west syncline of the South Wales

FIG 14. Granite country: view from Haytor looking towards Hound Tor, Dartmoor, July 1975.
Heather–western-gorse moorland and late-summer buff-tinged upland grasslands; tors crown
many of the summits.

coalfield with the high Old Red Sandstone ridge running through the Brecon Beacons along its northern edge. East–west Hercynian folding also dominates the structure of southernmost Ireland from Waterford to Kerry. Farther north in Britain, Hercynian folding (possibly of a slightly different age) rather surprisingly takes a north–south direction, as in the Malvern Hills and in the Pennines, where there are major faults of Hercynian age, though a good deal of movement along some of the fault lines is geologically much more recent. The collision between Africa and Europe that threw up the Alps some 20 million years ago had only relatively minor effects in Britain, but the gentle folding of the Jurassic and the chalk and younger rocks of southeast England created the structures underlying some of our best-known and best-loved landscapes in Purbeck, the Isle of Wight and the Downs and Weald of southeast England (Friend 2008).

Superficial deposits: glaciation
While the underlying 'solid' geology has big effects in determining the character of vegetation, the relationship is often blurred in detail by the presence of superficial deposits covering the bedrock. Sometimes these have developed *in situ* as residues from weathering over spans of geological time, as with the 'clay-with-flints' that often overlies the chalk. Often the superficial material has been transported from elsewhere, as with dune sands, or the wind-borne silt called *loess* that was deposited under dry conditions around the margins of the ice-sheets during the glaciations – quite widespread but always thin in south and east England, but forming massive deposits in parts of continental Europe and in China, where it is responsible for the colour of the Huang He, the 'Yellow River'. Much more important in Britain and Ireland is the material that was transported by the ice-sheets themselves and the streams flowing from their margins – boulder-clay, drumlins, moraines, outwash gravels and so on, often referred to collectively as glacial 'drift'. At its greatest extent, ice covered all of Britain and Ireland except for England south of a line from the Severn estuary to Essex. Glacial deposits cover a large part of the land north of that line.

GEOLOGY, CLIMATE AND THE DEVELOPMENT OF SOILS

Soil is a product of interacting processes, which include rock weathering, the production of organic matter by plants, downward leaching of soluble substances as water percolates through the soil, and return of nutrients to the surface as plants shed their leaves or die. In general, the soil that develops at a particular place depends on three factors: the nature of the *parent material* (more or less weathered bedrock or superficial deposits), the *climate* (especially the amount and distribution of rainfall, and the range of temperature round the seasons), and the nature of the *vegetation*.

The parent material largely determines the texture of the soil. The mineral particles in a soil vary greatly in size. For soil analysis, particles between 2 mm and 0.06 mm (60 μm) in size are regarded as 'sand', particles between 2 and 60 μm as 'silt', and particles less than 2 μm as 'clay' (Avery 1980; different conventional limits are sometimes used). If the proportions of sand, silt and clay are plotted on a triangular diagram (Fig. 15), the upper part of the figure is occupied by clay soils, heavy, 'cold', impermeable and easily waterlogged. The bottom left corner is occupied by sandy soils, light, 'hungry', free-draining and liable to drought. The silt soils of the bottom right corner lie between these two extremes, but they mostly occur in moist valley bottoms where river silt has provided the parent material. Soils near the middle of the diagram, 'loams', have a good balance of particle sizes and are free-draining but reasonably retentive of water. They occur over a wide range of parent materials, and generally make good agricultural or garden soils. Whatever the nature of the 'mineral skeleton', the organic-matter content of the soil also plays a very important role in soil structure, in water retention, and in determining the availability of plant nutrients.

An example of soil development on another western seaboard

The ground progressively exposed by the retreat of the glaciers around Glacier Bay in Alaska (at roughly the same latitude as northernmost Scotland) gives a nice illustration of the course of early soil formation in a cool, rainy oceanic climate not too unlike northern Scotland; summer temperatures in the two places are similar, but the winters in Alaska are a few degrees colder (Crocker & Major 1955). In this part of Alaska, the mosses *Racomitrium lanuginosum* and *R. canescens* are early colonists of newly exposed moraines, joined within a few years by scattered plants of the willowherb *Epilobium latifolium* and the mat-forming dwarf shrub *Dryas drummondii*. These are followed in turn by low-growing willows and the shrubby alder *Alnus crispa*, which after 40 years or so forms a dense continuous cover. Sitka spruce (*Picea sitchensis*) becomes established in this, and a century after the start of colonisation the alder has generally been shaded out by a continuous forest of spruce.

Soil development begins slowly. After a sluggish start for the first 20 years or so while the vegetation is still relatively sparse, organic matter increases

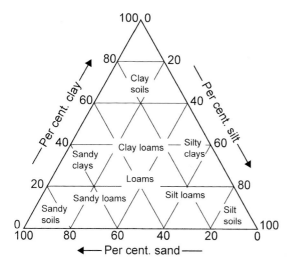

FIG 15. Triangular diagram of soil texture types, and their relation to the percentages of sand, silt and clay.

progressively, levelling off at about 4 kg/m² in the mineral soil under the spruce forest – with rather more organic matter than that remaining in the undecomposed litter of the forest floor, which continues to increase, though more slowly. The build-up of nitrogen in the soil roughly follows the incorporation of organic matter, but is particularly rapid under the alder, which (like our own *A. glutinosa*) has nitrogen-fixing root nodules. Calcium carbonate is quickly leached out in the early stages of the succession, from 5 to 9% in the newly exposed moraine to very low values after 30 years. The pH of the soil, *c.* 8.0 in the early stages of development, falls to *c.* 5.0 within 35–50 years; under old alder the surface pH is *c.* 4.2–4.6.

Glacier Bay is very wet, with a similar rainfall to western Scotland (Sitka spruce, introduced to Britain and Ireland, is now our commonest timber tree), and soil development there reflects that fact. The details will be different in other places, but the processes at work are similar, and soil development always needs at least decades and more often centuries (or even millennia) to come to completion.

Major soil types in Britain and Ireland
On typical well-drained soils developed under old woodland in the climate of lowland England and eastern Ireland (as on neighbouring parts of the Continent), leaching is less pronounced, surface leaf litter is broken down more quickly and organic matter is thoroughly incorporated into the soil by the combined activity of earthworms and the soil microorganisms as *mull* or mild

humus. If a sample pit is dug, no very obvious layers strike the eye, but on closer examination the upper parts of the soil appear rather greyer, and the lower parts somewhat more rusty brown in colour; deeper again signs of the rock structure of the parent material begin to become evident. Chemical analysis shows that the upper layer (the 'A horizon') is richer in organic matter and somewhat depleted in iron, aluminium and silicates, while the next layer (the 'B horizon') has less organic matter but shows at least some sign of the accumulation of these metals and silicate minerals; the 'C horizon' at the bottom contains little or no organic matter and is relatively unaffected by soil-forming processes. A soil of this kind is called a *brown earth* (Fig. 16a). These are typical soils of the temperate deciduous forests of western Europe. Brown earths are mildly to moderately acid (pH typically (4.5–)5.0–6.0, slightly lower in the A than in the B horizon).

In colder and wetter climates, or on free-draining and nutrient-poor parent materials, leaching dominates soil development. A layer of dark raw humus (*mor*) accumulates on the surface. Iron is leached almost completely from the A horizon, which is stained dark by humus at the top but often bleached to pale grey or almost white below (this bleached layer is sometimes called an E (for 'eluvial') horizon). Typically the B horizon forms a sharply defined layer, black to dark reddish brown from deposition of humic material, iron and aluminium oxides and clay particles that originated from higher up the profile. The B horizon is strikingly differentiated by its intense colour both from the leached A horizon above and from the little-altered C horizon below. *Podzol* soils of this kind are typical of well-drained sites in the northern coniferous forest zone from Scotland and Scandinavia right across northern Russia and Siberia (hence the Russian name), and are familiar under heathland in southern England, where striking examples can be seen in gravel and sand pits in Surrey, the New Forest and Dorset (Fig. 16b). Podzols are typically acid, with a pH in the A horizon commonly around 4.0–4.3, and rather higher in the B horizon. Soils of generally podzolic character occur very widely in the uplands of northern and western Britain and Ireland.

Often it is not possible to recognise distinct A and B ('eluvial' and 'illuvial') horizons in the soil, and a more or less humus-rich A horizon lies directly on top of the parent material. On hard acid rocks weathering is very slow, and a layer of highly organic soil may accumulate over the hard rock surface. A soil of this kind, which is generally both acid and drought-prone, is called a *ranker*. Rankers are very widespread, especially in western Britain and Ireland, but seldom extensive. Much more important are the *rendzina* soils developed over limestone, where the parent material contains too much calcium carbonate and too little other mineral material for the soil to be quickly decalcified by leaching. The humus-rich A horizon grades gradually down into the shattered surface of the parent limestone

FIG 16. Soil profiles. (a) A typical (but red-stained) brown-earth soil under pedunculate oakwood near Exeter, Devon. The soil is not divided into obvious horizons. (b) A dry-heath podzol on sandy soil in the New Forest, with typical distinct soil horizons. (c) A rendzina under chalk grassland, near Freshwater, Isle of Wight.

(Fig. 16c). Rendzina soils are generally neutral to slightly alkaline (pH typically 6.5–7.5) and freely drained. They occur very widely over limestones, where they are most typically developed on the scarp slopes and steep valley sides. Rendzinas are particularly extensive in England on the Jurassic limestones and the chalk; they are common on Carboniferous limestone in Ireland, and in England and Wales where the bedrock is not covered with superficial deposits.

There are two situations in which waterlogging, and consequent development of anaerobic conditions, are important in soil-forming processes. On impermeable clayey soils surface water from rain leads to waterlogging, which may be seasonal and patchy, but reduces and mobilises iron and manganese, leading to the mottled blue-grey and rusty brown soil colouring associated with 'gleying'. These soils are *surface-water gleys*. They occur very extensively over impermeable substrates (including glacial drift), and are especially prominent in upland and western districts, including much of Wales and western Scotland and Ireland. In valley-bottom soils the fluctuating water-table often lies near the surface for all or part of the year, leading to intermittently anaerobic conditions and gleying in the lower part of the profile, while the surface layers may remain freely drained and oxidising. *Groundwater gley* soils of this kind are typical of major river valleys and floodplains, and cover a particularly big tract of country around the Wash. As a final category, peats are 'soils' composed largely of organic matter. They will be considered in more detail in Chapters 13–15.

Of course soils do not all fall into clearly defined categories. All gradations exist, and soil scientists recognise many more types than the seven outlined above. However, these broad categories correspond with major groupings recognised in the 1983 quarter-inch maps of the Soil Survey of England and Wales.

PLANT NUTRIENTS AND TOXIC IONS: CALCICOLES AND CALCIFUGES

Essential elements for plant growth

Plants get water and nutrients from the soil. At least 16 chemical elements are essential for plant growth. Carbon, hydrogen and oxygen make up over 90% of the dry weight of the plant body, including cellulose, sugars, organic acids and oils and fats. Nitrogen (> 1%) is a necessary ingredient of proteins; sulphur (0.2%), although required in smaller amounts, is also essential. Phosphorus (0.2%) plays a key part in the structure of DNA and RNA, and in various other molecules crucial to plant metabolism. Potassium (1%) is important in regulating the ionic environment in plant cells. Calcium (0.2%) stabilises cell walls, and is important in controlling various cell functions, but its concentration in the cell cytoplasm is low. Magnesium (0.2%) is the central atom in the chlorophyll molecule, and has other functions related to enzyme activity. Iron (0.1%) is an essential constituent of the proteins that mediate oxidation and reduction reactions in plant cells. Manganese (0.05%) is a component of a few enzymes, and activates others. The remaining essential trace elements are present in only very small quantities. Boron affects pollen-tube growth (amongst other things) but its exact role is little understood. Zinc, copper, nickel and molybdenum are components of particular enzymes. The range of concentrations at which these nutrients are required is extraordinarily wide. For every one atom of molybdenum, plants require about 2000 atoms of iron, 60,000 atoms of phosphorus and a million atoms of nitrogen.

Calcicoles and calcifuges: toxic ions

Calcium, though not required in large amounts *as a nutrient*, has very important effects in soils and natural waters in determining pH and the uptake by plants

FIG 17. The Needles from above Alum Bay, Isle of Wight, September 1958. Heath in the foreground on podzolised Cenozoic sands and gravels with heathers and dwarf gorse; the chalk ridge culminating in the Needles is covered with downland turf on calcareous rendzina soils.

of other elements. At high pH (> 7.0) calcium is in abundant supply, and there is usually adequate nitrogen, in the form of nitrate. However, *calcifuge* plants adapted to acid soils are unable to take up sufficient iron and other metallic trace elements under these calcareous conditions and often suffer from 'lime chlorosis' – a problem familiar to gardeners trying to grow rhododendrons on ordinary garden soils. Conversely, at low pH iron and manganese become freely available, sometimes (especially under waterlogged conditions) so freely that (together with aluminium) they reach toxic concentrations. Inorganic nitrogen (generally in the form of ammonium ions) is often in short supply. *Calcicole* plants adapted to calcareous soils suffer from nutrient deficiencies on these acid soils, and are vulnerable to heavy-metal toxicity. Calcicole–calcifuge relationships underlie some of the most striking contrasts in our vegetation (Fig. 17). Many common species of mildly acid or neutral soils are neither markedly calcicole nor calcifuge. They often have a rather wide ecological range from moderately acid to moderately calcareous soils, but usually avoid the extremes.

Of course other elements may be toxic too, most often where spoil from mining has brought high concentrations of lead, zinc or copper to the surface. A number of plant species tolerant of high concentrations of heavy metals are often conspicuous on these sites. Alpine penny-cress (*Thlaspi caerulescens*) and spring sandwort (*Minuartia verna*) are particularly associated with lead- and zinc-rich mine waste in Mendip, north Wales and the Pennines (Chapter 11).

'NATURAL AREAS': BIOGEOGRAPHICAL TRENDS AND DIVISIONS

We are all aware of 'natural areas' in the sense of areas of country (often with evocative names) that have a character and unity of their own that sets them apart from their surroundings and places farther afield. Examples are Dartmoor, Gower, the Yorkshire Dales, the Burren, the Hebrides, the Fens and the Weald. They can be at any scale, we can invent new ones at will, and they can be an invaluable tool in thinking about landscapes, vegetation and biogeographical relationships. Thus Natural England, the statutory nature conservation body for England, defines 159 'National Character Areas' within eight regions.

Viewed as a whole, Britain falls into two very clear natural areas, roughly separated by a line from the mouth of the Exe to the mouth of the Tees, and reflected in the distributions of many species (Figs 18, 19). South and east of this line the countryside is dominated by relatively easily eroded Mesozoic and later rocks, often calcareous, and generally weathering to give deep soil parent materials. There are very few summits higher than 300 m, and the mean annual rainfall is almost everywhere less than 1000 mm. The soils are mostly fertile and make good farmland. By contrast, north and west of the Exe–Tees line old hard rocks predominate. This is a region of steep relief and high summits. Rainfall varies greatly but is usually more than 1000 mm a year, and over large areas of country exceeds 1600 mm. Soils are commonly thin and infertile. Surface-water gleys, podzolic soils and peats cover much of the landscape, which is largely given over to livestock farming (mainly sheep) and forestry, with rough grazing on the higher parts of the hills. Within this large northwestern area of Britain, a second important dividing line is the Highland boundary fault, which runs from near Helensburgh on the Firth of Clyde to Stonehaven on the Aberdeenshire coast. The Scottish Highlands to the north of it have a high mountain character, which even Snowdonia hardly approaches, and which looks towards the much bigger mountains across the North Sea in Norway, of which the Highlands are geologically a detached fragment.

Ireland is quite a different shape from Britain. It has been likened to a saucer, the midland plain occupying a low-lying flat expanse of Carboniferous limestone (largely drift and peat covered) in the centre, with a ring of mountains

FIG 18. Black horehound (*Ballota nigra*), a common hedgerow and roadside plant throughout 'lowland Britain', scattered in the Irish lowlands.

FIG 19. Bilberry (*Vaccinium myrtillus*), common throughout 'upland Britain' from Cornwall to the Shetlands, and much of Ireland; in southern England it indicates acid heathy soils.

forming a somewhat irregular upturned rim round the edge. In fact the southern third of Ireland is dotted with isolated groups of mountains, mostly rising to 800–900 m, made up of similar hard acid rocks to those of Wales just across the Irish Sea. The most extensive are the Wicklow Mountains south of Dublin, and the mountains of Cork and Kerry, which include the highest summits in Ireland. On the west coast, the mountains of Galway, Mayo and Donegal are geologically more comparable with the western Scottish Highlands, while the plateau basalt of the northeast matches the similar basalts of Mull and Skye in western Scotland. There is more than the Gaelic language in common between Ireland and western Scotland!

'Natural areas' conceived on a smaller scale are often an enlightening guide in making comparisons from one geographical region to another. It can be more helpful to compare corresponding facets of the landscape in different places than to attempt comparisons between complex landscapes as a whole. Some years ago I visited two intriguing fen and bog sites in deserted ox-bows of the River Tisza in northeastern Hungary – and one of my most vivid impressions was that there were almost no species that could not have been found in lowland England, underlining the essential similarity of the flora and vegetation across the north-European plain. The Breckland around Thetford on the Suffolk–Norfolk border is known for its assemblage of rare 'steppe' species – and on dry slopes of the Rhone valley in Switzerland I have seen a rich 'Breckland' flora, which could still be related to its context of traditional farming: a facet of the Swiss landscape discernible as just a tantalising glint in East Anglia. Dry rocky limestone hills across central Europe are particularly fascinating for anyone who knows our own western limestones and their plants. We cherish hoary rock-rose (*Helianthemum oelandicum*), goldilocks aster (*Aster linosyris*), spiked speedwell (*Veronica spicata*) and pasqueflower (*Pulsatilla vulgaris*) as rarities. On the Continent one can often see these species growing together in abundance, in habitats that are recognisably 'the same' as those they occupy with us, but more extensive. Conversely, some habitats are better developed with us than on the Continent, such as our oceanic woodlands, heaths and bogs, and the Atlantic coast. We appreciate Britain and Ireland all the better for seeing them in the context of our neighbouring continent. The similarities deepen our understanding – and throw into relief the differences that we ought to value.

CHAPTER 2

Prehistory: from Glaciation to the Iron Age

LIMATIC CHANGE IS A TOPICAL SUBJECT, but it is nothing new. Some 50 million years ago, when the London Clay was being laid down in the early Eocene, average temperatures were some 5–10 °C warmer than now, and more evenly spread from the equator to the poles. Rainy tropical climates extended to 45° N (fossil plants in the London Clay speak for at least a subtropical climate), and temperate forest extended to the poles. Antarctica was covered with forest, which shared many genera with the present flora of southern Chile and New Zealand. From that time climate gradually cooled. The first permanent ice-sheets were forming in Antarctica about 35 million years ago. By the mid-Miocene, around 12 million years ago, the Antarctic ice-sheet had spread almost to its present size. In the Pliocene, some 3–5 million years ago, climate was much as it is now, and the Greenland ice-cap began to form. The beginning of the Pleistocene, with its recurrent glaciations and interglacials, has been dated rather precisely to about 2.59 million years; it ended with the final retreat of the ice-sheets of the last glaciation and opening of the Holocene, the interglacial in which we live, about 11,500 years ago.

Fifty years ago it was generally accepted that there had been four major glaciations in Europe in the course of the last million years, and probably some more before that of which we knew very little. Over the last half-century, geochemical evidence from cores of deep ocean sediments and ice cores from Antarctica and Greenland has shown that the picture is both more complicated, and simpler. Several lines of evidence – oxygen isotope ratios from ocean and ice cores, temperature estimates from the ocean-sediment cores, analyses of the air entrapped in the ice cores – come together to show that, over the last million

Isotopes of the chemical elements

Atoms of the chemical elements are characterised by two properties, atomic number and atomic weight. The atomic number is the positive charge on the nucleus and the number of (negative) electrons orbiting it, and it determines the chemical properties of the element. The atomic weight is the mass of the nucleus compared with that of hydrogen, the lightest element. Most elements exist in forms with different atomic weights: these are called *isotopes*. Thus hydrogen has two stable isotopes, the common one ^1H with an atomic weight of 1, and a rare one ^2H with atomic weight 2 (sometimes called deuterium and abbreviated D; D_2O is 'heavy water'). Carbon has two stable isotopes, ^{12}C (common) and ^{13}C (rare), and so has oxygen, ^{16}O (common) and ^{18}O (rare). The isotopes of a particular element have exactly the same chemical properties, but the heavier isotopes generally take part in physical processes (e.g. evaporation) and chemical interactions (e.g. photosynthesis) slightly more sluggishly than their lighter counterparts. This leads to slight changes in the ratio of the two isotopes ('discrimination'), which can be measured by mass spectrometry and from which useful inferences can often be drawn. Measurements of ^{18}O discrimination are the basis of ocean-sediment and ice-core estimates of the relative amount of land-ice during the Pleistocene.

Some isotopes are unstable, and undergo radioactive decay more or less quickly. Carbon has an isotope, ^{14}C (carbon-14), which is unstable and decays with a half-life of 5730 years. Carbon-14 is continuously formed by the bombardment of atoms of the common isotope of nitrogen, ^{14}N, in the upper atmosphere by cosmic rays. It is quickly oxidised to $^{14}CO_2$, and exists at about 1 part in 1000 million of the total carbon in the atmosphere. A plant constantly takes in $^{14}CO_2$ by photosynthesis, so the amount of ^{14}C in its tissues is in proportion to that in the atmosphere. Once the plant dies, the ^{14}C decays and is not replenished, so the proportion remaining in organic material can be used as a measure of its age. This is the basis of radiometric C-14 dating. Radiometric dating demanded large sample sizes and long counting times, but more recently accelerator mass spectrometry (AMS) dating has become available. It is much more sensitive and can work with much smaller samples, because it relies on counting all the ^{14}C atoms directly, not just those that disintegrate within the counting interval. The half-life of ^{14}C is well suited to the timescale of Post-glacial events, but the interglacials are far beyond its range.

C-14 dating assumes that the proportion of ^{14}C in the air, and various other factors, have been constant. When ^{14}C dates were compared with long series of tree-rings and other sources of precise dates it was found that, beyond a few millennia into the past, radiocarbon dates could underestimate calendar dates by 10–15%. Calibration curves have now been constructed that give dates accurate to within a few decades over Post-glacial time.

years, the volume of land-ice has fluctuated rather regularly with a period of roughly 100,000 years, that growing ice-sheets on land are closely correlated with falling temperatures, lowered CO_2 concentration in the atmosphere and falling sea levels. The pattern of each peak was typically a slow, irregular build-up of ice, followed by an abrupt fall. Intriguingly, this periodicity also correlates with changes in the eccentricity of the earth's orbit round the sun, which appear to act as the pacemaker probably by giving rise to differences up to about ±10% in solar radiation during the northern-hemisphere summer.

What is perceived as a 'glaciation' depends on where you are. In Iceland (Fig. 20), Greenland, Svalbard or South Georgia glaciation is very much a present reality. From Stockholm, London or Chicago, evidence of the last glaciation is virtually on your doorstep (Fig. 21), and its end is not unimaginably far back in time – but we live our everyday lives safely distant from ice-sheets and glaciers, while in Rio de Janeiro or Mumbai it may all seem rather distant. If you live in Bangladesh or on a coral atoll in the Pacific your main concern is likely to be the variations in sea level that come with glacial–interglacial cycles.

FIG 20. The receding ice-front of Langjökull in Iceland, with fresh morainic material and outflow channels. The distance to the hill beyond the ice-front is about 15 km. Britain and Ireland might have looked like this towards the end of the last glaciation.

FIG 21. Glacial striae on a limestone surface newly exposed from under boulder-clay in the Burren, western Ireland. The striae are accentuated here because the limestone contains lumps of hard chert, leaving 'tails' as the ice has passed over them.

FIG 22. A view just below the Cabane de Valsorey, Grand Combin, Switzerland, at almost 3000 m. Well-developed vegetation on the moraines and outwash exposed since 1850 in the valley below, and a rich alpine flora on the sunny slope to the right. Moss campion (*Silene acaulis*) in flower. This is a 'full-glacial' flora in nearly our own latitude in a favourable spot.

TABLE 1. The Mid-Pleistocene to the present day: glacials and interglacials interglacials. This table takes in less than half the Pleistocene. Dates are given 'before present' (BP), conventionally taken as 1950. Odd Marine Isotope Stage (MIS) numbers indicate warm periods with ice minima; even numbers indicate colder periods and expansion of the ice-sheets.

Name		Approximate dates (BP)	MIS no.	Comments
Flandrian	Interglacial	0–11,500	1	Post-glacial (Holocene)
Devensian (Britain) Midlandian (Ireland)	Glacial	11,500–110,000	2–4, 5a–d	Ice covered much of Ireland and Britain north of a line from south Wales to north Yorkshire and the coast south to the Wash ('New Drift')
Ipswichian	Interglacial	110,000–130,000	5e	Longer than Post-glacial
Wolstonian	Glacial	130,000–? 375,000	6–?10	Possibly several glaciations with warm phases between them, ice covering a similar area to the most recent (Devensian) glaciation
Hoxnian (Britain) Gortian (Ireland)?	Interglacial	? 375,000–? 425,000	11	Formerly regarded as the longest ('Great') interglacial, but this is now doubtful
Anglian	Glacial	? 425,000–c.455,000	12	Generally taken as the 'glacial maximum'; ice covered most or all of Ireland, and Britain south to the Severn estuary, Oxford and south Essex ('Old Drift')
Cromerian complex	Alternating	c.455,000–c.620,000	13–21	Probably 4 interglacials and 3 glaciations

The succession of glacial and interglacial periods over the last 600,000 years is summarised in Table 1. The glacial periods hardly need explanation. We have models at hand in Greenland and Iceland, and we can see mountain glaciers in the Alps (Fig. 22). The interglacials that came between them were broadly like the present day in climate and vegetation. Each interglacial passed through a comparable pattern of development. All through the glacial maxima the boreal trees – birch (*Betula* spp.), pine (*Pinus sylvestris*), aspen (*Populus tremula*), shrubby willows (*Salix* spp.) – would have been widely, but thinly and patchily, distributed in favourable spots in the unglaciated country south of the ice-sheets. The trees of the 'mixed oak forest' – the oaks (*Quercus* spp.), wych elm (*Ulmus glabra*), alder (*Alnus glutinosa*), limes (*Tilia* spp.), beech (*Fagus*) and hornbeam (*Carpinus*) – would have been farther south, in the Mediterranean region and hardly reaching north of the Alps (Svenning *et al.* 2008). Initially, an open vegetation of cold-tolerant Arctic plants occupied the skeletal and unstable soils left by the receding ice. As the climate warmed, soils and plant cover became more stable, but the soils were still generally base-rich and unleached and came to bear a rich and diverse vegetation of dwarf shrubs and herbaceous plants. In the course of decades or centuries trees became established, first cold-tolerant and quickly dispersed trees such as birch and pine. These were followed by the trees of the mixed oak forest – the temperate deciduous forest, typically on brown-earth soils, which in various forms has dominated the North European Plain (including lowland Britain and Ireland) through successive interglacials, including our own. This phase was generally the climatic optimum of the interglacial, and it could be very prolonged. In the course of time, soils tended to become more leached, acid and peaty (blanket-bog growth is a likely possibility), and the interglacial entered a phase in which the mixed oak forest trees were progressively replaced by conifers, and heath subshrubs and other calcifuge plants increased. As the interglacial cooled to a close, the forest returned to pine, birch and a sparse glacial flora of Arctic plants as the advancing ice took over once again. The flora and the sequence of events is different in detail in each interglacial. In the Hoxnian of England, silver fir (*Abies procera*) and Norway spruce (*Picea abies*) played a significant role; these are continental trees not now native to Britain. In Ireland, the later stages of deposits of similar age have yielded very abundant pollen of the common rhododendron (*Rhododendron ponticum*; not now native to Ireland, but an invasive introduction), as well as seeds of the calcifuge Irish specialities St Dabeoc's heath (*Daboecia cantabrica*) and Mackay's heath (*Erica mackaiana*) (Fig. 225). The Ipswichian interglacial is notable for the abundance of hornbeam.

THE CLOSE OF THE DEVENSIAN GLACIATION AND THE LATE-GLACIAL

Twenty thousand years ago the last glaciation was at its peak. By 15,000 years ago the ice had retreated far enough to expose most of England and Ireland, and eastern Scotland, but the climate was still cold, and the vegetation was dominated by Arctic plants such as mountain avens (*Dryas octopetala*), crowberry (*Empetrum* spp.), dwarf willow (*Salix herbacea*) and mountain sorrel (*Oxyria digyna*)

FIG 23. Skogahliðar, Iceland. The southwest corner of Langjökull ice-cap can be seen on the skyline. *Empetrum–Racomitrium* heath and grassland, with patchy shoulder-high downy birch (*Betula pubescens* ssp. *tortuosa*) scrub, and lower patches of dwarf birch (*Betula nana*). The Pennines or the Scottish and Irish hills can be visualised looking like this during colder phases of the Late-glacial.

and mosses such as *Racomitrium* spp. and *Polytrichastrum alpinum*, with scattered patches of juniper (*Juniperus communis*), dwarf birch (*Betula nana*) and probably tree birches in favourable places; the view in Iceland gives some idea of how the British or Irish landscape may have looked at that time (Fig. 23). From this 'Older Dryas' period, by about 14,000 years ago the climate seems to have warmed to temperatures not much colder than the present day (Table 2). This relatively warm period, named the 'Allerød Interstadial' after the site in Denmark where it was first recognised, lasted for around 1000 years – a short time by geological standards but many human lifetimes, during which the climate must have felt as stable as now. It was brought to an abrupt end by a return to near-glacial conditions, the 'Younger Dryas' period, which lasted for about 600 years, with the mean temperature in the lowlands about 5 °C. The glaciers readvanced, and an ice-cap once again covered the Western Highlands from Loch Lomond to Wester Ross, with smaller masses of ice in the central Highlands. This cold stage seems to have ended at least as rapidly as it began; average temperatures rose by about

TABLE 2. The Late-glacial and Post-glacial stages and zones. The old Blytt–Sernander names are convenient labels for the main divisions of Post-glacial time and are still widely used. The dates ('before present', BP) are only a rough guide, to the nearest 500 years.

Name	Pollen zones	Approximate dates BP (1950)		Comments
		c-14 years	Calendar years	
Sub-Atlantic	VIII	2500–present	2500–present	Iron Age climatic deterioration; cooler and wetter, but varying through historical time. Present climate.
Sub-Boreal	VIIb	5000–2500	5000–2500	Becoming drier; starts warmer than now but cooling towards end.
Atlantic	VIIa	7500–5000	8000–5000	A little warmer than now, oceanic and wetter than Boreal. Climatic optimum.
Boreal	VI	9500–7500	10,500–8000	Warm, comparable with present day
	V			but drier and more continental.
Pre-Boreal	IV	10,500–9500	11,500–10,500	Warming rapidly
Upper Dryas	III		12,000–11,500	Cold; mean temperature c.5 °C
Allerød	II		13,500–12,000	Warm; mean temperature ? c.12 °C
Lower Dryas	I		14,000–13,500	Cold; mean temperature c.5 °C
Full glacial			–14,000	Cold; mean temperature < 5 °C

10 °C in only a few decades, and this date, about 11,500 years ago, is taken as the opening of Post-glacial time.

Vegetation and soils during the Late-glacial period

During the colder stages of the Late-glacial we may picture Britain and Ireland covered with a very open 'park tundra' of the kind sketched in the last paragraph. In the warmer Allerød phase tree birches spread to form copses or more extensive open woodland, especially in eastern England, and juniper was locally common in the north and west, but almost everywhere grass and tree pollen more or less matched the frequency of birch pollen, so our landscape must still have been quite open, perhaps rather like northern Scandinavia at the present day (Fig. 24). Lakes became more productive, and the sandy and clayey mineral deposits of the former (and succeeding) cold tundra landscape were replaced by calcareous marl or organic muds.

The raw soils left in this period of changing climate at the close of the last glaciation bore a flora whose richness and diversity surprised many people when it was first discovered. It included not only the expected Arctic and mountain

FIG 24. The tree-line above Storbakken, near Abisko, northern Sweden. The forest is mainly the northern form of the downy birch, giving way at a rather ragged limit to low scrub of arctic willows and dwarf birch. Most of the species here are also in the Scottish Highlands. Another 'Late-glacial' landscape; vegetation like this, or like the birch forest of the lower slopes, would have been widespread in Britain and Ireland in the Allerød period.

plants, but also calcicole and grassland plants, some of them with relatively southern distributions today, and a great diversity of species of other habitats including aquatics, marsh and rich-fen plants, and plants generally thought of as 'ruderal'. Early interpretations of these rich Late-glacial assemblages saw them as evidence of 'open habitats' left by the retreating ice, in which species of diverse ecology could flourish together. More realistic models can be seen around and above the tree-line in the central European mountains. There, a diversity of plant communities exist side-by-side, differentiated by slope, aspect, and soil moisture and pH – including heaths, calcareous grasslands, mires, tall-herb communities and patches of weedy vegetation. We have to remember that deglaciation did not happen overnight. The Late-glacial spanned 2000 years of fluctuating climate, through most of which sea level was lower and climate was more continental (and more stable) than it is now. And of course it had all happened before, at the ends of a succession of preceding glaciations.

Pollen analysis

Much of our knowledge of the history of British and Irish vegetation is drawn from the technique of pollen analysis. Most of our dominant forest trees, such as oak, birch, elm and alder, are wind-pollinated, and disperse large mounts of pollen into the air; so too are grasses and sedges and many other plants such as nettles and docks. This pollen rain (of which hay-fever is a by-product) falls year by year over the whole landscape including lakes and other wetlands. Pollen is remarkably resistant to decay, so lake sediments and peats preserve a continuous record of the pollen rain from their surroundings. Pollen grains are diverse in size, shape and ornamentation (Fig. 25), so after suitable preparation of samples for microscopy they can be identified to build up a *pollen spectrum* from the sample, which gives some picture of the vegetation at the time it was deposited. Much more information can be gained if pollen spectra from successive depths in the deposit are plotted together as a *pollen diagram* (Figs 26, 27). The interpretation of pollen diagrams is not straightforward. Different trees produce different amounts of pollen, and the pollen of different plants may be dispersed for different distances; thus pollen of forest trees will be dispersed farther than that of grasses and sedges. Pollen in peats is

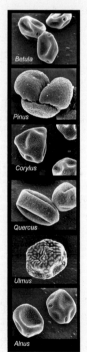

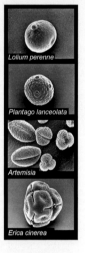

FIG 25. Some airborne pollen grains important in pollen analysis, × 340 approx. (a, left, top to bottom): Birch (*Betula*), with three thick-rimmed equatorial pores; pine (*Pinus*), a large and distinctive grain with two air-sacs; hazel (*Corylus*), three pores, without thickened rims; oak (*Quercus*), one of many grains with three longitudinal furrows, but differing in surface pattern; elm (*Ulmus*), about six equatorial pores, with distinctive surface ornamentation; alder (*Alnus*), usually four pores with bands of thickening in the pollen wall between the pores. (b, right, top to bottom): a typical grass pollen (*Lolium perenne*, perennial rye-grass), with a single pore and a smooth surface; ribwort plantain (*Plantago lanceolata*), a spherical grain with pores evenly scattered over the surface; mugwort (*Artemisia vulgaris*), one of a genus, all wind-pollinated, which includes coastal, alpine and steppe species. In bell heather (*Erica cinerea*) and common heather (*Calluna*) the 'grain' is in fact a *tetrad* of four grains, which fail to separate and are dispersed as a unit, as is usual in Ericaceae.

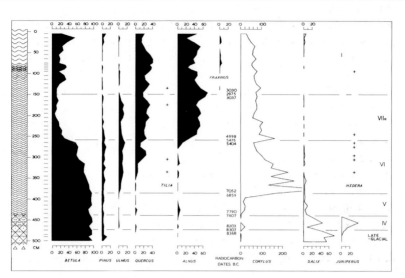

FIG 26. Pollen diagram from Scaleby Moss, northeast of Carlisle, the first to be radiocarbon-dated in Britain. The dates are in uncalibrated radiocarbon years BC; taking account of the calibration correction, the Post-glacial here began a little over 11,000 years ago. The solid black curves show the forest trees as a percentage of total tree pollen (AP). The open curves show the non-tree pollen types (NAP) as percentages of total tree pollen. From Godwin (1981).

likely to be influenced by the surrounding vegetation; pollen in lake sediments will generally give a wider picture. Comparison of pollen diagrams (and the modern pollen rain) from different sites and different regions is needed to disentangle many of these problems.

In the earlier days of pollen analysis it was customary to express pollen counts as a percentage of a conventional list of tree-pollen types ('arboreal pollen', AP). Hazel was excluded from 'tree pollen' mainly because it was seen as an understorey shrub, but also because its pollen is hard to distinguish from that of bog-myrtle (*Myrica gale*). Figures 26 and 27 use this convention, as do most pollen diagrams before about 1975. There was little against it when the emphasis was on changes and regional variations in Post-glacial forests, but it fails to give any sense of the relative sparseness of pollen in the Late-glacial, or of the quantitative impact of human activity, and hazel is an

(continued overleaf)

Pollen analysis (*continued*)

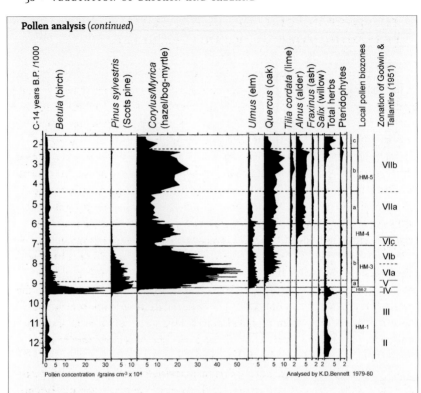

FIG 27. Pollen concentration diagram from Hockham Mere, a former lake in Norfolk, northeast of Thetford. Hazel, the most prolific pollen producer, reaches half-a-million pollen grains/cm^3 of sediment 8000–9000 years ago. This diagram shows a typical pollen sequence from southern England. From Bennett (1983).

embarrassment as a 'non-tree' pollen type when we know that it can be the dominant woody species in western parts of Britain and Ireland. For these reasons there has been an increasing tendency to present pollen diagrams showing either concentration of pollen per unit volume of sediment (Fig. 27), or annual influx of pollen per unit area of deposit. Obviously pollen influx diagrams only became possible in most places with the advent of C-14 dating.

THE POST-GLACIAL: CHANGING CLIMATE, SPREADING FORESTS, RISING SEA LEVELS

The spread of forest: the Pre-Boreal and Boreal periods

In the fast-warming climate of the beginning of the Post-glacial, the open tundra landscape quickly gave way to forest. In our area the first tree to form continuous woodland (and dominate the pollen rain) was birch, often accompanied by a peak in juniper (as at Scaleby Moss, Fig. 26). Probably most of this early Post-glacial birch was the northern form of downy birch (*Betula pubescens* ssp. *tortuosa*), which is common in the mountains of Scotland and Scandinavia at the present day. Birch is cold-tolerant, but what was probably more important for its rapid Post-glacial spread is that it is a light-demanding tree, and with its light wind-borne seeds it can rapidly colonise open country. This 'Pre-Boreal' phase of relatively pure birch forest lasted a few centuries, less in eastern England, more in the north and west, before the birch was joined by substantial quantities of other trees. Generally, the first of these to colonise in southern England was pine, followed by hazel (*Corylus avellana*), which must have been abundant in the forests at this time, often dominating the pollen rain. In more northern and western districts hazel often preceded pine, and in some places (as at Scaleby) pine never played a large part in the forest at all. The next tree to arrive was usually elm, probably wych elm, followed by the oaks.

Our climate during this 'Boreal' period of mixed forest (roughly from 9500 to 8000 years ago) was warm, but drier and more continental than now. Britain was for much of this time still a part of continental Europe, and the Irish Sea was a less formidable barrier than it afterwards became. Probably the succession of forest trees reflected migration and establishment rates much more than rising temperature. However, some late arrivals may indeed have been limited by temperature, notably small-leaved lime, which is known to require a warm late summer to set good seed. Open calcareous fens were widespread in the Boreal period and many rich-fen mosses were abundant, such as *Cinclidium stygium, Paludella squarrosa, Meesia tristicha, Helodium blandowii* and *Tomenthypnum nitens*, which have become rare or extinct in Britain and Ireland since that time. Ombrogenous bogs were uncommon. Late in this period climate showed signs of becoming more oceanic. Britain had by now become an island, the Irish Sea was wider, and these changes were bound to have climatic effects.

Changing sea levels

Every bit of ice locked up in an ice-cap or glacier on land is that much less water in the oceans, and when the last glaciation was at its height, ocean levels were about 120 m lower than now. As the ice melted and the water returned to the oceans, sea level rose worldwide. This is called the *eustatic* rise of sea level. At the start of the Post-glacial, sea level was probably still around 60 m below its present level but rising rapidly; by 8000 years ago it was only some 10 m below present sea level. The shallowest part of the North Sea, between Norfolk and Holland, is about 30 m deep. Locally another factor has influenced sea level around our coasts. Although the rocks of the earth's crust appear to us rigid, the crust is sufficiently flexible for heavy loads on it to come into *isostatic* equilibrium – given time, on a scale of millennia. Major loads such as continents and mountain ranges 'float' on the viscous mantle underneath. At the maximum of the last glaciation, Scotland was loaded with at least 1000 m of ice, which depressed the land beneath by several hundred metres. As the ice melted the surface gradually recovered to its former level, but not immediately; evidence for that is seen in the many 'raised beaches' round the Scottish coast (Fig. 11). In fact isostatic recovery is still taking place now. The Scottish Highlands between Glasgow and Inverness are rising at about 3 mm a year, while northernmost Scotland, the Western Isles, the northeast coast of Ireland and northern England

FIG 28. Changing sea levels. Sub-fossil pine stump on the shore, Streamstown, near Clifden, Co. Galway, September 1997. The seaweed-covered tree stump is evidently *in situ*. Erosion has exposed part of its roots, which were in mineral soil under the peat bed which engulfed it, in its turn to succumb to rising sea level. The top surface of the peat now bears upper-saltmarsh turf.

are hardly moving either up or down. Because the earth's crust has some degree of rigidity, south of this line England, Wales and Ireland are sinking at 1–2 mm a year. The results of the interaction between these two processes are seen in many coastal features in Scotland, in the drowned river valleys of southwest England, 'submerged forests' and apparently bizarre juxtapositions of vegetation on our shores (Fig. 28), and the interplay of sediment accumulation, peat growth and marine transgression seen in the Fens, the Norfolk Broads and the Somerset Levels (Chapter 13). More detail about land and sea-level changes around Scottish coasts is given by Friend (2012).

An eventful climatic change: the Boreal–Atlantic transition

About 8000 years ago, a major change took place in the vegetation of Britain and Ireland and neighbouring parts of western Europe, apparently everywhere at about the same time. The most obvious change in pollen diagrams was a sudden rise in the pollen of alder, which often roughly coincided with the beginning of ombrogenous bog growth. These changes were symptomatic of a switch to the warm, more oceanic climate that led Axel Blytt in Denmark and Rutger Sernander in Sweden to call this the 'Atlantic' period. Early in this period the sea had risen substantially to its present level. The next four millennia can lay claim to be the climatic optimum of the Post-glacial. The forests at this time were made up of a mix of high-forest trees: oak, wych elm, lime, some ash (*Fraxinus excelsior*), and abundant alder wherever the soil was wet enough. Hazel seems always to have been a common component of the forest, presumably as an understorey shrub.

There was a good deal of regional variation with geology and climate. Elm was particularly prominent in Ireland. Birch had declined to low levels generally but remained a major component in northern England and Scotland. The oaks made up a major proportion of the pollen rain in England and Wales – presumably mainly pedunculate oak (*Quercus robur*) on the moister and more fertile soils, and sessile oak (*Q. petraea*) on the poorer and drier acid soils and on dry limestone. Pine had dropped out completely from many lowland sites but continued to be prominent in the pollen diagrams in northern Scotland and in Ireland. Small-leaved lime is under-represented in pollen diagrams, because it is largely an insect-pollinated tree; it probably dominated forest at least locally on fertile base-rich brown-earth soils in lowland England north to the Scottish border. A notable feature at this time is the great expansion of holly and ivy, especially in western Britain and Ireland.

There must have been local openings in the forest, due to regeneration gaps, windthrow, and probably through browsing and grazing in places where animals congregated. It has been suggested that the primeval forest was more of the nature of wood-pasture, with extensive grassy areas created and kept open by forest herbivores. However, pollen analysis gives no support for this supposition, and positive evidence against it, so we can visualise the forest at this time as essentially unbroken (Mitchell 2005).

MAN COMES ONTO THE SCENE: NEOLITHIC FOREST CLEARANCE AND PREHISTORIC FARMING

Palaeolithic and Mesolithic cultures cannot have been without their impact on the tundra and forests, just as the Sami with their reindeer herds have played a significant part in shaping the seemingly 'natural' landscape of arctic Sweden, and the treeless prairies of North America owed a great deal to the traditional lifestyle of the plains 'Indians'. But about 5000 years ago a number of changes appear in pollen diagrams across western Europe, at about the same time, some of which can only be explained by Neolithic man having a much more specific and direct effect on the forest. The most widespread of these changes is a decline in elm pollen. Hazel typically increases, and grass pollen begins to appear again, often accompanied by pollen of weeds such as plantain (*Plantago* spp.), cereals (larger grains than wild grasses) and spores of bracken. Pollen diagrams show well the broad picture of long-term change from dominant forest to the relatively open countryside of the present day. In closer detail, the early stages often suggest that small areas of the forest were cleared, a crop grown for a few years, with the plot then abandoned and the forest allowed to revert; then the whole sequence was repeated elsewhere. This is the pattern of slash and burn shifting cultivation in the tropics today.

   How far the long-lasting forest clearance that was to come was the result of gradually changing practices of the indigenous Mesolithic hunter-gatherers, and how far it was influenced by migration from continental Europe, is still a matter of controversy among archaeologists. Early Neolithic farmers raised native pigs and cattle, but it is clear that at least some elements of the 'Neolithic package', including wheat, barley, sheep and goats, must have been introduced from farther afield. What is certain is that within a few centuries a settled Neolithic farming culture was established in both Britain and Ireland, with a substantial level of division of labour and social organisation, clearing forest for cereal cultivation and pasturing livestock, and producing pottery, polished stone axes and other skilled artefacts. The first long-barrows and causewayed camps on the English chalk and the first 'passage tombs' in Ireland were early Neolithic, and by 5000 years ago the population in both Britain and Ireland was sufficiently large and well organised to have conceived and built the winter solstice-aligned passage-tomb at Newgrange in Co. Meath, the extensive stone-walled Céide Fields in northern Mayo, the early circles at Stonehenge and Avebury in Wiltshire and innumerable dolmens and other megalithic structures in both islands. However, the greatest impact the Neolithic peoples had on our modern landscape and vegetation was the creation of open grassland and moorland where formerly there was forest.

What was the cause of the elm decline?

The earliest episodes of Neolithic forest clearance were localised. A small patch of forest seems to have been cleared of all trees indiscriminately, and after a few decades or a century or two reverted to forest. Early interpretations of the elm decline sought a climatic cause, but none could be demonstrated that was convincing. The epidemic of Dutch elm disease in the 1970s stimulated speculation that the Neolithic elm decline might have been at least in part due to disease, and a persuasive case can be argued that disease may have been partly responsible for the opening-up of the forest. The consensus view probably is that although elm disease may have played some part in 'the elm decline' in the pollen record, seen as an event, the long-term decrease in elm is a clear part of the package of changes that came with the growth of Neolithic agriculture. Our widespread native species, wych elm, continues to be quite common in our woods. The elms that were so prominent in the traditional agricultural landscape of lowland England and Wales in the twentieth century belonged to other species, small-leaved elm (*U. minor*) and English elm (*U. procera*), both probably introduced in prehistoric or Roman times for their agricultural value (Chapter 5). Elm foliage (even dried) provides good feed for livestock, and we can surmise that Neolithic people found it a valuable agricultural resource.

Continuing climatic change: the Bronze Age and the Sub-Boreal period

Blytt and Sernander were led to erect their climatic division of the Post-glacial by the changing character of the peat with time in the raised bogs of Denmark and southern Sweden. The climate continued warm, but was getting drier. Growth in the peat bogs was slower, and the peat was highly humified, darker and denser. Apart from elm the dominant forest trees do not seem to have changed greatly, but beech appears in the pollen record at a few sites.

Bronze tools came into use in Ireland and Britain over 4000 years ago, and gradually replaced stone for tools and weapons. This implies trade. Both southwest Ireland and southwest Britain had substantial and accessible deposits of copper ore but these were very localised, and tin was mined only around the granite masses of southwest England. The Bronze Age people were adept at working metal and produced fine pottery, and they continued and extended the forest clearance and farming begun by their predecessors. In some areas, deforestation had opened the way for the soil to leach and become podzolised, and many of our heathlands must date from this period. Some may have had a period of arable cultivation in their early history, but most probably originated as wood-pasture.

The later Neolithic and the Bronze Age left evidence of thriving populations in coastal and upland areas of both islands (Fig. 29). The 'reaves' of Dartmoor, though probably somewhat later, resemble the Céide fields in Mayo in marching in straight parallel lines across square kilometres of country almost regardless of topography – though the reaves do peter out on either side of the deep wooded valley of the Dart, to reappear on the other side. This was landscape planning on a massive scale. Some later fields conformed to (or used) the pattern of the reaves, but medieval field boundaries conspicuously did not. 'Celtic fields', the small squarish irregular fields whose outlines can still be seen in chalk

FIG 29. Bronze Age house ('hut circle') near Tavy Cleave, Dartmoor, Devon, January 1966. Purple moor-grass (*Molinia*) on hillside in foreground. Upland blanket-bog, heather-moor and hill grasslands in distance.

downland and moorland that has escaped the plough, mostly date from this period. On slopes these fields were commonly elongated along the contours, and the downward slip of the soil with repeated ploughing formed lynchets – but probably most of the lynchets so conspicuous on hillsides in chalk country date from medieval times (Figs 34, 37). Woodmanship must have been well established by this time. The trackways built across bogs in the Somerset Levels and elsewhere were not simply cut from the wildwood. They were sophisticated structures, which imply deliberate woodland management to produce the uniform hazel rods, poles and timber used to build them.

THE IRON AGE AND THE POST-GLACIAL CLIMATIC DETERIORATION

Iron-working reached our islands around 2700 years ago – about 700 BC. Iron was more plentiful and stronger than bronze, and must have revolutionised many aspects of life, not least farming. Iron-tipped ploughs could plough land more deeply and effectively than earlier wooden ones, iron axes could fell trees more efficiently, and the relative cheapness of iron brought metal implements within reach of many more of the population. The smith was a key man in the community, shoeing horses certainly, but making and repairing ploughs, harrows and hand tools, tools for the woodman, hinges and fastenings for doors and gates, and making ironwork and implements for domestic needs. No wonder 'Smith' or its equivalent is such a common surname across Europe. But the smith was only one craftsman among many, and the Iron Age Celts were renowned for their imaginative and skilled craftsmanship. It could be said that the pre-Roman Iron Age put together all the ingredients of the traditional British and Irish rural scene except the church – and maybe pre-Christian festivities in some measure did duty for that.

The Iron Age had an urgent need for iron tools, because about 600 BC the climate became cooler and wetter. Dartmoor has a plethora of Bronze Age remains, but little from the Iron Age – and it is still a pretty inhospitable place on a wet day in winter. Raised bogs began to grow again with renewed vigour; the change is often marked in the peat by flooding horizons with abundant papery remains of the rhizomes of Rannoch rush (*Scheuchzeria palustris*, Fig. 211), and much of the thickness of peat in our blanket bogs dates from this time. Non-tree pollen increased. Oak and alder remained the most abundant trees in the pollen record. Elm declined still further, and small-leaved lime disappeared from many sites. A notable novelty was the regular occurrence of significant amounts of beech pollen, but mainly in the southeast, and not in quantity to rival the established trees. Hornbeam was another tree that appeared in small quantities at this time. The climate became gradually warmer and drier through Roman times and the 'Dark Ages' until in medieval England many monasteries were making their own wine. But that is trespassing on the territory of the next chapter.

Iron Age Britons lived in tribal groups ruled over by a chieftain. By the time of the Roman invasions the indigenous population had been reinforced by Gauls displaced from the Continent. The population had become more organised, sophisticated and to some degree Romanised. Coinage began to be used, and people had begun to congregate into towns. But tribal rivalries still ran deep. Iron Age people built innumerable hill-forts. Some of these were reinforced in the last century BC into what must at the time have been formidable defensive structures. Their earthworks are still impressive, and sites such as Maiden Castle and Hod and Hambledon Hills in Dorset are now among our finest chalk grassland localities, with rich calcicole floras. Julius Caesar in 55 BC found the Britons skilled in using war chariots. He would also have found a prosperous population, productive farming of cereals and livestock, and managed woodlands in settled parts of the country – but still with remnants of the wildwood in the more inhospitable parts, and large expanses of fen and bog.

CHAPTER 3

History: from the Romans to Modern Times

T HE ROMANS WERE LITERATE, and from this period onwards we have written records to supplement what we can learn about our history from other sources. But those other sources remain important. Written records inevitably reflect the point of view of the writer; and what is seen as important and worth recording at the time may not be so with historical hindsight. Where purely factual information is concerned (e.g. accounts, estate records, birth, death and marriage records) we can always be grateful for written records. Where there is an element of value judgement we have to be cautious. The Roman Empire was as aware of its civilising mission as was the British Empire in its heyday, or the United States now. The Romans were disparaging of 'barbarians', just as our grandparents' generation was disparaging of (and sorry for) the 'natives' in our Empire. We should not judge our ancestors of two millennia ago uncritically by the Romans' view of them.

'History' as it is mostly taught comes in neatly packaged compartments: the Roman period, the Dark Ages, the Anglo-Saxon period, the Norman conquest and so on. This is largely the history of ruling elites. Social history and the history of the landscape is a more gradual and continuous process, driven by economic pressures and technological change, and tempered by the natural conservatism of human societies. Major catastrophes like the Black Death and periods of warfare leave their mark, and can trigger or delay change in unforeseen ways, but their effect in the long run is often less than would have been forecast at the time.

ROMAN BRITAIN

In the late summer of 55 BC Julius Caesar mounted a reconnaissance expedition to the south coast of Britain, and a more serious attempt at invasion the following year, but it was not until AD 43 that Claudius initiated the conquest of the southern part of Britain. The southeast was quickly overrun, and within five years the Romans had probably reached the Severn and the Trent. There followed a period of consolidation, punctuated by the rising of the Britons led by Boudicca, Queen of the Iceni, in AD 60. By the early and mid-70s, the Romans had conquered south Wales, and penetrated as far north as York. During his governorship (AD 77–84), Agricola completed the conquest of Wales, and then quickly marched northwards to the Scottish border and the Clyde–Forth valley, culminating in the defeat of the Caledonian army in Perthshire (AD 83). Truly a *Blitzkrieg*. That was the farthest point the Romans reached, and it is one thing to win battles, and quite another to hold and govern territory. Permanent Roman settlement extended only to Hadrian's Wall from the Tyne to the Solway, begun in AD 122.

The Romans' influence upon the British landscape may be compared with Britain's upon India. The Romans brought roads, planned towns, villas, fen drainage, trade, and peace and prosperity. The Roman occupation may have had little effect on the ordinary countryside, or on everyday life, just as the British Raj brought railways, public works and unified administration but left the Indian countryside and the essentials of Indian life much as they ever were. Although Britain has a wealth of Roman remains, clear effects of the Roman occupation on our fields, lanes and woods are very localised.

Many hundred 'Roman villas' are known in England. These were essentially large working farms (sometimes with other, more industrial, rural activities), which must have employed substantial numbers of servants and labourers. They were sometimes built near the main Roman roads and towns, but often they seem to have relied on the minor road network, which was probably little if at all less dense than it is in country districts now. But all this took place against a predominating background of traditional native farming (Fig. 30), which went unrecorded because it was too ordinary for comment – and of which the archaeological evidence is largely buried under our modern network of field boundaries and lanes. The straight 'Roman roads' that we know were only a small part of the road network in Roman times. They were the motorways of the age, built primarily for military use, and their upkeep was paid for by the state.

The coming of the Romans coincided with a time of falling sea level on the East Anglian coast. The Romans brought with them centuries of experience of

FIG 30. Outlines of small semi-regular fields, probably of Iron-Age origin ('Celtic fields'), on the hillside below Malham Cove, Yorkshire. April 1974.

drainage engineering from the Mediterranean, and for the first time made the silt fens round the head of the Wash habitable. In the course of this work they dug the Car Dyke, a remarkable waterway stretching for some 140 km between Cambridge and Lincoln as a cut-off channel intercepting all the minor Fenland rivers, and embanked the seaward margin of the silt fens between King's Lynn and Skegness. Settlement of the newly drained fen seems to have been haphazard, and did not last for more than a century or two.

Roman Britain was probably somewhat, but not greatly, more wooded than now. Woods would have taken their place along with fields in the ordinary farmed countryside. The larger river valleys were probably unwooded. So most likely were the shallow soils of the chalklands, and many areas that now are moorland. These would have been used for grazing. Remnants of the wildwood were probably confined to remote places. It is important to remember that for the Romans, as for the Iron Age Celts they conquered and the Anglo-Saxons that followed them, woods would have been a valuable resource. The wood not only supplied timber from the large trees. It provided (renewably, from the underwood) for the energy supply of the community, for cooking, for keeping warm in winter, for industry (iron smelting and working, pottery and brick and tile making). The luxuries the more opulent Romans loved – hot baths, underfloor central heating – all depended on a reliable local supply of brushwood. There was no other source of energy.

The villas declined after the Roman legions left early in the fifth century. Some Roman ways persisted for a time, including by then Christianity. Over most of England the Roman way of life succumbed to fifth-century Anglo-Saxon invaders, but this was not marked by an immediate collapse into chaos. The Dark Ages were principally 'dark' in that they left few written records. Neither pollen analysis nor archaeology records any sudden increase of wildwood after the Romans left.

ANGLO-SAXON TIMES: THE EARLY MEDIEVAL PERIOD

The Anglo-Saxons found England in many ways not radically different from what had greeted the Romans four or five centuries previously. The Anglo-Saxons founded many settlements. They seem to have liked more compact villages than their Celtic predecessors. There was probably more displacement of population with the Anglo-Saxon influx. Both place-name and blood-group evidence show Anglo-Saxon influence predominating over most of England and southern and eastern Scotland, with predominantly Celtic populations remaining in Cornwall, Wales, an area behind the Fens (which shares St Neots and St Ives with Cornwall) and in the Pennine dales, north and west Scotland, and of course Ireland. Scandinavian settlement during the Anglo-Saxon period added another linguistic (and blood-group) element.

Consequently, Britain has more complexity in its place-names than many European countries. Even in Anglo-Saxon country, names of rivers are often of Celtic origin: Avon, from *afon* in Welsh or *abhainn* in Gaelic; Usk, Esk, Exe and Axe from Gaelic *uisce*, water (the same root as 'whisky'); Derwent meaning a river where oaks grew (Welsh *derwen*, oak); and names ending in -caster, -cester or -chester were towns by Roman forts (Latin *castra*). Most English place-names are from Anglo-Saxon roots, but in the north and east of England, settled by Danes from the ninth century onwards, unmistakably Scandinavian place-names are common, such as Whitby, Scunthorpe and Derby. Tenth-century Norwegian settlement in northwest England is marked by names such as Kendal and the many 'Kirkbys' and '-thwaites', and by the 'fells' and 'gills' of the Pennines and Lake District. In the west, Celtic place-names predominate, and Norman French was a later addition to the mix.

What can be gleaned from Anglo-Saxon charters indicates that the distribution of woodland did not differ greatly from what is recorded in Domesday Book, after the Norman conquest. Pasture can be taken for granted, as can the enclosed-field farming landscape. Some specific references to (hay) meadow can be inferred from Anglo-Saxon charters. Opinion is divided as to when open-field farming became established. There is no evidence for its existence in the Roman period. Open fields, cultivated in common, seem to have been an early-medieval invention that spread across Europe wherever conditions were suitable, which was mostly in flatter country. The perceived benefits are unclear. Possibly economies of scale in ploughing, making the best use of arable grazing, and saving the upkeep of fences all had a part. Place-name evidence suggests that it was already established by late Anglo-Saxon times; no doubt practice varied in different places. There are no certain mentions of it in Domesday Book (possibly because it was too commonplace to record), though a few entries could be interpreted as referring to open fields. Classic open-field farming reached its peak under the Normans and Plantagenets, several centuries later.

In the 500 years between their coming to Britain and the Norman conquest, the Anglo-Saxons (and Scandinavians) probably increased the area of farmland, managed the woodland more intensively, founded villages and made many minor changes, but they did not radically alter the English landscape.

THE NORMAN CONQUEST OF ENGLAND: THE LATER MEDIEVAL PERIOD

The Normans brought to Britain organising ability, technological prowess, the Norman-French language, and big ideas. One of those was Domesday Book, the great survey of England commissioned by William the Conqueror and completed in 1086. For landscape history, with this survey the written word comes into its own.

Woods and meadows

From the Domesday record it has been calculated that in 1086 about 15% of England was wooded. The most heavily wooded areas were the Weald (70%), followed by Worcestershire (40%). Buckingham and Derbyshire were about 26% wooded, West Yorkshire, Gloucester and Hampshire were about average (14–16%), Devon and East Yorkshire 4%, and the Isle of Ely 1%. The Fens and the Breckland were almost bereft of woodland, as was a belt running from East Yorkshire to Wiltshire, and some tracts between the Severn and the Welsh border. Woodland remained a valuable resource, yielding a greater return than arable, but the period from 1100 to 1300 was a time of rapid population growth and the area of

FIG 31. Coppice produce near Sixpenny Handley, Dorset, May 1976. Hazel rods stacked on the right, a hurdle in course of making in the centre, finished hurdles stacked on the left. Bean poles and pea sticks were also on sale. Hurdles were just one of the traditional products of the coppice. A vitally important product was firewood, either as wood, or burnt to form charcoal. In earlier times 'coal' generally meant charcoal.

farmland expanded at the expense of woodland, so that by the mid fourteenth century the proportion of woodland in England had fallen to about 10%. Medieval woodmanship came to a high pitch of development during this period. Woods supplied not only timber, but smaller wood for hurdles, wattle-and-daub building, posts and firewood. The underwood often brought in a greater income than the timber; 'wood' meant the products of the underwood, not the big trees (Fig. 31). Woods had a major role as renewable biomass plantations.

Meadows are the best-recorded land use in the Domesday survey, in spite of covering less than 2% of England. Hay meadows were a vital ingredient of medieval farming (indeed of all historical periods). In the summer months stock could graze the permanent pastures, but there was a period in winter when (even if the ground was not snowbound) grass growth was too poor to provide adequate forage for livestock, and they had to be fed on hay. This was especially so for the oxen (and latterly horses) that drew the ploughs, which worked hardest in the winter months, and meadows were sometimes valued in terms of the number of ploughs or oxen they would support. In well-wooded places, foliage

FIG 32. Fen produce: reed (*Phragmites*) and sedge (*Cladium*) for thatching. Fenside, Norfolk Broads, August 1970. Fenland areas also provided hay, summer grazing, fish and wildfowl.

(e.g. elm and hazel) cut from woods and dried sometimes supplemented hay as winter fodder. Meadows increased greatly during the Norman period, and became some of the most valuable land. Pastures enormously exceeded meadows in area, but are less thoroughly recorded in the Domesday survey. The recorded numbers of livestock suggest that pasture must have amounted to a quarter or a third of the landscape.

The early Norman period was a time of congenial climate, and continuing fall in sea level. This gave a filip to continued settlement of the northern part of the Fens, both seaward over the upper saltmarshes and landward into the peat fens. In the thirteenth century, the 'Roman Bank' was built (or rebuilt) into a massive and elaborate earthwork protecting the northern part of the Fens. Fenland became the most agriculturally prosperous region of England, and its produce complemented that of the surrounding upland (Fig. 32). Much the same was true of the Somerset Levels.

Arable farming: the open fields

The three centuries that followed the Norman Conquest saw the heyday of medieval open-field arable farming. This was largely superimposed on land that had formerly been farmed as enclosed fields. In its classic form, in open

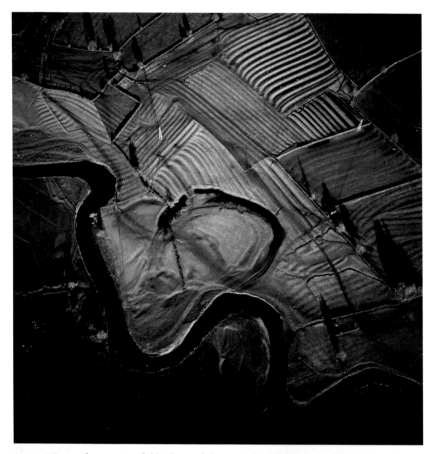

FIG 33. Pre-enclosure open-field ridge-and-furrow in the Dove valley near Uttoxeter, Staffordshire. The classic reversed S-shape of the ridges is well shown in the low afternoon sun. North is at the top left-hand corner. The river has evidently changed its course repeatedly since the last ridge-and-furrow ploughing. Two electricity pylons in the upper half of the picture give the scale. Since this picture was taken, the ox-bow in the centre has been obliterated by modern cultivation (© Adrian Warren & Dae Sasitorn/www.lastrefuge.co.uk).

country, the open-field system generally combined some or all of the following characteristics. The arable land of a village was divided into strips, about 220 × 11 yards (200 × 10 m), distributed either regularly or at random among the participating farmers. The strips were combined into furlongs, and these into fields; the same crop was grown by all the farmers in any one field. Hedges were few. Often the village had three large fields; two would be sown in rotation (winter wheat or rye might alternate with a spring crop of oats, beans or peas), while the third was left to lie fallow (the details varied from place to place and from time to time). After harvest, the animals of the village were turned loose on the stubble and weeds of the cultivated strips, and the fallow. Individual farmers shared some of the labour of cultivating one another's crops. The way in which strips were ploughed created a ridge-and-furrow pattern.

The characteristic shapes of medieval ridge-and-furrow stemmed from the constraints of ploughing narrow strips with teams of eight oxen, with a mouldboard that turned the soil to the right. It was difficult to plough straight to the end of the strip, and easier to begin the turn 25 m or so before the headland, which created the elongated reversed S-shape of the ridges (Fig. 33). In the Midlands and north of England large areas of country show evidence of ridge-and furrow left by open-field cultivation; beautiful examples are illustrated by Beresford & St Joseph (1979). The open-field system was in its heyday in the thirteenth and early fourteenth centuries, when it covered at least 14% of England, some land in Wales, and some Norman lands in eastern Ireland. But even in its heyday, there was land being newly brought into open-field cultivation, and open fields being enclosed. Thus in the Yorkshire Dales and in east Devon there was substantial late-medieval enclosure of former open

FIG 34. Medieval lynchets below Pikedaw, west of Malham, Yorkshire, April 1986. The lynchets formed part of the strip-cultivated 'West Field'. The open fields at Malham were enclosed in late medieval times.

(or 'subdivided') fields, along with a shift away from arable to a more livestock-based economy (Figs 34, 35). An important aspect of the open fields was that they largely underlie (and are an expression of) Oliver Rackham's distinction between 'ancient countryside' and 'planned countryside', of which more will be said later in this chapter. The open-field system coexisted with traditional farming in 'planned countryside', which had numerous pockets of enclosed fields, as well as in 'ancient countryside', where farming in enclosed fields remained overwhelmingly predominant.

By no means all land was suitable for working by large teams of oxen so, especially in hilly areas, smaller-scale methods were widely used (Fig. 36a), including 'lazy-beds' of the kind used until modern times in the Hebrides and western Ireland (Fig. 36b). On steep slopes, cultivation strips were elongated parallel with the contours. Ploughing led to soil creep down-slope, in due course creating lynchets, steep banks separating one cultivated strip from the next. Most of the lynchets so conspicuous in chalk country date from medieval times (Fig. 37).

FIG 35. Medieval enclosure at Axminster, Devon. Documentary evidence shows than in east Devon open fields ('subdivided arable') were mostly enclosed in the thirteenth and fourteenth centuries, leading to narrow strip-shaped closes such as this group north of Axminster, contrasting with the meadows along the river, and the irregular-shaped fields away from the town. Field boundaries as they were in the 1880s from the first edition of the Ordnance Survey six-inch map. Black dots show hedgerow and woodland trees. A sample of 'ancient countryside'. The straight line cutting across the map at bottom right is the railway. From Fox (1972); reproduced with the permission of the Devonshire Association.

FIG 36. (a) Medieval village below Hound Tor, Dartmoor, about 330 m above sea level, February 1980. The remains of the thirteenth-century stone buildings in the middle distance overlie a succession of turf-walled houses, possibly dating back to Saxon times. Oats were grown. The ridges of a cultivated field, visible on the slope beyond, are only about 2 metres wide and are more akin to lazy-beds than to classic Midland ridge-and-furrow. The site was abandoned in the late fourteenth century, probably because of deteriorating climate (Beresford 1979). (b) Ridge-and-furrow left by modern lazy-beds, near Roundstone, Co. Galway. Despite their name, lazy-bed cultivation was hard work!

An animal that came into Britain after the Norman Conquest and was to have a major impact on our landscape is the rabbit (*Oryctolagus cuniculus*). Originally introduced in the twelfth century as a delicacy, by the following century it had

FIG 37. Medieval lynchets near Worth Matravers, Dorset, May 1996. The lynchets on either side of the Winspit valley were in open-field cultivation from the thirteenth century till enclosure at the end of the eighteenth. Since then they have been used as pasture for livestock.

become an important article of commerce. When first brought in, rabbits were seen as needing careful tending, but they soon showed themselves hardy and impossible to confine, and have become an established part of our countryside and country tradition. More is said about them in Chapter 11.

The Black Death

The catastrophic plague that swept through Britain and Ireland in 1348–1350 recurred at intervals in succeeding years. It is thought that the initial outbreak killed 30–45% of the population. The death rate in later outbreaks of the plague may have been lower, but seems to have borne most heavily on children and young men. The population in the 1370s was probably not more than half what it had been in the early years of the century. The Black Death at one stroke relieved population pressure and land-hunger, and created a scarcity of labour. There were not enough able-bodied men to cultivate all the open fields – and there were fewer mouths to feed. This had two major effects on our countryside. It led to widespread abandonment of open-field arable cultivation in favour of pasture and less labour-intensive sheep farming. And it gave respite to the woods, which had been hard-pressed in the thirteenth and early fourteenth centuries.

FROM THE SIXTEENTH CENTURY TO THE NINETEENTH CENTURY

The accession of Henry VII in 1485 has conventionally been taken in England as the opening of Modern History. This date is only one of a number that mark the passing of the medieval period, and the transition to an age much easier to relate to our own ('the past is a foreign country: they do things differently there'!). It was the time of the High Renaissance in Florence; in 1492 Columbus sailed to America; in 1517 Martin Luther nailed his 'Ninety-Five Theses' to the church door in Wittenberg, sparking the Protestant Reformation. It was a time of exhilarating new ideas, of questioning old certainties, and of new horizons and opportunities. In the five centuries since that time, power has shifted progressively from the Church and the monarch to trade, commerce and secular authority.

By the close of the medieval period the main outlines of the present-day landscape in England and Wales were already apparent. The distribution of woodland today is not greatly different from that in 1500, and much of the network of minor roads and field boundaries goes back to Norman times and beyond. In England the open-field landscape of the (mainly) flatter parts of country, with compact villages, few woods, and very few old hedges, formed an irregular wedge from east Yorkshire and the Norfolk coast to the chalk country of Dorset – the future 'planned countryside'. The rest of England was occupied by 'ancient countryside' of scattered farms and irregular fields enclosed by medieval (or older) hedges, with more diverse settlements. Much the same distinction was observed by sixteenth- and seventeenth-century writers between 'champion' country and 'several' or 'woodland' country. Subsequent changes radically altered farming practices, field boundaries and land tenure in the 'planned countryside' that had been in open-field cultivation, but had much less effect on the 'ancient countryside'.

Scotland and Ireland from the sixteenth to the eighteenth century
The Scottish Highlands and islands, and Ireland, were different. Both regions were poor, and were seen as land to be exploited by (sometimes unscrupulous) landlords, and (not without cause) as unruly and a haven for enemies of the state by those in authority in London or Edinburgh (or Dublin). Physical inaccessibility and religious and language differences did nothing to further understanding on either side. The consequence was continuing poverty and continuing suspicion, violence, resentment and neglect. Neither the Irish nor the Highlanders shared in the relative prosperity of the English or the Scottish lowlands in the sixteenth and seventeenth centuries. Ireland had long been a

place where Norman barons could acquire land and do as they pleased out of reach of the King of England's authority. Henry VIII had such trouble with his rebellious Anglo-Irish and Irish subjects that in 1534 he laid down that all lands in Ireland were to be surrendered to the Crown, and then re-granted. When Elizabeth I came to the throne, Catholicism in Ireland represented a dangerous threat, and she responded by making her father's edict a reality and applying it with ruthless severity. A last rising by Hugh O'Neill, Earl of Tyrone was defeated in 1601 at the Battle of Kinsale. This led to the 'plantation' of Ulster with (largely) Scottish presbyterian settlers from 1610 onwards. Ireland was the victim again later in the seventeenth century, when the bitter divisions of the Civil War in England spilled across the Irish Sea. And when William of Orange came to the English throne in 1688, to replace the deposed Catholic James II, it initiated a train of events that this is not the place to recount in detail. This was an inauspicious background for a new century, exacerbating rather than healing the alienation between predominantly Catholic Ireland and Protestant Britain.

Medieval Ireland had less woodland than England, but was relatively well wooded compared with later years, and had a coppicing tradition. Woodland probably covered 2–3% of the land area at the time of the Civil Survey in the 1650s. This was mostly recorded as 'shrubby' (most likely hazel?), or oakwood. Only a tenth of the wood recorded in the 1650s was still there two centuries later; most was replaced by ordinary farmland. There is little doubt that the primary cause of this destruction was the expansion of agriculture to meet the needs of an expanding population. With an abundant supply of peat as domestic fuel, woods had little value, so they were grubbed up for farming. Woods were preserved longest in districts with an active iron industry, notably western Co. Waterford. Ironworks remained productive for a century or longer on local woodland; they were probably relatively small enterprises, supplying a local market, and fuelled by a renewable regularly harvested supply of underwood.

The Highlands' problems were different, and their troubles came later. For most of the seventeenth century there was a Stuart king in London, but after 1688 that tie of loyalty to the British Crown was broken, culminating in the Jacobite risings of 1715 and 1745, and the eventual crushing defeat of 'Bonnie Prince Charlie' at Culloden in 1746. These uprisings spurred the government in London 1724 to appoint Major-General George Wade to build barracks, bridges and roads for control of the region. Between 1725 and 1737 Wade directed the construction of some 400 km of road and 40 bridges, linking the permanent garrisons in the Highlands, and raised militias to supplement the garrisons. The net effect of the Jacobite rebellions was to open up the Highlands and strengthen the grip of the authorities on the region.

From the scanty evidence, there was little woodland in the Scottish Lowlands or Southern Uplands in medieval times, and of what there may have been, little trace remains. Ancient woodland is virtually confined to the Highlands. The oakwoods of the west Highlands share a similar history with the western oakwoods of England, Wales and Ireland, and similarly supported iron working and tanning. The native pine woods of Deeside and Speyside were long a source of timber. There may well have been substantially more pine forest than there is now, but we would probably be wrong to assume that the Highlands were extensively covered with pine forest within historical time. Birch was no doubt always there.

The draining of the Fens, 1637 onwards

In 1600, Parliament passed 'An Act for the recovering of many hundred thousand Acres of Marshes', but nothing much came of it until, under the sponsorship of the 4th and 5th Earls of Bedford and the advice of the Dutch engineer Cornelius Vermuyden, the 'Old Bedford River' was created in 1637, and a parallel channel, the 'New Bedford River' or 'Hundred Foot Drain', in 1651. These were enormous straight drains, which diverted the waters of the Great Ouse, which had flooded the Great Level every winter, directly northeastwards until they rejoined the old course of the river at Denver. The Ouse Washes between the two Bedford Rivers provided a vast reservoir for flood-water in winter. The water of the minor Fenland rivers (Cam, Lark, Little Ouse and Wissey), which formerly discharged directly into the Great Ouse, now only joined the main flow at Denver. The Bedford Rivers were only the largest of a complex system of drainage works. At first the enterprise was a great success, but within a few years the consequences of shrinkage and wastage of the peat began to tell. Once the peat was no longer waterlogged, the organic matter in the fertile black peat soil oxidised and wasted away, lowering the surface of the land. This led over the years to demands for ever-deeper drainage. At first this was achieved by windmills, and from about 1820 these were replaced by steam-powered pumping stations, till 1851 with scoop-wheels, then for a century with centrifugal pumps, replaced successively from the 1940s onwards by diesel and electric pumps.

Some drainage works in the Somerset Levels were undertaken in medieval times, mainly by Glastonbury, Athelney and Muchelney Abbeys and their tenants, winning pasture and meadow from the marshes. Serious interest in draining the Levels re-emerged in the seventeenth century, when Vermuyden undertook some minor works but could not persuade the commoners to a more general drainage scheme. That had to wait till 1795, when the 17 km King's Sedgemoor Drain (Fig. 38) was completed. The Drain, which was upgraded in 1972, operates

FIG 38. King's Sedgemoor Drain, completed in 1795, looking towards Bawdrip and Knowle from near Chedzoy, Somerset, September 2010.

FIG 39. The steam pumping-engine at Westonzoyland, Somerset, installed in 1862 to replace an earlier (1831) beam-engine and scoop-wheel. The two vertical cylinders drive a horizontal crankshaft, while a bevel-geared flywheel driving the vertical impeller-shaft of the centrifugal pump at the bottom of the pump-well. The machine could raise 100 tonnes of water a minute a height of 1.2 m, and worked regularly until 1951 when it was replaced by a diesel pump. Similar pumping engines were used in the Cambridgeshire Fens.

entirely by gravity, but flood-water from many of the Somerset moors is pumped, originally by steam (Fig. 39), now by electric pumps. The Huntspill River, cut early in the Second World War, drained the Levels between Glastonbury and the sea, and added to the efficacy of King's Sedgemoor Drain. The Somerset Levels, largely used as pasture (with a thriving osier industry in the past), have largely escaped the problems of peat wastage that beset the Cambridgeshire Fens.

The agricultural revolution in England, 1700–1850

In 1701 Jethro Tull invented a horse-drawn seed drill, which would sow seed more precisely and less wastefully than the hand-drilling customary at the time. He had various other inventions (or re-inventions) to his credit, including a horse-hoe, and in 1731 he published his book *The New Horse Hoeing Husbandry*. His methods were not generally adopted for many years, but he came to be recognised as a major pioneer of mechanised farming. Charles, 2nd Viscount Townshend ('Turnip Townshend') introduced into England the four-field crop rotation practised by farmers in the Waasland region of the Netherlands – wheat, barley, turnips and clover. This did away with the fallow of the traditional medieval three-field rotation. The root crop provided winter feed for livestock and made it easier to hoe between the rows, so helping to keep weeds in check, and including a legume in the rotation countered the depletion of soil nitrogen by the other crops. Notable advances were also made in animal breeding during the eighteenth century. These innovations were slow to catch on, and they were not widely adopted until the Napoleonic Wars and the Industrial Revolution had radically changed the economic climate in the first half of the nineteenth century. The Royal Agricultural Society of England was founded in 1838, with the motto 'Practice with Science'. John Bennet Lawes established an experimental farm on his estate at Rothamsted in 1842. The science of plant nutrition and the first artificial fertilisers were developing at about the same time.

The Enclosure Acts, 1761–1845

Open-field farming on the medieval pattern was increasingly an anachronism, and by the opening of the eighteenth century change had become inevitable. Since the heyday of the open fields, there had been some local enclosure, but it became a political issue in the late seventeenth century, with traditionalists (including the Church) opposing enclosure, and progressive landowners and farmers advocating it. In total there had been about 4½ million acres (nearly 2 million hectares) of open field. Around 150,000 ha were enclosed before 1760. Between 1761 and 1844 more than 2500 Enclosure Acts were passed, dealing with nearly a million hectares. The General Inclosure Act of 1845 dealt with most of

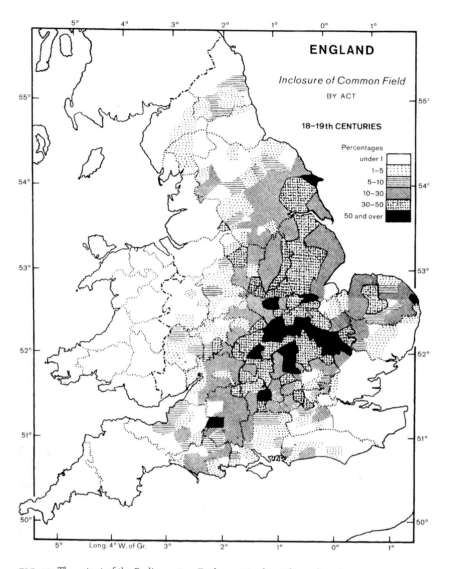

FIG 40. The extent of the Parliamentary Enclosures in the eighteenth and nineteenth centuries. From Gonner (1912). This map gives a rough idea of the distribution of 'ancient countryside' (white) and 'planned countryside' (shaded), but the two form a finer mosaic than is shown here, particularly in the areas with lighter shading. See also the maps in Barnes & Williamson (2006).

the remainder; another 80,000 ha were enclosed after this Act (Fig. 40). Over the same period, land in Ireland, on which the only hedges were along townland boundaries and fields of 50–100 ha were not unusual, was subdivided into fields of which few exceeded 6 ha.

Significant agricultural improvement could not have taken place without enclosure, but the whole process was disruptive of traditional rural life, displacing many country people to the expanding industrial towns, and it was expensive in legal fees. The old open fields, which often ran to a hundred hectares or more, needed to be subdivided and fenced. The enclosure commissioners tried to form square or squarish fields, usually 2–4 ha on smaller farms. The field might be up to 20 ha or so on large farms, but these large fields were often subdivided later. The new fields were commonly divided by hawthorn (*Crataegus monogyna*) hedges (Fig. 41), often with ash or elm saplings at intervals. These grew up into our familiar lowland landscape of hedged fields with hedgerow trees, but the hedges are nothing like as rich in species as the much older hedges of 'ancient countryside'.

In England, large areas of heath and upland common were also enclosed during the nineteenth century. Some of the heathland was taken into arable cultivation, most extensively in Norfolk and the chalk country, but much remained uncultivated, particularly on the poorer and more intractable soils, and in the uplands. There, the Enclosure divisions stand out by their straightness,

FIG 41. 'Planned countryside': former unfenced common divided into large fields with straight hedges on the chalk north of Lulworth, Dorset.

and in country with suitable walling-stone often take the form of dry-stone walls, not hedges.

The Highland Clearances: eighteenth and nineteenth centuries

The troubles of the Highlands did not end with the repressive measures taken by the government in the wake of the Jacobite risings. The 'Clearances' are an emotive subject for Scots. But they were seen as 'improvements', and moving with the times, by many landlords, who sometimes had little scruple for the people they displaced. In the 1730s, McLeod of McLeod is said to have been experimenting with sheep in Skye. Many on the mainland followed him. Tenants were 'encouraged' (often forcibly) to move off land judged suitable for sheep and were accommodated in poor crofts on the coast, where they were left to fend for themselves as best they could. The 'Year of the Sheep', 1792, is remembered as a notable year of enforced migration. The people were not only moved to the coast; many were put on ships to Nova Scotia, and many of these seem to have been Catholics. In 1807, Elizabeth Gordon, 19th Countess of Sutherland, wrote of her husband Lord Stafford (later Duke of Sutherland) 'he is seized as much as I am with the rage of improvements, and we both turn our attention with the greatest of energy to turnips.' The same year, 90 families were compelled to leave their crops in the ground and move all their possessions 30 km to the coast, and that was only beginning. The duke was obviously an ambitious, energetic and forward-thinking man, but he is still remembered by many Highlanders with a bitterness rivalling the hatred many Irish people still feel for Cromwell.

To the landlords, the clearances did not necessarily mean depopulation. Apart from fishing, the crofters' settlements on the coast were a source of cheap labour for the kelp industry, which was profitable at the time; indeed at least until 1820 emigration was actively discouraged. Tim Robinson (1986) has given a vivid account of the kelp industry in the Aran Islands (Co. Galway). Attitudes changed during the 1820s with the collapse of the demand for kelp, and many highlanders moved to the growing cities of Glasgow, Edinburgh and Dundee, or emigrated to Canada, Australia and New Zealand. The Highland potato famine from 1846 to 1857 caused renewed hardship, and gave further impetus to depopulation and (sometimes enforced) emigration.

During the nineteenth century, the Highlands became a playground for the rich – deer-stalking, grouse-shooting, and salmon and trout fishing. The Balmoral estate was bought by Prince Albert for Queen Victoria in 1852, adding a royal cachet to the Highlands. This brought some employment to the region, but did nothing for the crofters in the west.

The Irish potato famine 1845–1852

The story of the Irish potato famine – 'The Great Famine' – has been told in detail many times; all that is needed here is to record the bare facts. Largely owing to repressive laws against Catholics, only reformed in 1793, early-nineteenth-century Ireland was backward and poverty-stricken. There had been sporadic failures of the potato crop from the early eighteenth century, and more frequent failures during the early decades of the nineteenth. In the late 1830s and early 1840s many districts suffered severe losses. Maybe a more virulent form of potato blight (*Phytophthora infestans*) was introduced from America in 1844. At all events, the infection spread rapidly, and by mid-August 1845 had reached Holland, Belgium, northern France and southern England. On 13 September the *Gardeners' Chronicle* stopped 'the Press with very great regret to announce that the potato Murrain has unequivocally declared itself in Ireland.' The blight destroyed at least a third of the crop in 1845. Three-quarters of the crop was lost in 1846, and that winter a third of a million destitute people were employed in public works. Seed potatoes were scarce in 1847 and few were sown so, despite an average crop, hunger continued. Yields in 1848 were only two-thirds of normal.

After a period of very rapid growth, the population of Ireland was just over 8 million in 1841, and about half of that a century later. Probably about a million people died of starvation or disease as a direct result of the famine. The rest emigrated. Emigration from Ireland was nothing new. Between the defeat of Napoleon and the famine probably at least a million emigrated. By 1854, between 1½ and 2 million more had left Ireland. Emigration was to Britain, the United States, Canada and Australia. Already by 1850 the Irish made up a quarter of the population in Boston, New York, Philadelphia and Baltimore (Maryland). Emigration continued throughout the Victorian period and beyond, but now it was driven by the quest for greater opportunity and reward, rather than dire necessity.

The Irish nationalist John Mitchel wrote, 'The Almighty, indeed, sent the potato blight, but the English created the Famine.' He was sentenced to 14 years' transportation to Bermuda for his pains, but there was a lot of truth in what he wrote. The Famine, and all that had gone before, left Ireland a bruised, backward and rather isolated country, with very little woodland, but wonderful bogs (now sadly depleted). In the west, evidence of depopulation remains very obvious on the ground. I remember Dr Harry Godwin (as he then was) returning from the 1949 International Phytogeographical Excursion in Ireland, saying that his pervading impression was of a 'cow-blasted landscape'. In the last half-century things have changed – most would say for the better.

Industrialisation

The industrial revolution went hand-in-hand with the agricultural revolution. Growing industrial towns absorbed the population displaced from the country by enclosure and agricultural improvement, but at the same time provided the market for the expanding productivity of farming in England, where enclosure and the drive for 'improvement' combined to encourage widespread ploughing-up of pasture to increase the area of arable. In a manner of speaking, this enabled England (and lowland Scotland) to absorb (or export) its population problems. The British Empire provided both space for surplus population, and a source of cheap food. Industry put farm machinery within reach of even small farmers, and increasingly made chemical fertilisers available to supplement the practices of traditional husbandry. Steam power was harnessed to agricultural use with traction engines, which beside their use for transport could power threshing machines. With the invention of the internal combustion engine, mechanical assistance in farming was to become all-pervasive.

The pioneer industrialists were proud of their achievements, and often lived within sight of their mills and factories. But mass industrialisation created a drab and brutal environment, and unhealthy living conditions for the mass of people in the growing industrial towns and the coal-mining areas that fed their industries. Many at the time (and later) deplored industrialisation and its consequences. Anna Seward (1747–1809) in her sonnet *to Colebrookdale* wrote of the desecration of a beautiful valley by '... umbered fires on all thy hills ... with columns large, of black sulphureous smoke, that spread their veils, like funeral crape upon the sylvan robe, of thy romantic rocks.' And almost two centuries later Hoskins (1955) wrote with feeling about Nottingham's transformation from 'one of the most beautiful towns in England' to 'a squalid mess', through a combination of industrial growth, enclosure and self-centred greed. These things are the seamy side of Britain's historic rise to economic pre-eminence during the nineteenth century.

The industrial revolution had mainly indirect effects on our landscape and vegetation. One direct effect was atmospheric pollution, especially smoke and sulphur dioxide pollution around the major industrial towns – and as most people still used coal fires to keep warm at home, small towns were not immune. In my schooldays, a day in London left your shirt collar and cuffs with a dark ring of grime, and you could smell sulphur dioxide at the main-line railway stations. When I first knew the industrial north of England, the towns looked like a wall-to-wall Lowry painting, and the buildings were black. That all changed with cleaner air; the incidence of bronchitis has fallen, and epiphytic lichens and mosses now grow in places where they have not been seen for a century.

Canals and railways

The Exeter canal, built in 1564–7 to bypass the weirs on the Exe and enable barges to reach the port of Exeter, was the first modern canal in Britain; it had locks (an innovation) and a towpath. James Brindley's canal (1760) built for the Duke of Bridgewater to carry coal from the coal-mines at Worsley to Manchester was the pioneer that others followed. The eighteenth-century canals were impressive engineering achievements, with locks, cuttings and tunnels. Brindley's canal included an aqueduct over the River Irwell, and Telford's Pontcysyllte aqueduct carrying the Ellesmere Canal over the River Dee is a striking sight from the A5 road near the Welsh border. The canals vastly cheapened the transport of heavy materials over long distances (and incidentally brought stretches of water to country that formerly lacked them). The railways completed the job the canals had begun, of making cheap coal widely accessible. This ended most people's reliance on wood as the staple source of energy. Coal was convenient, compact, generally available and less expensive than the alternatives. With hindsight, one could say that it had the deeper psychological effect of severing our tie to the land and the seasons. If you ran out of coal, all you had to do was to order (or mine) more. Ocean transport by steamship had the same effect with food and other commodities. There is nothing wrong in that, as long as you recognise that all resources are finite.

THE TWENTIETH CENTURY

The last was an eventful century: two world wars and the rise and fall of the Soviet Union, the development of powered flight, the general adoption of electricity for lighting and power, and of internal-combustion-engined vehicles for transport on land (and sea), the development of telephones, radio and television, the development of information technology and the Internet, and the establishment of the European Community, to list only a few of the more notable. Perhaps it is surprising that the twentieth century did not have *more* effect on our landscape and flora and fauna.

Agricultural pressures in the twentieth century

In 1860, C. C. Babington, Professor of Botany at Cambridge, wrote, 'Until recently (within 60 years) most of the chalk district was open and covered with a beautiful coating of turf, profusely decorated with *Anemone Pulsatilla* [pasqueflower] and *Astragalus Hypoglottis* [purple milk-vetch], and other interesting plants. It is now converted into arable land.' This was the period of rapid population growth in the industrial cities, and enclosure and agricultural improvement were in full swing. He might have added that many ancient woods went the same way as the chalk turf, but many woods found a new role as fox covert and pheasant preserve. After Babington's time, the countryside regained a degree of stability, and in general did not change greatly over the decades until 1939. Britain was dependent on imports for a substantial part of its food in the 1930s. During the Second World War, the German U-boat campaign exploited that dependence, and although nobody went hungry, rationing was severe and there was little choice and few luxuries – and both sides devoted a substantial part of their war effort to 'the Battle of the Atlantic'. After the war it was deliberate government policy to make Britain self-sufficient in food. Ploughing was encouraged; in particular, large areas of the flatter chalk country were converted to arable. Looking at air photographs taken in the 1940s, with their wealth of archaeological detail still preserved beneath the old turf, it is hard not to regret that this phase of post-war ploughing was not done more sensitively. For decades after the war, the Ministry of Agriculture, Fisheries and Food was King, to the intense frustration of other interests in the land!

Many of the most notable twentieth-century changes in the British and Irish landscape had either natural or economic causes. The collapse of the rabbit population through myxomatosis was probably started by a deliberate introduction (Chapter 11), but would very likely have happened sooner or later anyway, once the virus was in Europe. Dutch elm disease has to be seen as a natural phenomenon, and the outbreak in the 1960s was certainly not the first (Chapters 2 and 5). In farming, ready availability of artificial fertilisers and pressure to maximise yields spelt the death of the traditional species-rich permanent grasslands (Chapter 9) and that, combined with the newly available selective weedkillers, largely eliminated arable weeds (Chapter 10). New breeds, economic pressure to maximise yield, and the ready availability of pelleted livestock feeds, combined to change farmers' attitude to livestock, and to a decline in casual grazing, so many heaths and rough grasslands went ungrazed. And there was no longer an economic demand for the products of the traditional woods, with the result that the burden of managing (and conserving) these traditional elements in our landscape has fallen on the statutory conservation and amenity bodies, and on private organisations such as the National Trust, the Royal Society for the Protection of Birds and the County Wildlife Trusts.

Forestry

Forestry is not specifically a twentieth-century phenomenon; many landowners in the eighteenth and nineteenth centuries established plantations of various sorts. Plantation forestry has nothing to do with woodmanship as traditionally practised. It is more akin to growing an arable crop. Woodmanship had to do with managing a renewable resource, as has much Continental forestry. The Forestry Commission was established in 1919, with the aim of making good the timber felled during the 1914–18 war, and increasing the stock of timber to provide a strategic reserve. After the 1939–45 war planting was continued on a larger scale, and subsidies were made available for planting trees. Quick-growing conifers were the preferred choice, particularly Sitka spruce and Douglas fir (*Pseudotsuga menziesii*) – or, on soils too poor for those species, Scots or lodgepole pine (*Pinus contorta*). The conifer plantations have their place in the landscape, like other crops, provided they are economic. But the foresters' remit was to grow timber, and the old deciduous woods did not accord with that primary aim.

The subsidies that were intended to promote the provision of a strategic reserve of timber had by the 1980s outlived their original aim, and become a burden of no benefit to the taxpayer, either economic or strategic. However, they remained potentially a profitable subsidised investment, and so created a powerful vested interest in their continuance. This worked against the interests of nature conservation. Matters came to a head over proposals to drain and plant wide areas of the Flow Country of Caithness and Sutherland (Fig. 42). The foresters believed they were working for progress and improvement (as did

FIG 42. Clash of interests: 'improvement' or 'vandalism'? Peatland prepared for forestry, Flow Country near Forsinard, Sutherland, September 1989.

progressive landowners in earlier centuries). A nearby hotel saw forestry as an unwanted intrusion into a wild landscape, which they and their guests valued. Landscape and nature-conservation interests were opposed to the replacement of a near-unique landscape and habitat by yet more conifer plantations, especially as the real economic value of the planting was questionable. It made a *cause célèbre* at the time, and had political repercussions including the break-up of the then all-UK Nature Conservancy Council.

Many of the general public objected to conifers as such when their true objection was to plantation forestry single-mindedly focused on timber production – and they, like many foresters, failed to understand the nature of the old deciduous woods they *did* like. Happily, understanding of traditional woods, their value to amenity and biodiversity, and their role as historical documents, has increased. Many foresters are interested in woods and their management, of whatever kind. The Forestry Commission now has a broader remit to include amenity. The subsidies that caused so much damage for no useful return to the taxpayer have gone. But the tendency to confuse 'woodland' and 'forest' (in the forester's sense) lingers on.

The growth of ecology

The German biologist Häckel coined the word *ecology* in 1866 for the study of the interactions of organisms with their physical environment and with other organisms, but the development of the science of ecology to its present prominence has been largely a twentieth-century phenomenon. Early studies in ecology were largely of individual plant and animal populations and communities (Chapter 4), but from the 1940s and 1950s interest expanded to the worldwide scale, with the writings of Charles Elton, Eugene and Howard Odum, Rachel Carson and many others. Public awareness of ecological issues was alerted by the finding of pesticide residues and industrial chemicals such as DDT, dioxins and CFCs worldwide, and in places and organisms far from their intended use. A good deal of ecology is scattered through the pages of this book. Ecology can sometimes set firm limits, and make categorical predictions, but often (like weather forecasting) it can only express probabilities. Nevertheless, if we ignore such predictions and they turn out to be right, we have only ourselves to blame. The 'muck and mystery' brigade, masquerading under the banner of 'ecology', do the science of ecology and us all a disservice. Ecology is too serious for that. We ignore it at our peril.

CHAPTER 4

Plant Communities

P LANTS IN THE WILD seldom grow in isolation; they interact with other plants and with their physical environment in all manner of ways. Plant ecology as a systematic study is only a little over a century old. In the first half of the twentieth century, plant-ecological thinking in the English-speaking countries was heavily influenced by the ideas of the great American ecologist Frederic E. Clements (1874–1945). Clements (1916, 1936) envisaged vegetation – plant communities – in terms of 'climatic climaxes', stable communities in equilibrium with the climate of particular regions (e.g. temperate deciduous forests, boreal coniferous forests, tundra), and 'succession' towards these climaxes. Succession might begin from many starting points (e.g. an abandoned field, bare rock, a lake), and he called the course of succession in a particular instance a 'sere' – a 'xerosere' from bare dry ground, a 'hydrosere' from open water. Clements likened the climax to 'a complex organism inseparably connected with its climate and often continental in extent', and developed an elaborate intellectual framework to describe it – and a terminology to match. Another American ecologist, Henry A. Gleason (1882–1975), who had begun his career in the Clementsian tradition, increasingly found that tradition both frustrating and inadequate, and argued for 'the individualistic concept of the plant association', seeing the plant community as simply the sum of the tolerances of the individual species (Gleason 1917, 1939). Both points of view have merits. Gleason's ideas made little impact at the time, but were immensely influential in the later development of ecology.

It was perhaps natural that North American ecologists, in a continent relatively newly colonised by Europeans, should have seen vegetation in terms of climatic climaxes and succession, but American ideas were influential in

Britain too. Tansley (1939) largely uses Clements's terminology and successional concepts. A notable early contribution from Europe was the elucidation by the German agricultural botanist and peatland ecologist Carl Albert Weber (1856–1931) of the succession leading to the raised bogs of the North German Plain (1902). However, in the long-settled landscape of continental Europe the emphasis tended to fall more on description and classification of plant communities. This work became particularly associated with the name of the Swiss ecologist Josias Braun-Blanquet (1884–1980) – although Braun-Blanquet's own *Plant Sociology* (1927, 1932 and many subsequent editions) is a wide-ranging textbook of all aspects of plant ecology. Over the decades, Continental 'phytosociologists' have built up a body of description and classification from which we can draw many useful insights into our own vegetation.

A less controversial division exists between *autecology*, the study of the responses of individual plant species to their environment, and *synecology*, which deals with the composition and functioning of vegetation – of entire plant communities – whatever view one takes of their nature.

PLANT LIFE-FORMS AND LIFE STRATEGIES

Plants vary greatly in shape and functional architecture. The Danish botanist Christen C. Raunkiaer (1860–1938) a century ago devised a scheme of 'life-forms' based on the position of the buds by which the plant passes the unfavourable season – winter in northern Europe, summer in arid climates (Raunkiaer 1907, 1934). In our climate, most plants start growth in spring from winter buds at about the soil surface – they are *hemicryptophytes* in Raunkiaer's terminology (H in Fig. 43) – either with a rosette at the base or with evenly leafy stems. Some plants have their perennating buds on underground rhizomes (Grh), or in corms or bulbs (Gb). These *geophytes* are commoner in Mediterranean and semi-arid climates where the soil surface becomes not only dry but intolerably hot in the summer sun. Nevertheless, plants as diverse as bluebell, ramsons (*Allium ursinum*), bracken and many of our orchids are geophytes. Subshrubs, *chamaephytes*, have their perennating buds at the tips of the shoots, up to about 25 cm above the soil surface; they may be entirely woody (Chw) like the heathers, or a proportion of the shoots may die back in the unfavourable season (Chh). They are prominent in the Mediterranean region where the ground surface becomes too hot in summer, and in the Arctic where the soil is frozen in winter and growth can begin when the surface is thawed and shoots are warmed by the first spring sunshine. *Therophytes* evade the unfavourable season by perennating as seeds; we have many annuals

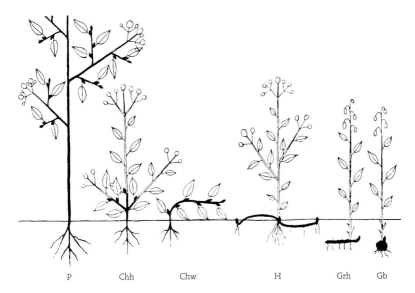

FIG 43. Raunkiaer life-forms. Light outlines show the parts of the plant that die in the unfavourable season, solid black the parts that remain alive. The diagrams represent (left to right) a *phanerophyte* (P), herbaceous (Chh) and woody *chamaephytes* (Chw), a *hemicryptophyte* (H), and rhizomatous (Grh) and bulbous *geophytes* (Gb). An important life-form omitted from the diagram is the *therophytes*, which pass the unfavourable season as seeds; further life-forms are *helophytes* and *hydrophytes*, with perennating buds under mud and water, respectively. Simplified from Raunkiaer (1934).

in our climate, but they become more prominent in the Mediterranean. Shrubs and trees with their buds more than 25 cm above the ground are *phanerophytes* (P); shrubs less than 2 m high may be distinguished as *nanophanerophytes*. The humid tropics, with no unfavourable season, have a preponderance of phanerophytes. Plants of all life-forms have diversified in the course of evolution. Subtropical and temperate shrubs and trees evolved protected perennating buds to survive summer drought or winter cold, and many went beyond that to become deciduous, losing their leaves in winter – or in the dry season. Most trees at our latitude are deciduous, but in humid temperate climates only a little warmer than ours many trees are evergreen with laurel-like leaves, as in Madeira and the Canary Islands, and New Zealand.

Plants interact also with animals, and a very important innovation was the evolution of the grasses, which probably took place hand-in-hand with the evolution of the large grazing mammals in the early Cenozoic era. The essential feature of grass architecture in relation to grazing is that the growing-points of the shoot, and the growing-points of the individual leaves, are at the base, not at the apex as in most other plants. Various other groups of monocotyledons have the same growth-habit as the grasses, notably the sedges (Cyperaceae).

Raunkiaer's life-form concept is illuminating, but tells only a part of the story. If habitats are grouped in relation to disturbance and 'stress' (anything that limits productivity, e.g. drought, salt, lack of nutrients or light), there are four possibilities. Stable but unstressed habitats are inevitably competitive, bearing vigorous closed vegetation. Stable but stressed habitats are tenable to specialised species that can tolerate the unfavourable conditions. Disturbed but otherwise favourable habitats are open to species that can establish, grow and set seed quickly. No plant can cope with the fourth possibility, the habitat that is both disturbed and unfavourable. Professor Philip Grime (1979) suggested that plants could be seen in terms of three 'strategies', *competitors*, *stress-tolerators* and *ruderals*.

This threefold C-S-R model has the merit that it can be set out as a triangular diagram like the conventional diagrams for the mechanical composition of soils (Chapter 1). The sides of the C-S-R triangle can usefully be thought of as the inverse of the opposite vertices – the left-hand side as (at least potentially) 'productive', limited by neither water, nutrients or light, the right-hand side as 'stable', and yield along the base as limited, by either habitat constraints or disturbance. Figure 44 shows the most-frequent occurrence of some species from the Sheffield region in vegetation types classified by the C-S-R strategies of their component species. Stinging nettle (*Urtica dioica*), meadowsweet (*Filipendula ulmaria*), rosebay willowherb (*Chamerion angustifolium*) and bracken (*Pteridium*

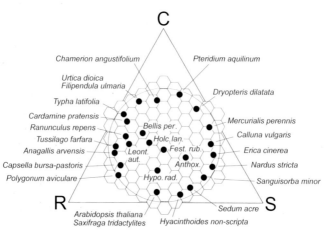

FIG 44. C-S-R diagram for some common species from the Sheffield region. For further explanation see text. Based on Grime, Hodgson & Hunt (2007).

aquilinum) grow in communities predominantly of C-strategists. Bell heather (*Erica cinerea*), mat-grass (*Nardus stricta*), salad burnet (*Sanguisorba minor*), biting stonecrop (*Sedum acre*) and bluebell (*Hyacinthoides non-scripta*) characterise vegetation of habitats imposing various kind of 'stress'. Scarlet pimpernel (*Anagallis arvensis*), shepherd's-purse (*Capsella bursa-pastoris*) and knotgrass (*Polygonum aviculare*) are species of weed and ruderal communities. Bulrush (*Typha latifolia*), cuckooflower (*Cardamine pratensis*), creeping buttercup (*Ranunculus repens*) and colt's-foot (*Tussilago farfara*) compromise between C and R strategies, and dog's mercury (*Mercurialis perennis*) and heather (*Calluna vulgaris*) are dominants of communities that are stable but stressed in different ways. Thale-cress (*Arabidopsis thaliana*) and rue-leaved saxifrage (*Saxifraga tridactylites*) are small annuals of seasonally-desiccated wall-tops and similar places. Common grassland species occupy the centre of the diagram – here, red fescue (*Festuca rubra*), Yorkshire-fog (*Holcus lanatus*), sweet vernal-grass (*Anthoxanthum odoratum*), daisy (*Bellis perennis*), cat's-ear (*Hypochaeris radicata*) and autumn hawkbit (*Leontodon autumnalis*).

The C-S-R concept is elegant, and a productive framework for thought, but, like Raunkiaer's, it tells only a part of the story and should not be pressed beyond its limitations. In particular, stress is many-faceted, and 'one man's meat is another's poison.'

A related concept comes from studies of growth. Both individual organisms and populations tend to follow S-shaped (sigmoid) growth curves. Growth begins slowly, then gathers pace and passes though a rapid phase until it slows again with the approach of the carrying capacity of the habitat (or adulthood). Often the growth curve roughly fits a logistic relationship, of which one formulation is the equation $dN/dt = rN(1 - N/K)$, where dN/dt is the rate of change of N with time, N is the number of individuals at time t, r is the *intrinsic rate of increase*, and K is the *carrying capacity* of the habitat (Fig. 45). Clearly, starting with an empty

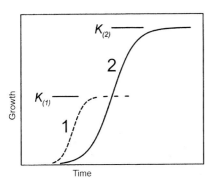

FIG 45. Two logistic growth curves. Species 1 will win in the short term because of its greater intrinsic rate of increase (*r*), but will be outcompeted in the long term by the slower-growing species 2. Weedy species are adapted to disturbed habitats in which growth-rate and pre-emption of resources are paramount: they are *r*-selected. For dominant species of stable communities, selection pressure is for share of the carrying capacity (*K*) of the habitat, and retention of nutrient capital: they are *K*-selected.

habitat, the species with the highest intrinsic rate of increase will win in the short term, but as the carrying capacity is approached competition becomes progressively more important in determining the fraction of the habitat a particular species occupies in the long term. Species adapted to open habitats can be thought of as *r-selected*: they are primarily adapted to *acquire* resources (Grime's 'ruderals', many of them annuals, biennials or short-lived perennials, belong here). Species adapted to stable, closed habitats are *K-selected*: their primary adaptation is to *retain* resources (generally perennials, these embrace Grime's 'competitors' and 'stress-tolerators'). Evidently there is a trade-off between these two adaptive patterns; compromise is possible, but no species can have the best of both worlds.

HABITAT TOLERANCES AND PREFERENCES OF SPECIES, ECOLOGICAL NICHES AND COMPETITION

The basic needs of plants for water and nutrients from the soil, and light and a tolerable microclimate from their above-ground environment, have already been sketched out in Chapter 1. No two plant species respond to the factors of the habitat in exactly the same way, which is one reason why so many species can grow together in the same community. Each has its own *ecological niche*, which may be likened in a general way to its 'job' in the community. It is often useful to think of a plant's ecological niche in terms of its preferred place along gradients of habitat factors such as soil pH, moisture, light and temperature. Professor Heinz Ellenberg (1988) gave scores on a 10-point scale for seven habitat factors: light, temperature, continentality, moisture, soil-reaction (pH), nitrogen and salinity, for all the species in the central European flora, and Dr Mark Hill and his colleagues have adapted these (and summarised a great deal of other information) for the British and Irish flora (Hill *et al.* 2004). Plants compete with one another for resources both above and below ground, and generally show a broader tolerance of habitat factors in laboratory experiments (or in gardens), in which they are free of competition, than they do in the wild; the *potential niche* is generally broader than the *realised niche*. It is in fact uncommon for plants to be limited in a simple way by a single environmental factor – if only because most factors vary so much, and interact. Nevertheless, it is often possible to infer likely governing factors from the distribution limits of species. Some examples were given in Chapter 1.

Two salient points may be made about plant competition. One is that in single-species stands seedling densities are often immensely greater than can ever come to maturity. Growth is invariably accompanied by self-thinning, and yield of the stand is in general proportional to $(1/\text{density})^{-3/2}$, flattening off as maximum size is approached. This relationship also holds in the thinning tables used by foresters to maximise yield from plantations. The other point is that if individuals of two species are competing at a range of densities, three possibilities can be envisaged. If the two species are competing for the same mix of resources, the one with the more vigorous growth (and uptake) will over generations (or seasons) progressively oust the other. If the two species are *not* competing for exactly the same resources, they will progressively converge on a stable equilibrium between them. This must be a common situation in species-rich communities. The third possibility is that one or both species is actively inimical to the other (e.g. by producing toxic root exudates, or slow-to-decay leaf litter), and in such a case the weaker competitor will be eliminated at an accelerating rate. This sort of process is probably often implicated in plant succession – and (arguably) underlies the success of *Sphagnum*.

PLANT COMMUNITIES

The structure of plant communities

Plant communities, except the very simplest, are not just haphazard aggregations of individuals; they have at least some degree of *structure*. This is most obviously shown by *layering*. A population of weeds is a one-layered community, and so are many grasslands. Some grasslands and many heaths are two-layered, with a *field-layer* of the dominant flowering plants, and a *ground-layer* of mosses, liverworts and lichens. Most forests have three or more layers; an oakwood often has a *tree-layer*, a *shrub-layer* of (mainly) hazel, a herbaceous field-layer (of e.g. primrose, wood anemone), and a *ground-layer* of woodland mosses (Fig. 46). The lower layers must be tolerant of the shade cast by the layers above, but benefit from the shelter they provide. Each layer generally has a *dominant* species; sometimes two or more species share dominance, but almost always the bulk of the herbage is made up of relatively few species. Plants may also be layered below ground. T. W. Woodhead (1906) described how in the oakwoods around Huddersfield creeping soft-grass (*Holcus mollis*) roots in the surface layers of the soil, the rhizomes of bracken are deeper, and the bulbs of bluebell are deepest of all. In chalk and limestone heaths, and in the Burren grasslands, calcifuges (e.g. bell heather) are rooted in the leached surface layers, while the calcicoles root in the lime-rich soil underneath (Chapter 11). It is tempting to suppose that in richly diverse communities the species must be at least to some degree complementary

FIG 46. A traditional coppice-with-standards wood near Sixpenny Handley, Dorset, April 1974. This is a three-layered community, with a rather open tree-layer of large oaks, a shrub-layer of hazel, which would have been regularly coppiced, and a field-layer of wood anemone (*Anemone nemorosa*).

to one another, but in species-rich road verges near Bibury in Gloucestershire monitored for almost half a century by the late Professor Arthur Willis without showing significant change, Thompson and his colleagues (2010) could find little evidence for this. The resolution of this enigma awaits new insights.

Plants are never perfectly evenly distributed on the ground (except in plantations or orchards), neither are they distributed truly at random. Some degree of clumping is almost always present. This may arise from irregularities of seed distribution, from underlying patterns in the soil or bedrock, or from the growth-form of the dominant or other prominent plants (Fig. 47). Bracken can form striking circles in the course of invading grassland; the disturbed patch (rabbit burrow or whatever) where the spores germinated and growth began can often be made out in the centre. White clover (*Trifolium repens*), another species that spreads by rhizome growth, often forms circles in newly sown lawns in a similar way (Fig. 48). As the community ages, the circles break up into an all-over pattern, but patchiness persists at a scale related to the rhizome systems of the plants. Patchiness is also created as individual dominant trees and shrubs age, die and are replaced. As there are many causes of pattern and patchiness, some degree of patchiness is virtually universal, and appears at every scale.

FIG 47. Daisies (*Bellis perennis*) in lawns. (Left) A relatively newly sown lawn, probably 3–5 years old, with the daisies still conspicuously clumped around the points of initial colonisation. (Right) An old-established lawn; the initial colonisation pattern has disappeared, but an irregular finer-scale pattern persists, related to the size of the daisy plants. Exeter University campus, May 1976.

FIG 48. (Left) Bracken ring on a hillside near Aberayron, Cardiganshire, September 1956; the rhizomes have grown radially from the point of establishment, and in this case have died away behind the vigorous invading front. (Right) Rings of white clover in a recently seeded lawn; clover is still present in the interior of the rings, but is less vigorous. Exeter University campus, June 1973.

Some numerical relationships: how many species can we expect in a given area?

One of the questions asked by the early plant ecologists was 'what is the *minimal area* of sample we must take for it to be representative of the community?' It became customary to plot species–area curves, which were found to be initially steep at small areas, but flattened off as larger and larger areas were examined. The degree of flattening that satisfied the investigator was taken as the minimal area. If species-number is plotted against area on a *logarithmic* (as against a linear) scale, it was found that the flattening was illusory, and that the number of species generally continued to increase far beyond any practicable area for routine quadrat samples. Species recording from squares of the Ordnance Survey national grid ('square-bashing') renewed interest in species–area relations, but at a larger scale. How many species should be expected from a complete survey of a 10 km (or 2 km) square? In preparation for a *Flora of Hertfordshire*, John Dony (1963) surveyed some available quadrat data, along with the number of species recorded from various larger defined areas, up to the total flora of Britain and Ireland. Plotted on a graph of log(species number) against log(area), all the (somewhat variable) sets of data lie remarkably close to a straight line (Fig. 49). This gave him an estimate of the number of species to be expected in a 2 × 2 km square ('tetrad'). Species–area relationships have a voluminous literature,

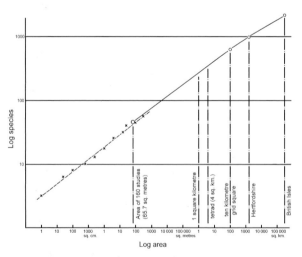

FIG 49. Plant species numbers from a wide range of defined areas, plotted on a logarithmic scale on both axes. Included are quadrat data, starting at 1 cm², from Hopkins (1955, crosses and broken line), the median number of species recorded by Dony in 160 samples (66.7 m²) from various habitats, the number of species recorded for the BSBI maps scheme (1955–62) in Hertfordshire (1631 km²) and the British Isles (310,600 km²). Separate lines link the higher points on the graph. Slightly simplified from Dony (1963)

of which the consensus is that, overall, the double logarithmic relationship generally gives the best approximation to reality, even though data from particular places can depart quite widely from it (Connor & McCoy 1979). This implies that, generally, 'minimal area' as an intrinsic property of the community is an illusion, and that we are at liberty to choose any scale we like to describe vegetation – and that our human scale, a metre or two, is as valid as any other. We may adjust the scale on which we look at vegetation upwards or downwards when dealing with organisms very different in scale from ourselves, such as forest trees, or bryophytes and lichens.

The distribution of commonness and rarity

It is characteristic of many communities of plants and animals that they are made up of relatively few common species; and large numbers of species that are less common or rare. The maps in the *New Atlas of the British and Irish Flora* illustrate this in a general way. It is instructive to consider the species that are most-nearly ubiquitous in Britain and Ireland (Table 3). Predictably, the list contains many species abundant in ordinary farm grassland: the common pasture grasses, buttercups (Robert Browning's 'the little children's dower'), sorrel, tormentil, several common clovers and vetches, self-heal, plantains, common daisy, yarrow and a clutch of common yellow composites. (St Patrick might have had several clovers at his feet as a symbol of the Trinity, and white clover (*seamaìr bhán*) is as likely as any other; the Irish diminutive *seamróg* (shamrock) could have referred to any small clover – and the legend does not appear in written form until many centuries after his death.) The table also contains a group of plants of wetter places, such as lesser spearwort, meadowsweet, silverweed, cuckooflower, wild angelica, marsh bedstraw, marsh thistle, colt's-foot, several common rushes, glaucous sedge and yellow iris, a versatile group with their headquarters in shady places, such as bracken, male and broad-buckler ferns, common dog-violet, primrose, brambles, broad-leaved willowherb, herb-robert, ivy, cleavers and honeysuckle, and some common weeds such as stinging nettle, chickweed, docks, dandelions and annual meadow-grass. The 'near-misses' that just fail to make 85% of the possible 10 km squares tell much the same story. Notably few woody plants appear in the list – common sallow, hawthorn and common gorse. The only near-ubiquitous tree is the 'non-native' sycamore, which has been with us for centuries. We shall now never know whether the odd winged seed has blown across the Channel and established with us, and sycamore could surely now be granted UK and Irish citizenship for its long residence and impeccable EU credentials! None of our long-established forest trees makes 85% of the grid squares or more. Those that come nearest are listed in the last part of Table 3.

TABLE 3. The most-nearly ubiquitous plants in Britain and Ireland. Species are listed in the order of the *New Atlas of the British and Irish Flora*. The numerals show the number of 10 km squares from which the species was recorded in 1987–99 in Britain/Ireland.

Species	No. of 10 km squares	Species	No. of 10 km squares
IN > 85% OF 10 KM SQUARES IN BRITAIN AND IRELAND			
Field horsetail *Equisetum arvense*	2564/874	[Sycamore *Acer pseudoplatanus*]	2495/905
Bracken *Pteridium aquilinum*	2581/933	Herb-robert *Geranium robertianum*	2450/980
Male-fern *Dryopteris filix-mas*	2484/877	Ivy *Hedera helix*	2435/935
Broad buckler-fern *Dryopteris dilatata*	2592/886	Wild angelica *Angelica sylvestris*	2646/938
Meadow buttercup *Ranunculus acris*	2739/934	Hogweed *Heracleum sphondylium*	2624/911
Creeping buttercup *Ranunculus repens*	2738/931	Self-heal *Prunella vulgaris*	2742/931
Lesser spearwort *Ranunculus flammula*	2450/893	Greater plantain *Plantago major*	2702/951
Stinging nettle *Urtica dioica*	2721/963	Ribwort plantain *Plantago lanceolata*	2784/967
Common chickweed *Stellaria media*	2671/896	Thyme-leaved speedwell *Veronica serpyllifolia*	2512/804
Common mouse-ear *Cerastium fontanum*	2784/945	Germander speedwell *Veronica chamaedrys*	2539/926
Procumbent pearlwort *Sagina procumbens*	2726/885	Marsh bedstraw *Galium palustre*	2512/892
Sheep's sorrel *Rumex acetosella*	2590/786	Cleavers *Galium aparine*	2584/927
Common sorrel *Rumex acetosa*	2763/934	Honeysuckle *Lonicera periclymenum*	2491/893
Curled dock *Rumex crispus*	2577/983	Spear thistle *Cirsium vulgare*	2746/944
Common dog-violet *Viola riviniana*	2690/930	Marsh thistle *Cirsium palustre*	2614/929
Common sallow *Salix cinerea* s.l.	2462/904	Field thistle *Cirsium arvense*	2682/932
Cuckooflower *Cardamine pratensis*	2635/894	Common knapweed *Centaurea nigra*	2575/943
Primrose *Primula vulgaris*	2495/879	Cat's-ear *Hypochaeris radicata*	2661/940
Meadowsweet *Filipendula ulmaria*	2605/948	Autumn hawkbit *Leontodon autumnalis*	2705/874
Brambles *Rubus fruticosus* s.l.	2485/945	Prickly sow-thistle *Sonchus asper*	2465/911
Silverweed *Potentilla anserina*	2571/953	Dandelion *Taraxacum officinale*	2536/939
Tormentil *Potentilla erecta*	2558/914	Common daisy *Bellis perennis*	2762/962
Hawthorn *Crataegus monogyna*	2380/931	Yarrow *Achillea millefolium*	2731/922
Bird's-foot trefoil *Lotus corniculatus*	2538/899	Ragwort *Senecio jacobaea*	2662/949
Tufted vetch *Vicia cracca*	2512/887	Groundsel *Senecio vulgaris*	2426/876
Bush vetch *Vicia sepium*	2418/903	Colt's-foot *Tussilago farfara*	2471/812
Meadow vetchling *Lathyrus pratensis*	2538/899	Toad rush *Juncus bufonius*	2608/862
White clover *Trifolium repens*	2777/955	Jointed rush *Juncus articulatus*	2654/902
Lesser trefoil *Trifolium dubium*	2378/874	Soft-rush *Juncus effusus*	2729/947
Red clover *Trifolium pratense*	2652/950	Glaucous sedge *Carex flacca*	2512/898
Common gorse *Ulex europaeus*	2415/927	Red fescue *Festuca rubra* s.l.	2748/936
Broad-leaved willowherb *Epilobium montanum*	2477/814	Perennial rye-grass *Lolium perenne*	2640/923

continued overleaf

IN > 85% OF 10 KM SQUARES IN BRITAIN AND IRELAND

Crested dog's-tail *Cynosurus cristatus*	2670/917	Tufted hair-grass *Deschampsia cespitosa*	2576/764
Annual meadow-grass *Poa annua*	2737/953	Yorkshire-fog *Holcus lanatus*	2767/952
Smooth meadow-grass *Poa pratensis* s.l.	2647/822	Sweet vernal-grass *Anthoxanthum odoratum*	2737/933
Rough meadow-grass *Poa trivialis*	2579/806	Common bent *Agrostis capillaris*	2656/832
Cock's-foot *Dactylis glomerata*	2642/962	Creeping bent *Agrostis stolonifera*	2730/942
False oat-grass *Arrhenatherum elatius*	2594/904	Yellow iris *Iris pseudacorus*	2416/917

'NEAR MISSES' (SELECTED)

Marsh-marigold *Caltha palustris*	2496/725	Devil's-bit scabious *Succisa pratensis*	2395/869
Lesser celandine *Ranunculus ficaria*	2478/732	Nipplewort *Lapsana communis*	2306/826
Sticky mouse-ear *Cerastium glomeratum*	2433/744	Field wood-rush *Luzula campestris*	2510/742
Knotgrasses *Polygonum aviculare* s.l.	2435/831	Mouse-ear hawkweed *Pilosella officinarum*	2443/792
Wavy bitter-cress *Cardamine flexuosa*	2433/815	Carnation sedge *Carex panicea*	2220/829
Rosebay willowherb *Chamenerion angustifolium*	2488/635	Common sedge *Carex nigra*	2322/790
Fairy flax *Linum catharticum*	2276/763	Sheep's fescue *Festuca ovina*	2507/546
Pignut *Conopodium majus*	2349/686	Floating sweet-grass *Glyceria fluitans*	2489/780
Cow parsley *Anthriscus sylvestris*	2341/758	Meadow foxtail *Alopecurus pratensis*	2225/670
Hedge woundwort *Stachys sylvatica*	2360/721	Marsh foxtail *Alopecurus geniculatus*	2366/718
Foxglove *Digitalis purpurea*	2460/747	Timothy *Phleum pratense* s.l.	2257/601
Brooklime *Veronica beccabunga*	2185/831	Purple moor-grass *Molinia caerulea*	1970/839
Heath bedstraw *Galium saxatile*	2309/684		

COMMON TREES AND SHRUBS IN < 85% OF SQUARES (IN DESCENDING ORDER OF ABUNDANCE: TOTAL 10 KM SQUARES IN PARENTHESES)

Ash *Fraxinus excelsior*	2371/903 (3274)	Goat willow *Salix caprea*	2241/515 (2756)
Alder *Alnus glutinosa*	2383/858 (3241)	Beech *Fagus sylvatica*	2258/705 (2723)
Elder *Sambucus nigra*	2328/858 (3186)	Pedunculate oak *Quercus robur*	2138/532 (2670)
Hazel *Corylus avellana*	2341/820 (3161)	Wych elm *Ulmus glabra*	2135/941 (2626)
Blackthorn *Prunus spinosa*	2209/872 (3081)	Silver birch *Betula pendula*	2064/227 (2291)
Rowan *Sorbus aucuparia*	2383/693 (3076)	Aspen *Populus tremula*	1914/270 (2184)
Holly *Ilex aquifolium*	2218/817 (3035)	Sessile oak *Quercus petraea*	1499/444 (1943)
Downy birch *Betula pubescens*	2197/763 (2960)		

However, the distribution of commonness and rarity depends greatly on how you view the data. The species lists from the 3859 10 km squares of the *Atlas* do not accurately reflect the abundance of different plants, because a widely distributed but a minor ingredient of a common vegetation type (e.g. common mouse-ear) may appear as abundant as a widespread dominant grass (e.g. red fescue) – indeed more abundant than a forest tree or other plant that makes a far greater impact on our landscape (e.g. oak, heather). A minor but evenly scattered species, beyond a certain abundance, will inevitably be found in almost every grid square.

The engineer, glass-technologist, and spare-time ecologist and conservationist Frank W. Preston (1896–1989) suggested that, in general, species-abundance follows a bell-shaped 'normal' distribution on a logarithmic scale – i.e. a scale in which equal distance along the x-axis represents multiplication by a constant factor (e.g. doubling) (Preston 1948, 1962). He pursued the mathematical consequences of his theory for the structure and properties of animal and plant communities, including species–area relationships. Mathematically inclined readers will find his papers fascinating. He tested his ideas mainly on data from birds. Most species-lists from quadrat samples of vegetation (or from mapping projects) are unsuited to testing whether Preston's hypothesis holds for plants for the reason outlined above, but weight or percentage cover are appropriate. In Figure 50 a small set of data from one of the communities described in Chapter 9 is plotted in this way.

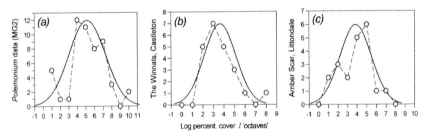

FIG 50. Number of species plotted against the logarithm of the number of 'hits' of 100 random pins in native localities of Jacob's-ladder (*Polemonium caeruleum*) in Britain (omitting *Polemonium* itself); most of the sites are false oat-grass (*Arrhenatherum*) grasslands (MG2). (a) All data from 13 native sites in Britain; (b) The Winnats, Castleton, Derbyshire; (c) Arnber Scar, Littondale, Yorkshire. The curves are normal distributions with the same mean and standard deviation as the points. The frequencies are calculated in intervals centred on 1, 2, 4, 8, 16 ... 'hits' – each figure twice the one before ('octaves'). The figures for the 13 sites are summed for graph (a). This is a small dataset, so random 'noise' has a large effect, but the graphs support Preston's hypothesis rather than otherwise. Data from Pigott (1958).

A log-normal (or similar) distribution implies that most species are moderately abundant, with few species very abundant, and few species very rare. Some characteristics of both common and rare species can be recognised. Dominant species in cool-temperate climates tend to be wind-pollinated (forest trees, grasses), and often have seeds with no particular adaptations to wide dispersal. Common and gregarious species are often insect-pollinated, but not specialised to particular pollinators, and often conform to one of a few common flower forms (e.g. yellow disc, white disc with yellow centre, massed small white flowers). Less-common plants often have flowers adapted to particular pollinators (bumblebees, butterflies, night-flying moths), and rely on the flower-constancy of their pollinators and the precise placing of pollen on their bodies for effective pollination. Many of the precise pollination systems in orchids could only have evolved in relatively rare species. But many plants that are 'rare' with us are common elsewhere – *Diapensia lapponica* as its name implies in Lapland (and elsewhere in northern circumpolar regions), hoary rock-rose (*Helianthemum oelandicum*) on limestone in the central and south European mountains (and of course on Öland). Plants that are rare everywhere are rare indeed!

SOME DYNAMIC ASPECTS OF PLANT COMMUNITIES

Potential natural vegetation

When Lewis and Clark set out upon their memorable journey across the continent of North America (1803–6), they were the first to traverse the great climaxes from deciduous woods in the east through the vast expanse of prairie and plain to the majestic coniferous forest of the northwest. At this time the oak–hickory woodland beyond the Appalachians was almost untouched by the ax except in the neighborhood of a few straggling pioneer settlements, and west of the Mississippi hardly an acre of prairie had known the plow

(Clements 1936)

Climatic climaxes were (and are) close to the American consciousness (though Clements' generation may have underestimated the impact of the native Americans). In Europe it was different. Virtually the only truly 'natural' vegetation was on high mountains, and in remote boreal forests and bogs. Human influence was all-pervasive in habitable country, and had been for centuries or millennia. We too tend to underestimate the influence of our ancestors in moulding the landscape we live in, and which we are prone to regard as 'natural'.

The 'potential natural vegetation' of Europe has long been a preoccupation of Continental ecologists. John Cross (2006) has published a map of the potential natural vegetation of Ireland. For Britain the data are more scattered, but from the pollen in peats and lake sediments (Chapter 2) we can reconstruct some picture of what Britain might have been like in the absence of man. Deep fertile soils in the lowlands of England would probably be dominated by (pedunculate) oak forest with wych elm and locally a substantial proportion of small-leaved lime; alder would be abundant in wetter places and beech in drier sites. On the acid soils of the rainier north and west acid (sessile) oakwood would probably be dominant, giving way to birch and pine in the Scottish Highlands. Oakwood of one sort or another would cover most of the Irish lowlands – sessile oak with an understorey of bluebells on the more acid soils, pedunculate oak with wych elm, ash and hazel on the more calcareous soils. As in Britain, the hard rocks of the rainy uplands and west would be dominated by acid woodland of sessile oak. In both islands, the forest cover would be broken by expanses of ombrogenous bog (Chapter 15), and treeless vegetation on the highest summits (Chapter 17).

Plant succession and the acquisition and cycling of nutrients by vegetation
The concepts of 'climax' and 'potential natural vegetation' imply that a piece of ground in any other state will tend to develop towards the climax – in other words to undergo vegetational *succession*. A distinction is often drawn between *primary successions*, from bare ground (as at Glacier Bay, Chapter 1) or open water (Chapter 13), and *secondary successions*, from some kind of pre-existing vegetation. The same kinds of processes are involved whatever the starting point. Individual plants germinate, establish, grow, compete and die. In so doing, they take up nutrients, photosynthesise and build up organic matter, which is incorporated in their leaves and roots. Some part of the vegetation is eaten by grazing animals – farm livestock, insects – but most is usually returned to the soil by decay organisms. Nature is good at recycling. Succession can be thought of as the balance between upgrade and downgrade processes, and the climax as the point at which these are in balance. Generally, the total mass of living material in the plant community rises in the course of succession, as do the quantities of organic matter and nutrients stored in the vegetation and the soil. An important caveat is that successions often pass through stages in which all the dominant plants are young and vigorous together; these stages may be more productive than the climax. Succession takes time. The dominant plants become established relatively quickly, but many years may pass before the less-common K-selected species (which may be very characteristic of the mature community) colonise and establish. Other things being equal, long-established plant communities are generally the richest in species.

Plant nutrients are added to the soil by weathering of soil minerals, and brought in by rain and as airborne dust. Animals may bring nutrients in (seabird colonies) or lead to their net loss (grazers). Nutrients are constantly lost from the soil in the drainage water, so in rainy climates soils tend to leach and become acid. Over time, the gains and losses must come into balance; change will only cease when a sustainable equilibrium is reached. It could be argued the 'climax' is a will-o'-the-wisp, because these processes need centuries or millennia to approach equilibrium, and on that timescale climate itself changes.

All of this presupposes that the plant community (and the ecosystem as whole) can 'do its own thing' without outside interference. But forest fires and volcanic eruptions have always happened, and grasslands are where herds of grazing mammals evolved. This was long before our ancestors began to use fire as a management tool, and before tools and domesticated animals gave them increasing control over the landscape. Many of our plant communities are maintained by some combination of grazing, cutting or burning – and secondary successions from these are very common (Chapter 8). Primary successions to forest do not generally go through a grassland stage; woody plants are often among the early colonists.

Even a 'climax' community is far from static. Quite apart from the constant turnover of organic matter and cycling of nutrients, plants, even forest trees, do not live for ever. In the course of centuries there must be a constant turnover of trees in the canopy. We can visualise two processes by which this can happen. In a mixed-age forest, individual trees age and die. Their canopy thins, they may lose branches, and they become liable to windthrow, leaving a gap in which regeneration can take place. Seedlings of many climax forest trees (e.g. oak, beech) commonly exist for many years in a suppressed state in the understorey. An opening in the leafy canopy lets in light and gives them the opportunity to grow to maturity. Many conifers are prone to forest fire and windthrow, leaving wide areas for regeneration at one time. These generally form extensive even-aged stands. Some conifers indeed are fire-adapted to the extent that the cones remain closed until the stimulus of fire causes the cone-scales to gape apart and release the seeds. Australian *Eucalyptus* and *Banksia* species behave similarly. Trees with this pattern of regeneration generally have light, easily-dispersed seeds, as do early-successional trees such as birch, ash and the maples.

It is a general principle that a plant cannot occupy the same piece of ground indefinitely. Either it has a limited life span, and must sooner or later give up its place to others, or it must move on, usually year by year. The 'herbaceous perennial' growth form is potentially immortal, because the plant in effect renews itself every year from lateral buds, so it is never in precisely the same place two years in succession. Some plants produce new shoots only millimetres from the old, so forming tight clumps; some produce extensive (above-ground) runners or (underground) rhizomes, appearing as looser patches or scattered shoots.

SPATIAL VARIATION AND BOUNDARIES IN VEGETATION

We all recognise some broad plant communities, such as oakwood, heath, chalk down, saltmarsh. But can boundaries be drawn between plant communities, or is variation in vegetation continuous? This was a matter of some controversy in the 1950s, with the American ecologists John Curtis and Robert Whittaker arguing strongly for a continuum view of vegetation (heir to Gleason's 'individualistic concept of the plant association'). Duncan Poore came to a similar conclusion from his work in the Scottish Highlands. He coined the non-committal term *nodum* for a plant community of any rank, and used it for commonly recurring combinations in the continuous field of variation. Most plant ecologists now would accept that variation in vegetation is essentially continuous.

However, recognisable boundaries can arise in vegetation for two reasons. Some are imposed by the physical environment, such as spring-lines, abrupt changes in soil depth, or geological boundaries affecting the physical and chemical properties of the soil. Often the effect of these physical boundaries has been sharpened by land use – most of all where they are followed by a fence or hedge (Fig. 51). Boundaries also arise from the meeting of communities of different growth or life-form. Gregarious tussocky dominants, such as purple moor-grass (*Molinia caerulea*) and black bog-rush (*Schoenus nigricans*) often grow in rather sharply delimited patches, and their growth-form dictates the range of microhabitats available to the associated species. Cycling of organic matter and nutrients is radically different in a wood and a grassland. The transition between a grassland with scattered hawthorns and a hawthorn scrub which is in all essentials a wood can take place in remarkably few years, and within a few metres on the ground. Woodland herbs need the shade and shelter of the dominant trees; tree-seedlings spend their early life in the ground flora, with which they may compete, upon which they may depend for shelter and support – and the ground flora is attuned to the annual leaf-fall and the yearly flush of nutrients that it brings. The contrast between heath and chalk grassland depends much more on the tolerances of the individual species to pH and mineral nutrients in the soil (i.e. on their *autecology*), but these are worked out in the context of competition and other interactions with neighbouring plants.

FIG 51. Looking east along the scarp of the Mid-Craven Fault, near Malham, Yorkshire, August 1963. Limestone grassland dominated by sheep's fescue (*Festuca ovina*) and *Sesleria* covers the steep Carboniferous limestone slopes to the left, with enclosed meadow/pasture with red fescue (*Festuca rubra*) and yellow oat-grass (*Trisetum flavescens*) on the deeper soils of the gently sloping fields below. The slopes to the right of the road are on base-poor Bowland Shales; the enclosed fields near the road are damp acid pasture, giving way to mat-grass (*Nardus*) and heath rush (*Juncus squarrosus*) moorland above. The influence on the vegetation of the geological and topographical boundaries has been sharpened by land-use.

Even if no discontinuities existed, when we talk about anything as complex and diverse as vegetation we are compelled to use words – and words (if they are to be useful) need definitions, and definitions imply limits. A good analogy is colour. Colour varies continuously, but it can be precisely specified by three measurements: hue, saturation and brightness in one system, the density of yellow, magenta and cyan printing inks in another, but always just three. (The human eye has receptors sensitive to the primary colours red, green and blue.) Yet when we talk about colour we use words, and do not worry about the fact that there are no sharp boundaries in the spectrum. We can do the same with vegetation; much of the ensuing chapters can be thought of as describing and defining 'colours' in the landscape. When we actually encounter recognisable boundaries in vegetation, that is a bonus!

CHAPTER 5

Woodlands: Introduction

HOW NATURAL ARE OUR WOODS?

W HEN THE ICE RETREATED and the climate warmed, Britain and Ireland became covered with forest, along with the rest of northwest Europe, and we are used to regarding woodland as the natural state of the landscape. But natural woodland – the wildwood – was a very different thing from most of the woods we see around us now, which are at best 'semi-natural'. The species of which they are made up are mostly native wildwood species, but the structure and detailed composition of the woods have been heavily influenced over the millennia by man. Truly natural woods of long standing are rare in our islands, and mostly in inaccessible places, such as sea-cliffs (Fig. 52), rocky slopes high on our hills, and islands in lakes in western Scotland and Ireland. Natural woods are untidy places compared with most of the woods we are accustomed to!

The earliest Neolithic farmers were interested only in cleared land to grow crops, and they cleared the forest to waste. As more land was cleared and prehistoric society became more sophisticated, Neolithic and Bronze Age people 'domesticated' areas of the wildwood, to provide useful products they needed, such as straight hazel rods for making hurdles and wattle-and-daub building, and longer and stouter poles for more substantial structural uses, and an accessible supply of firewood. The wood generally provided both 'timber' from the large trees, and 'wood' from the coppiced underwood. The underwood was at least as valuable in the traditional economy as the timber, often more so. Farm

FIG 52. (a) Sessile oak (*Quercus petraea*) woodland on exposed cliff slopes at the Dizzard, south of Bude on the north coast of Cornwall. (b) Interior of the wood: bracken, brambles and great wood-rush, with primroses and wood anemone in the foreground. April 1981.

animals were kept out of woods. Some areas of woodland were used as wood-pasture, and these in general had no underwood. We should remember that the survival of these traditional woods depends on the fact that they were useful to our ancestors – they formed a part of the economic fabric of their societies.

TRADITIONAL WOODLAND MANAGEMENT

In traditional woodland, a proportion of trees (from seedlings or saplings on the floor of the wood) were allowed to grow to maturity to provide *timber*. The desired species (most usually oak) would be chosen, and the trees formed a more or less open canopy. Individual trees would be felled when needed. Because of their wide spacing, the crowns of these trees were wide and richly branched.

FIG 53. Spring flowers in a hazel coppice near East Anstey on the Devon–Somerset border, May 1984. Primroses, moschatel (*Adoxa moschatellina*), lesser celandine (*Ranunculus ficaria*) and early-purple orchid (*Orchis mascula*).

The main use of timber was for house and farm building and other local needs. (Contrary to popular supposition, shipbuilding was a relatively minor and localised use, and the demand for merchant ships greatly outweighed that for the Navy.) The space between the trees was occupied by *underwood*.

The underwood – coppiced hazel, or sometimes ash, wych elm or small-leaved lime – was generally the principal crop of the wood. It was cut to the ground on average every five years, though longer intervals between coppicings were sometimes used to get larger poles. Hazel is the traditional material of hurdles, and the coppice also provided stakes and poles, and firewood. Large quantities of underwood were turned into charcoal for smelting and iron working.

The years immediately following coppicing produced spectacular displays of woodland flowers (Fig. 53), until the coppice shrubs grew up and the canopy closed before the next coppicing. Many coppices are now derelict, and decades have elapsed since they were last cut. The spring woodland flowers are still there, but much sparser.

In parts of the country where there was a thriving leather industry, mainly in the west of Britain, oak bark for tanning was in great demand. This led to large

FIG 54. Sessile oak coppice, Stoke Woods, Exeter, May 1972; patchy field-layer of great wood-rush, with a scatter of the mosses *Dicranum majus* and *Polytrichastrum formosum*. The two trunks nearest the camera spring from the same stool, whose outline can be clearly made out. This coppice was singled a few years after the photograph was taken.

FIG 55. Oak coppice woodland (here pedunculate oak) clothing the slopes of the Dart valley upstream of Bench Tor, February 2000. Birch and common gorse on the well-drained heathy ground near the camera, foxy red-hued bracken on deeper soils of the slopes beyond.

tracts of oakwood (largely sessile oak) being treated as underwood and coppiced at 20–30-year intervals (Figs 54, 55). The bark was stripped for the tanneries, and much of the wood was burnt for charcoal. Coppicing in most of these oakwoods ceased with the Second World War, and this coppice is now derelict. Some has been 'singled' (regrowth from the coppice stool reduced to the one best trunk), so converting it to 'stored coppice'. Nevertheless, many of these woods and the valleys in which they lie are beautiful and species-rich, particularly in bryophytes and lichens (Chapter 7). No doubt any woody species other than oak were weeded out, but some birch and rowan are usually present. If the primary aim was to grow timber, the dominant trees might be managed as high forest, but woods that appear to be this often turn out to be singled coppice.

In wood-pasture the mature trees are out of reach of grazing, but seedling or sapling trees get eaten, especially those of the more palatable species. If an ongoing crop is wanted from the trees, it can be got by *pollarding* the trees. The trunk is cut off somewhere between 2 and 4 m from the ground, and shoots from the top of the cut trunk in the same way as from a coppice stool, but out of reach of the livestock. Pollard willows (Fig. 56) used to be a familiar sight on riverbanks in southern England, and old pollard oaks and beeches are a common feature of former wood-pastures and old parkland.

FIG 56. Pollard willows near Oath, West Sedgemoor, Somerset, Sept. 2010. The 'mop-head' trees produced a crop of young shoots for basketry and other uses out of reach of grazing cattle.

Particularly from the eighteenth century onwards, landowners created *plantations*, often but by no means always of exotic trees such as sycamore (*Acer pseudoplatanus*), sweet chestnut (*Castanea sativa*), Norway spruce (*Picea abies*) and European larch (*Larix decidua*). Native Scots pine was widely planted, especially in Scotland. These plantings were sometimes for timber, but often for other purposes. Sweet chestnut might be planted with a view to a crop of chestnuts, or for timber, or for coppicing to produce split-chestnut paling. Sycamore was often planted for shelter, and does surprisingly well even in exposed country up to altitudes of 400 m or more. The eighteenth-century landowners were very conscious of 'picturesque' landscape, not only in laying out their own estates, but in planting clumps of trees at focal points in the wider landscape – usually beech in chalk country, and pine in heathland (Chobham Clump near Woking was a much-loved landmark on the Surrey skyline in my childhood).

INDICATORS OF ANCIENT WOODLAND AND ECOLOGICAL CONTINUITY

Growing understanding of medieval woodland management, and awareness of the biodiversity of old woodland, has led to an interest in indicator species of 'ancient woodland'. It became apparent that a great many of our familiar woodland plants have been remarkably slow to colonise any woods planted since 1600. That year is taken as a defining date for ancient woodland, because tree-planting was not yet fashionable and any wood existing at the close of the sixteenth century had almost certainly been woodland for many centuries already. Indicators of ancient woodland vary somewhat from one part of England

TABLE 4. Species listed as indicators of ancient (pre-1600) woodland, in at least half (> 7) of 13 lists from different parts of England, collated by K. J. Kirby. Figures in parentheses give the number of lists in which the species is cited.

Species	No. of lists	Species	No. of lists
(A) SPECIES INCLUDED IN MORE THAN THREE-QUARTERS OF LISTS (≥ 10 LISTS)			
Adoxa moschatellina	(10)	Lathraea squamaria	(11)
Anemone nemorosa	(11)	Luzula pilosa	(12)
Carex pallescens	(10)	Luzula sylvatica	(11)
Carex pendula	(10)	Lysimachia nemorum	(10)
Carex remota	(10)	Melampyrum pratense	(10)
Carex strigosa	(10)	Melica uniflora	(12)
Carex sylvatica	(10)	Milium effusum	(11)
Chrysosplenium oppositifolium	(11)	Neottia nidus-avis	(10)
Convallaria majalis	(10)	Paris quadrifolia	(12)
Galium odoratum	(13)	Veronica montana	(10)
Lamiastrum galeobdolon	(11)		
(B) SPECIES INCLUDED IN BETWEEN HALF AND THREE-QUARTERS OF LISTS (7–9 LISTS)			
Allium ursinum	(9)	Hyacinthoides non-scripta	(8)
Aquilegia vulgaris	(8)	Hypericum pulchrum	(8)
Bromopsis ramosa	(7)	Lathyrus linifolius	(9)
Campanula trachelium	(8)	Narcissus pseudonarcissus	(7)
Carex laevigata	(9)	Orchis mascula	(9)
Conopodium majus	(8)	Oxalis acetosella	(9)
Dipsacus pilosus	(7)	Platanthera chlorantha	(7)
Dryopteris affinis	(7)	Poa nemoralis	(8)
Elymus caninus	(7)	Polystichum aculeatum	(8)
Epipactis helleborine	(8)	Potentilla sterilis	(7)
Equisetum sylvaticum	(8)	Primula vulgaris	(9)
Euphorbia amygdaloides	(9)	Ranunculus auricomus	(9)
Geum rivale	(8)	Sanicula europaea	(8)
Helleborus viridis	(7)	Vicia sylvatica	(8)

to another, but Table 4 gives the species regarded as ancient-woodland indicators in more than half of 13 lists collated by Keith Kirby in 2005. Most of these are quite common woodland plants. What is important about the list is that no one species, or no select group of species, is especially diagnostic. It cannot be said that if a particular rare species occurs, that wood is ancient!

Lichenologists realised that certain uncommon lichen species are restricted to big old trees, such as grow in old parkland or wood-pastures. This led the late

Dr Francis Rose to devise 'indices of ecological continuity', based on select lists of lichens. He envisaged his 'revised index of ecological continuity' as a tool for screening ancient parkland and wood-pastures, which could be widely used, and his 'new index of ecological continuity' as providing a finer-scale grading of sites, but placing greater demands on the identification skills of the observer (Rose 1976, 1992).

Indicator species of 'ancient woodland' or of 'ecological continuity' inevitably suffer from the limitation that they are only applicable in a particular area. Indicators of ancient woodland in the west of Britain and Ireland would be very different from lists appropriate to East Anglia or southeast England, and would have to include a high proportion of bryophytes and lichens.

WOODLAND TREES OF BRITAIN AND IRELAND

The oaks (*Quercus* spp.)

The two native oaks are the commonest dominant trees of established woodland in Britain, and on acid rocks in Ireland. They cast a moderately deep shade, but are somewhat more light-demanding than beech. The pedunculate oak (*Quercus robur*) is the dominant tree on the deeper and better brown-earth soils on clays and clay-loams with soil pH from around neutral to mildly acid. It can tolerate some degree of waterlogging, but is at its best on moist but well-drained lowland soils. Sessile oak (*Quercus petraea*) tends to grow on poorer, more sharply drained and more acid soils, but while this generalisation usually holds true in southeast England, it is fraught with qualifications and exceptions elsewhere. While the oakwoods on the valley sides around Dartmoor (on acid Carboniferous shales and grits) are of sessile oak, with pedunculate oak on the wetter valley floor, woods on steep valley sides on the granite, as well as the three small oakwoods high on the moor, are of pedunculate oak. Oakwoods on hard limestones are commonly of sessile oak, so there are more subtle soil factors than pH differentiating the two species. Both flower at about the time the leaves expand and are wind-pollinated. The crop of acorns is variable from year to year, especially at higher altitudes. The acorns germinate where they fall. Some are collected by jays, which often drop them, so oak seedlings can be found quite far from woods. The normal life span of the oaks (corresponding to our 'threescore years and ten') is probably about 300–400 years, but this figure may be even more variable than the human life span. Most of the other dominant trees of European broad-leaved forests probably attain similar ages.

Ash (*Fraxinus excelsior*)

This is the commonest dominant of chalk and limestone woodland, and probably the commonest woodland dominant in Ireland. It is a quick-growing, light-demanding tree, which casts a relatively light shade. It is notably late to come into leaf in spring, and the dark masses of wind-pollinated flowers appear at about the time the leaf-buds burst, ripening in late summer to bunches of wind-dispersed 'ash keys'. Ash is a very effective pioneer, and its seedlings can be almost as much nuisance in the garden as those of sycamore, but it is palatable to livestock so needs a degree of respite from grazing to get established. Perhaps surprisingly for a tree that is so often associated with dry chalk and limestone, ash is tolerant of moist soils and is a common ingredient of fen woods. Unless coppiced, it is a rather short-lived tree, rarely lasting more than two centuries.

Wych elm (*Ulmus glabra*)

This is our native elm, and the only one that is a true woodland tree. It is a bigger tree than ash, and casts a deeper shade. It mainly occurs on base-rich soils, and is often common in woods on limestone, but never more than very locally dominant. It flowers in late winter on the bare twigs, producing the bunches of discoid winged seeds ('samaras') in summer. It is eaten avidly by livestock.

Beech (*Fagus sylvatica*)

Beech is a widespread forest dominant in central Europe, but was a late Post-glacial arrival in Britain, where its natural range is limited to an area from southeast England and the Chilterns to the limestone valleys of southeast Wales. It is widely tolerant of our climate outside those limits, in plantings and beech hedges high on Exmoor and Dartmoor (where it self-sows freely), and north to Aberdeenshire and throughout Ireland. It casts the deepest shade, and is the most shade-tolerant of our common forest trees, and is an uncompromising dominant on soils ranging from the rendzinas of the beech 'hangers' on the chalk to the acid brown earths and podzols of the plateau of the Chilterns and the heathlands of the New Forest. Like oak, it flowers as the leaves expand in spring, and the triangular beech nuts are shed in autumn and germinate readily where they drop.

Hornbeam (*Carpinus betulus*)

Hornbeam was another late arrival in Britain. It is widespread on the Continent, where it is rather southern and does not reach the Atlantic coast, and is a tree of rather base-rich woods, avoiding the most acid soils. As a native with us, it

FIG 57. Hornbeam coppice, Hog Wood, near Ifold, West Sussex, April 1985.

is confined to the southeast, with its headquarters from Kent and East Sussex to Hertfordshire and Essex. Here it grows on somewhat moist silty sands and sandy clays, usually in company with oak, and is nearly always coppiced (Fig. 57). Hornbeam roadside hedges were part of my Middlesex childhood. It belongs to the same family as birch, but the winged fruits are very different. It is often planted, mostly for garden hedging, because it stands clipping well.

The limes (*Tilia* spp.)

The lime we are most used to seeing is the hybrid, *Tilia × europaea*, which is widely planted in parks and as a roadside tree. We also have both parent species as native woodland trees. Small-leaved lime (*T. cordata*) is a stately tall tree and was probably a major constituent of the wildwood over much of lowland England. It survives as a (usually) minor ingredient of ancient woodland in localities widely scattered over its former area, sometimes as a canopy tree, but more often as coppice. Unlike most of our other dominant trees, it bears its semi-erect clusters of flowers in summer, and is pollinated by both insects and wind. Large-leaved lime (*T. platyphyllos*) is much more local, usually in rocky woodland on limestone, as in the Derbyshire dales and the Wye valley; on the chalk it forms a sparse fringe at the base of the escarpment woods of the western South Downs (Abraham & Rose 2000). Apart from its larger leaves, it is easily recognised by its pendulous flower-clusters and fruits. Both limes are long-lived trees (old coppice stools can live for many centuries), and both are palatable to livestock. Small-leaved lime is limited northwards by its need for high-enough summer temperature to ripen viable seed (Pigott & Huntley 1981). In recent summers it has produced abundant seed (and seedlings) in northwest England.

The tree birches (*Betula* spp.)

The birches are elegant fast-growing but relatively short-lived trees; a birch a century old is a veteran. They are light-demanding trees, and cast only a light shade. Both our species produce their catkins in spring at about the time the leaves first expand, and are wind-pollinated. The ripe female catkins break up to release the tiny winged seeds, which are spread far and wide by wind, making the birches rapid and efficient colonisers. Our two species are closely related, but the silver birch (*B. pendula*) favours better drained and more base-rich soils, so it is the predominant species on dry heathland, and on chalk and limestone. Downy birch (*B. pubescens*) favours damper and more acid soils, so it is the species usually encountered in the damp oceanic west, or on acid peat. In upland areas of Scotland it is represented by ssp. *tortuosa*, which occurs too in Norway, northern Sweden and Iceland.

Scots pine (*Pinus sylvestris*)

This is one of the world's most widely distributed trees, extending from Scotland to Siberia and from Lapland to the mountains of the Mediterranean. With us, Scots pine is now native only in the Scottish Highlands, but early in the Post-glacial it was ubiquitous in Britain and Ireland, and it seems to have hung on locally almost into historical times in some areas of dry limestone (such as the Burren) or heath. On the Continent it is a tree of poor soils and exposed places. In Scotland its headquarters is in the eastern Highlands, especially Deeside and Speyside, where, with its windborne seeds, it vigorously colonises heather moorland unless prevented by burning or heavy grazing. Scots pine has been exploited for timber for centuries in the areas that now have the greatest concentration of pine forest. Pine is widely scattered elsewhere in the Highlands, but seldom in large stands (Chapter 7).

Hazel (*Corylus avellana*)

We do not usually regard hazel as a 'tree' because we most often see it as an understorey shrub, often coppiced. Our species, left to itself, forms a large shrub or untidy small tree up to 5 or 6 m (or more) tall. It is one of our most widely distributed woody plants. It forms a regular part of the understorey of most of our woods, and extends far outside the confines of woods to dominate large areas of scrub (woods in all but name) in coastal districts of western Ireland and western Scotland – as also in western Norway. Hazel was immensely abundant in the pollen rain in the Boreal period some 9000 years ago. Hazel coppices well, perhaps a legacy of an evolutionary history of recovery from browsing. The yellow 'lamb's tail' male catkins are a familiar sight in late winter. The heavy hazel nut is the seed of a plant attuned to the stability of established woodland or scrub (by comparison with light airborne seeds of the pioneer birches and ash), but observation of colonisation of limestone pavement suggests they serve hazel well enough (Chapter 6), no doubt with some help from small mammals.

Alder (*Alnus glutinosa*)
Alder dominated the pollen record in Britain and Ireland following the 'Boreal–Atlantic transition' about 8000 years ago. That was a time when the North Sea basin was filling, and the climate switched from rather continental to much more oceanic, and the indications are that the forests became much wetter. Alder must have dominated large tracts of country around the lakes and bogs whose deposits were sampled by the twentieth-century pollen-analysts. Alder is generally associated with moving water, either where water is upwelling or draining laterally through the soil; a widespread and important habitat is flushes or seepages in woods mainly of other trees (Fig. 86), or on the banks of rivers or streams (Fig. 58b), where it is commonly cut back at intervals to keep the banks clear for fishing and prevent excessive shading. Alders have root-nodules with symbiotic microorganisms that fix atmospheric nitrogen, so they are important in building up the nutrient capital of the wet woods in which they grow. Alder produces its catkins very early in spring, and is wind-pollinated. The ripe female catkins do not disintegrate like those of birch, but their scales gape apart to release the seeds and persist as centimetre-long black 'cones' all the following winter.

The sallows and willows (*Salix* spp.)
The genus *Salix* includes numerous species, and must be the most diverse in stature in our flora, ranging from tall riverside trees to the tiny arctic-alpine dwarf willow (*S. herbacea*), hardly a centimetre high. All are woody, and have simple leaves and catkins (male and female on separate plants). The species that concern us here are trees and shrubs of wet places. Colloquially, we generally distinguish 'willows', tall mostly riverside trees, with long leaves and catkins, from 'sallows' ('sallies'), large spreading shrubs with ovate leaves and short 'pussy willow' catkins, which form patches of scrub on wet ground anywhere. The flowers of willows produce nectar, and the female catkins are pollinated by insects, and by wind. The two common riverside willows, white willow (*S. alba*, Fig. 58a), with its grey-green foliage and 'church-tower' outline contrasting with the deeper-green and more spreading crown of crack-willow (*S. fragilis*), were probably both introduced in prehistoric times, as was osier (*S. viminalis*) and almond willow (*S. triandra*), smaller trees of river- and stream-sides and

FIG 58. Riverside trees. (a) White willow, Hinton St Mary, Dorset, August 1972. (b) Alders beside the River Otter, Otterton, Devon, March 2011. This early spring photograph shows the characteristic straight main trunk reaching high into the crown

osier beds – 'sally gardens' in Ireland. All four (and many hybrids) were used for basketry, and often coppiced or pollarded to produce a supply of pliable young shoots. There are three common sallows that are natives of long standing. Grey sallow (*Salix cinerea*) is an exceedingly common shrub of damp or wet soils ranging from calcareous fens to acid peaty gleys. It is more tolerant than alder of stagnant conditions, and its profuse fluffy wind-borne seeds make it a ready pioneer. In some places it covers many hectares of country (as in parts of Cornwall); in others it grows as patches or more isolated groups or individuals where it has happened to find a congenial moist-enough spot. Goat willow (*Salix caprea*) is less gregarious, and favours better-drained and more calcareous soils. It is a rather regular minor ingredient of woodland on chalk or limestone. In the north of England and Scotland, several other 'sallows' occur. Bay willow (*S. pentandra*) comes farthest south of these; tea-leaved willow (*S. phylicifolia*) and dark-leaved willow (*S. myrsinifolia*) are more northern. Finally, a very common and often only waist-high shrub of damp heath and moorland is eared sallow (*S. aurita*), with roundish grey-green wrinkled leaves and large stipules.

Minor woodland trees

Field maple (*Acer campestre*) is a defining small tree of lowland calcareous woodland in Britain, almost always present, but never even approaching dominance; it does not reach Ireland as a native tree. Its small lobed leaves colour an attractive yellow in autumn. Wild cherry (*Prunus avium*) is thinly scattered through lowland oak, ash and beech woods. It is unnoticeable except when it is in flower before the oak canopy is expanded, but the bark with its horizontally elongated lenticels is distinctive. Bird cherry (*Prunus padus*) with its racemes of white flowers is more confined to base-rich soils and is generally more upland and northern, but it occurs in fen carr in East Anglia. Aspen (*Populus tremula*) grows in woodland on a wide variety of soils, and up to high altitudes, but it is capricious in its occurrence. The catkins are borne before the leaves expand in spring; pollination is by wind. It readily suckers, so forming patches, and readily colonises bare ground by wind-borne seed. Holly (*Ilex aquifolium*) is very widespread as a minor shrub-layer component in all kinds of woodland, perhaps most characteristically in beechwoods on acid soils, where holly is often conspicuous in the shrub-layer and may form a subsidiary tree-layer under the beech canopy. It is often prominent in western oakwoods, as it famously is at Killarney, Co. Kerry. It is one of the few species that will grow amongst rhododendron. Holly occasionally forms small 'woods' on its own, as in various places in the New Forest. Perhaps surprisingly, young growth of holly is very susceptible to grazing. Strawberry-tree (*Arbutus unedo*) shares the evergreen habit of holly, but its undoubtedly native occurrence in moist rocky woodland around the lakes at Killarney (with an outlier in Co. Sligo) is far distant from its mainly west-Mediterranean distribution, so it is a rare but special tree with us. Yew (*Taxus baccata*) is widely but thinly distributed in lowland oak and oak–ash–field-maple woodland, and in beechwoods on acid soils. It is more constant in escarpment beechwoods, where it locally forms an almost continuous shrub-layer beneath the beech. Very locally it dominates woods on its own, notably on the chalk in southeast England, but also on Carboniferous limestone in the Wye valley, at the head of Morecambe Bay and near Killarney (Chapter 6). Rowan (*Sorbus aucuparia*) is very widely distributed as a minor ingredient in a variety of woods, particularly on acid soils, and in the uplands. It is often associated with sessile oak and downy birch. Well-grown rowans can be found on mountain ledges far outside woods. Its white insect-pollinated flowers appear in early summer and are followed by the bird-dispersed orange-red berries. Whitebeam (*S. aria*) is confined in Britain to an area from Kent to Dorset, the Wye Valley, the Cotswolds and the Chilterns, and nearly confined to calcareous soils. Like rowan (but unlike the apomictic *Sorbus* species considered in Chapter 8), it is a sexually reproducing tree, with white insect-pollinated flowers and red berries in late summer. It is occasional in oak–ash–field-maple woodland and beech and yew woodland on chalk and limestone, but probably only where it has been able to colonise open chalk and limestone grassland, or gaps in the wood. Many apparently suitable woods are without it. Wild service-tree (*S. torminalis*) is our third sexually reproducing member of the genus. It is an inconspicuous tree with brown fruits, probably most noticeable when the leaves fall in autumn, and their orange-yellow colour and characteristic shape stand out against the subdued brown of the oak leaves. It is widely but rather patchily distributed in woods on both calcareous and mildly acid soils.

SOME OTHER TREES IMPORTANT IN THE LANDSCAPE

Elms (*Ulmus* spp.)

The native wych elm is not our most familiar elm. That distinction must go to the hedgerow elms of the traditional farming landscape. Anyone whose memory extends back to the 1970s will remember the English elms (*Ulmus procera*) that were so much a feature of the lowland English landscape. The English elm is a magnificent tree, reaching 40 m high with a trunk 2 m in diameter (Fig. 59). It had many virtues as a hedgerow tree, giving shade, shelter, and foliage palatable to stock. Its timber is resistant to rotting when wet, and to crushing, which led to its use for water pipes, pier and wharf timbers, and lock gates. It was too liable to warp and shrink for much use in building. The English elm is female-sterile, but it is easy to propagate from suckers. Genetic analysis (Gil *et al.* 2004) has shown that probably all 'English elms' are part of a single clone, and genetically identical with trees in Spain and Italy, and with the Atinian elm used (pollarded) by the Romans to train vines. The Roman writer Columella records its introduction to Spain. There is no record of its introduction to England, but it is likely that it came with the Romans. When the fields were hedged after the Enclosures in the eighteenth and nineteenth centuries, English elm was widely planted as the preferred hedgerow tree. In East Anglia, Kent, the east Midlands and in the southwest forms of the small-leaved elm (*U. minor*) were predominant. All of these trees can propagate by suckering, which is useful for making hedges, and renders them virtually indestructible, and all were probably early introductions.

The elm disease caused by the fungus *Ophiostoma ulmi* and spread by the elm-bark beetle *Scolytus scolytus* (Figs 60, 61) has been around for a long time. There were outbreaks at intervals from the early nineteenth century, and there is no reason to suppose they were the first. The causal pathogenic fungus was first

LEFT: **FIG 59.** English elm, showing symptoms of Dutch elm disease, Fiddleford, Dorset, August 1975.

ABOVE: **FIG 60.** Elm-bark beetle (*Scolytus scolytus*), Exeter, December 1975. Greatly enlarged; the beetle is about 5 mm long.

BELOW: **FIG 61.** Elm-bark beetle. (a) Larval galleries under the bark. The primary burrow made by the beetle in which the eggs are laid is at the top right; the burrows of the individual larvae radiate from it. (b) The fully fed larvae are *c.* 5 mm long.

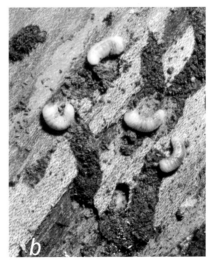

isolated in the Netherlands in 1921, hence the name 'Dutch elm disease'. About the mid 1960s a new virulent form of the disease appeared in both Europe and North America, and it has devastated elms in both areas. The first outbreak in England was at Tewkesbury, Gloucestershire, and by the mid 1980s most of lowland England was affected. English elm proved very susceptible to the disease, and almost 100% of mature trees succumbed. However, the disease did not kill the roots, which continued to send up suckers. After 20 years' growth the trunks had become large enough to suit the beetles, and the trees were reinfected. Other elms were not as susceptible. The various clones of small-leaved elm (*Ulmus minor*) showed varying degrees of resistance. The distinctively conical 'Cornish elm' (*Ulmus minor* ssp. *angustifolia*) proved susceptible when in the minority in English elm country, but largely escaped in Cornwall, where it is still a common feature of the hedgerows. Wych elms, perhaps because of their relative isolation, though susceptible, were only locally infected (Peterken & Mountford 1998).

Sycamore (*Acer pseudoplatanus*)

Sycamore was introduced into Britain in the sixteenth century. It was first recorded from the wild in 1632, and it was widely planted from the eighteenth century onwards. It is a native of continental Europe, where it occurs almost throughout Germany and is common in limestone woods in the hilly south of the country. It grows well throughout Britain and Ireland, and regenerates from seed (Fig. 62). It is anathema to many conservationists, but it seems to me that it

FIG 62. Sycamore (*Acer pseudoplatanus*) by Great Close Scar, Malham, Yorks., at nearly 400 m altitude. This species is widely planted in the uplands for shade and shelter – here with a ring of dry-stone walling to protect the recently planted tree from grazing sheep.

is not as black as it is painted, and it actually does little harm. If it was bent on reducing our islands to a coast-to-coast monoculture of sycamore it would have done so long before now! It finds its optimum with us in base-rich woodland in the hilly west, a distribution recalling the 'borderland woods' of Peterken (2008), and much the same niche as in its native home. As an opportunist and coloniser it has a lot in common with ash, but it casts (and will withstand) deeper shade. In general it does not take over established woodland, and it remains surprisingly uncommon in our lowland woods. A lot of conservation effort is wasted on trying to eliminate sycamore, which could be devoted to worthier causes, with more prospect of success.

WOODS AS HABITATS

Woods are sheltered, shady and cool in summer, and even after leaf-fall in autumn the leafless trees do something to temper the winter storms. Woods, in short, have a microclimate that is different from treeless places, and that microclimate varies round the seasons.

The microclimate of woods

Something has already been said about climate and microclimate in Chapter 1. Woods are complicated microclimatically because they are structurally complex, and the crowns of the trees are high above the ground. Microclimate is determined first by *radiative* exchanges of energy with the wider environment – radiation from the sun and sky during the day, radiation to the sky and other surroundings during the night – and second by *convective* exchanges of matter and energy with the air. Wind, blowing over vegetation, will probably be cooling it on a sunny day, and warming it on a cold clear night. At the same time it will be removing water vapour from the surface (so speeding evaporation when conditions are suitable) and imposing drag on the stems and leaves as it passes.

In summer the biggest energy exchanges, and most evaporation and drag, take place in the upper layers of the tree canopy. The drag of the tree crowns slows the wind. The leafy canopy absorbs most of the incident radiation; the light intensity reaching the ground may be only 1% of that above the canopy in a shady wood or forest. In a multi-layered wood each successive layer influences the climate of the layer(s) below, so it follows that the ground vegetation experiences a wholly different climate from the canopy, and the shrub-layer experiences a different climate from either. A plant on the ground in a dense wood is deeply shaded, windspeed is very low, therefore humidity is high –

and evaporation is low because of the combined effect of high humidity, low windspeed and low net radiation income. Obviously some woods are more shady and sheltered than others.

In winter there is no leafy canopy and energy exchanges take place on the branches and trunks of the trees, and on the ground. There will still be a gradient of windspeed from the uppermost twigs to the ground but it will be much less steep. Radiative heat exchange will still be concentrated to some extent in the dense twiggery of the canopy, but it will be much more diffuse, and much

Changes in
light intensity

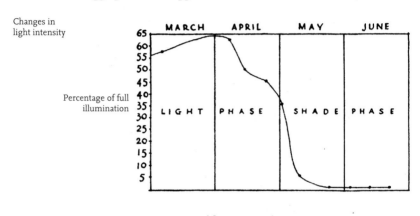

Vegetative and flowering periods

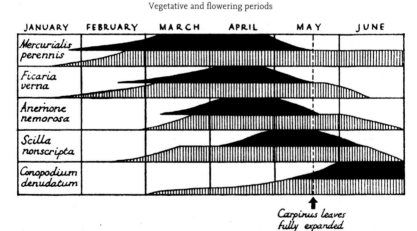

Carpinus leaves
fully expanded

FIG 63. Vegetative and flowering periods of woodland herbs in relation to light under the tree canopy in oak–hornbeam woods in Hertfordshire. From Tansley (1939), based on Salisbury (1916). The leaves of lesser celandine, wood anemone and bluebell die by the end of June, but dog's mercury and pignut remain green all summer.

more tempered by convective heat exchange with the air than in summer (or in a grassland). Consequently, the twigs will track air temperature closely, and will not face the hazard of frequent ground frosts on clear nights, or benefit from the warming effect of the early spring sunshine. Except when the trunks are wet with rain, evaporation will be largely from the soil and plants on the floor of the wood. Through the depths of winter the radiation income is low even above the canopy. As radiation increases in spring, air temperature begins to rise, and daytime ground temperatures rise faster than the temperature of the air and of the buds of the trees. Radiation income in April matches that in August, but the air is still generally cool. Conditions on the woodland floor are congenial for growth and flowering of the woodland plants. For most of them this is the

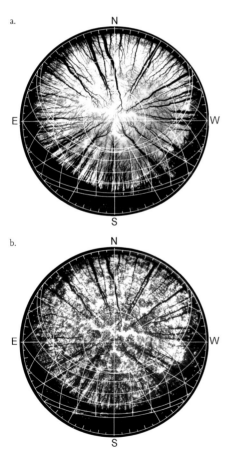

FIG 64. Hemispherical ('fish-eye') photographs of the canopy of the sessile-oak coppice shown in Fig. 54: (a) early in April, and (b) some weeks later with the leaves expanded. The edge of the picture is the horizon, the middle is directly overhead. The two concentric circles mark 30° and 60° above the horizon. The curved lines from east to west are the tracks of the sun across the sky on the 22nd of each month, and the lines crossing them are the hours of the day from solar noon, when the sun is due south. At the equinoxes in March and September the sun rises due east at 0600 h and sets due west at 1800 h solar time. As in a star map, with north at the top east is on the left. The wood slopes at roughly 20° to the north-northwest, and receives no direct sun from early November to the end of January.

peak flowering season, for instance of lesser celandine (*Ranunculus ficaria*), wood anemone (*Anemone nemorosa*) and bluebell (*Hyacinthoides non-scripta*, Fig. 63). For many, the vegetative season lasts only a few weeks longer, and the leaves have died down by the end of June. By late April or early May, air temperature is at last high enough for the buds of the trees to burst and the woodland canopy to expand (Fig. 64). In July the wood takes on an entirely different aspect. Foliage of dog's mercury (*Mercurialis perennis*) carpets the woodland floor throughout the summer, with ferns, shade-tolerant herbs such as pignut (*Conopodium majus*), and tall woodland grasses such as hairy brome (*Bromopsis ramosa*), giant fescue (*Festuca gigantea*) and wood millet (*Milium effusum*).

ECOLOGICAL FACTORS IN THE DISTRIBUTION OF WOODLAND

Woods vary from place to place with climate and soil. We might ask two rather different but related questions. Suppose that man had *not* come on the scene, what would the wildwood cover of our islands look like now? In other words, what is the potential natural vegetation? Or we can look at our present-day woods, and ask how their composition and distribution relate to climate, topography and soil – regardless of the fact that most our woods are far from being 'natural'.

Potential natural vegetation

We can look for evidence of the potential natural vegetation of Britain and Ireland in the pollen record and the record of plant macrofossils preserved in peat or lake deposits, or we can extrapolate from the distribution and behaviour of individual species and their interactions and the composition of plant communities at the present day. That the uncertainties of inference are substantial goes without saying!

Regional differences in the character of the forest cover over Britain and Ireland can be mapped from the evidence of pollen analysis. This shows that the pre-clearance wildwood can be divided into five broad regions. In most of northern Scotland beyond the Great Glen the pollen indicates dominant birch (with some pine locally), and perhaps an open unwooded fringe in the extreme north and west. The Highlands east of the Great Glen were a region of dominant pine, and there was a long-persistent outlier of pine in the west of Ireland. The Highlands south of Rannoch Moor, the Isle of Mull, and the Scottish lowlands were predominantly oak–hazel territory, as were the Southern Uplands, the

Lake District, the Pennines, most of Wales and Devon. In Ireland two blocks
of country on acid rocks are similarly oak–hazel country, the mountains of
Galway, Mayo and northern Ireland, and the granite, Ordovician and Old Red
Sandstone mountains from Wicklow to Cork and Kerry. Over much of Ireland
(predominantly on limestone or calcareous boulder-clay) the pollen shows elm
and hazel as major trees in the forest, and the same is true of southwest Wales
and Cornwall. In England, the Midlands and the south and east form a coherent
region dominated by mixed oak forest in which small-leaved lime was abundant
and locally dominant. However, we must remember that this was the climatic
optimum of the Post-glacial, when temperatures were several degrees warmer
than now. Oak, elm, lime and ash grew abundantly together in the mixed oak
forest that then covered much of central Europe, but several key trees in that
assemblage, including small-leaved lime, have retreated southwards and become
more local as the climate has cooled

But, as ever, 'the devil is in the detail', and these five broad regions conceal a
great deal of local variation in detail with topography and soil. John Cross (2006)
has published a detailed potential natural vegetation map for Ireland. It agrees
broadly with the pre-clearance regions defined by pollen analysis, but takes
account of the (at least largely natural) expansion of ombrogenous bogs since
that time, and envisages more oak–hazel forest on acid rocks in both the north
and the south of Ireland. All such maps invite questions (which is what they
should do!). Was there indeed a zone of birch forest on the higher parts of the
Wicklow Mountains and McGillycuddy's Reeks? As far as I know, no-one has so
far attempted a similar exercise for the whole of Britain. It would be interesting
to do so, but not an enterprise to be undertaken lightly. And should naturalised
introductions like sycamore, rhododendron and (in Ireland) beech be included
in our assessment of 'potential natural vegetation'?

The distribution of woodland types at the present day

We are on much surer ground here, but immediately come up against the
problem that, although the extremes may be very different, they intergrade, and
it is extraordinarily difficult to draw clear boundaries between woodland 'types'.
Our woods form a continuum (like colour), but as with colour, we need to define
some reference points if we are to talk intelligibly about them. Continental
ecologists have the same difficulty in drawing clear boundaries (Wilmanns 1978).

Woods vary with soil type, and there is also climatic variation with altitude
and latitude and with degree of oceanicity or continentality, and there are
distinctive woods on permanently wet soils. The next two chapters broadly
follow the scheme adopted by John Rodwell (1991a) in *British Plant Communities*

	RENDZINAS AND CALCAREOUS BROWN-EARTHS	BROWN-EARTHS OF LOW BASE STATUS	RANKERS, BROWN PODZOLIC SOILS AND PODZOLS
COLD NORTHERN UPLANDS AND SUB-ALPINE ZONE	[Salix lapponum scrub (W20)]	Scots pine–Hylocomium splendens woodland (W18) (Caledonian pine forest)	
		Juniper–wood-sorrel woodland (W19)	
COOL AND WET NORTH-WESTERN SUB-MONTANE ZONE	Ash–rowan–dog's mercury woodland (W9)	Oak–birch–wood-sorrel woodland (W11)	Oak–birch–Dicranum woodland (mossy western oakwood) (W17)
WARM AND DRY SOUTH-EASTERN LOWLAND ZONE	Ash–field-maple–dog's-mercury woodland (W8)	Oak–bramble–bracken–bluebell woodland (W10)	Oak–birch–wavy hair-grass woodland (W16)
	Beech–dog's mercury woods ('hangers') (W12)	Plateau beechwoods with bramble (W14)	Beech–wavy hair-grass woods (heathy beechwoods) (W15)
WET WOODS	Sallow carr and alder woods on fen peats and wet mineral soils (W1-3 and W5-7)		Wet birch woods mainly on raised-bog peat (W4)

FIG 65. Habitat relationships of the major types of British woodlands, based on Rodwell (1991a), with the NVC code-numbers used there. The Irish ash–hazel limestone woodland (Corylo-Fraxinetum) shares some of the character of the British northern and southern ash woods (W8 and W9) and the Irish mossy oakwoods (Blechno-Quercetum) are equivalent to the western oak–birch woods in Britain (W17). Equivalent wet woods probably occur in both islands, but need further study.

(Fig. 65). Soils fall into three major categories. The leached, acid soils (podzols and rankers) developed over old hard rocks or decalcified sands and gravels, are poor in nutrients and have a low pH (usually < 4.5), and a layer of acid humus accumulates on the surface. Brown-earth soils vary from mildly acid to around neutral (pH 4.5–6.5), organic matter is broken down at the surface and the humus incorporated into the soil by earthworms, and essential plant nutrients are usually in adequate supply; these are optimal soils for tree growth. The more calcareous brown earths intergrade with the rendzina soils over chalk and limestone, which vary from around neutral to weakly alkaline (pH often 6.5–7.5); calcium is abundant and the soils are usually reasonably nutrient-rich, but calcifuge species experience phosphate and trace-element deficiencies on these soils. The altitudinal-geographical divisions of the first column of Figure 65 are not hard-and-fast; they have very fuzzy boundaries. Thus woods of quite upland nature can occur at relatively low altitudes in southern England on poor acid soils, and favourable sites in the north and west can feel surprisingly southern in character. Juniper–wood-sorrel scrub grows at a wide range of altitudes from the Burren and the Dales of the north of England to the treeline in the Cairngorms. Some woodland types are in effect defined by the native range of individual tree species, Scots pine and beech are obvious examples. Field maple is a defining species of lowland ashwood in England and Wales, but hardly reaches Scotland and is absent from Ireland.

CHAPTER 6

Woods on the Better Soils, and Wet Woods

T HIS CHAPTER IS ABOUT THE WOODS that probably most of us first think of: oak, ash and beech woods on deep brown-earth soils of the lowlands, or on shallower soils over chalk and limestone, with abundant woodland flowers in the field-layer, and it includes most woody vegetation in wet places. However, the line between this chapter and the next is not sharp. There is intergradation and overlap, and some woods might equally well have found a home in either chapter.

LOWLAND WOODS ON WELL-DRAINED SOILS

Oakwoods and ashwoods

Woods on deep mildly acid brown-earth soils throughout lowland Britain, and more locally in Ireland, are dominated by pedunculate oak, often with an understorey of hazel and scattered hawthorn. Less often, sessile oak is dominant. Silver birch may sometimes be prominent, and various other trees may occur, including holly, rowan, wild service-tree, wild cherry, ash, yew, hornbeam and small-leaved lime. Most of these occur as scattered individuals, but some can be locally dominant, for instance hornbeam and small-leaved lime. Under the trees, the most constant plants are bracken, brambles and honeysuckle (Fig. 66c), and ivy is prominent in places. Bluebell is the most constant of the field-layer herbs, often staging spectacular displays of blue in early May (Fig. 1). The two male-ferns (*Dryopteris filix-mas* and *D. affinis* agg.) and broad buckler-fern (*D. dilatata*) are locally prominent, as is creeping soft-grass (*Holcus mollis*), which forms an

FIG 66. Oakwood plants. (a) Wood anemone (*Anemone nemorosa*); (b) wood-sorrel (*Oxalis acetosella*); (c) honeysuckle (*Lonicera periclymenum*); (d) common dog-violet (*Viola riviniana*).

understorey to the bracken (Fig. 67). Many other characteristic woodland plants are more scattered, but when in flower in spring they add their own patches of colour to the woodland floor before the bracken unfurls its fronds. The most widespread include wood anemone (Fig. 66a), wood-sorrel (Fig. 66b), pignut (*Conopodium majus*) and common dog-violet (Fig. 66d). Mosses are seldom prominent, but *Isothecium myosuroides* (Fig. 89a) often forms a green zone around the bases of the trunks, and *Kindbergia praelonga*, *Mnium hornum* and the beautiful *Thuidium tamariscinum* are frequent on stumps or on the ground (Fig. 68).

This *Quercus robur–Pteridium aquilinum–Rubus fruticosus* woodland[W10] is very variable. It is the heir to much of the wildwood, the forest that covered our islands before Neolithic man began forest clearance. Even before man was a

FIG 67. Sessile oak with bluebells at the peak of their flowering season, with creeping soft-grass (*Holcus mollis*) and bracken; a classic mix (Woodhead 1906). Bulbarrow Hill, Dorset, May 1977.

significant influence, it must have varied from place to place with climate and soil. In the woods as we see them now, the effects of millennia of human activity are superimposed upon this underlying natural variation. Much coppice-with-standards woodland is essentially this community (Figs 69, 70). The tree-layer would have been subject to selection, often for pure oak (and may bear the signs of systematic and intensive management, often followed by neglect), and the neat regular understorey of hazel coppice is an obvious artefact. Where the main crop taken was small timber for charcoal and bark for tanning, as was often the case in the west, the oaks themselves were coppiced, leading to a wood of similar species-composition, but with little or no hazel, and of different appearance and structure. A 'natural' wood might be composed of much the same species, but would be different again, and an altogether more untidy affair.

On more calcareous soils we encounter a different kind of woodland. Oak may still be dominant, but it is usually accompanied by ash, and it may be completely replaced by that tree. Hazel is constant, and in many places hawthorn is too. Field maple is a widespread and characteristic constituent. Wych elm and sycamore are frequent and occasionally dominant – the first a long-standing native, the other introduced a few centuries ago. Other trees in smaller quantity or more locally prominent include hornbeam, small-leaved lime, silver birch, goat willow, aspen, elder (*Sambucus nigra*) and holly. This *Fraxinus–Acer campestre–*

FIG 68. Mosses of oakwoods. (a) *Mnium hornum* often carpets banks and stumps; it produces big pendulous capsules, which mature in early spring. (b) *Thuidium tamariscinum*, a frequent and striking species on banks and the floor of the wood. (c) *Kindbergia praelonga*, common in many woodland situations. (d) *Atrichum undulatum*, on the ground, superficially like *Mnium hornum*, but with a band of green lamellae over the midribs of the leaves and very different cylindrical capsules.

Mercurialis woodland[W8] is conspicuously rich in species (Fig. 71). The shrub-layer is enlivened by a group of five calcicolous shrubs, dogwood (*Cornus sanguinea*), spindle (*Euonymus europaeus*), wayfaring-tree (*Viburnum lantana*), buckthorn (*Rhamnus cathartica*) and wild privet (*Ligustrum vulgare*) – and the more widely distributed blackthorn (*Prunus spinosa*) and guelder-rose (*Viburnum opulus*). The field-layer is rich too, in summer often dominated by a green carpet of dog's

LEFT: **FIG 69.** Yellowham Wood, near Dorchester, April 1969. A traditional coppice-with-standards wood: pedunculate oak with an understorey of hazel. An area coppiced a year or two previously. The cut stools of hazel have started to re-grow; wood anemones and primroses in flower. Fig. 70 shows how much the hazel has grown five years later. See also Fig. 46.

BELOW: **FIG 70.** Yellowham Wood, April 1974. The nearer part was coppiced six or seven years previously.

FIG 71. Small-leaved lime wood on Devonian limestone, Chudleigh Rocks, Devon, May 1972. Other woody species include wych elm, field maple, ash and hazel; field-layer dominated by ramsons (*Allium ursinum*).

FIG 72. Plants of base-rich woods. (a) Dog's mercury (*Mercurialis perennis*); (b) sanicle (*Sanicula europaea*); (c) yellow archangel (*Lamiastrum galeobdolon*); (d) oxlip (*Primula elatior*).

mercury (Fig. 72a), but in spring often colourful with a profusion of woodland flowers: bluebells, primrose, ground-ivy (*Glechoma hederacea*), yellow archangel (Fig. 72c), lesser celandine, herb-robert (*Geranium robertianum*), lords-and-ladies (*Arum maculatum*) and ramsons (*Allium ursinum*). Later-flowering species include sanicle (Fig. 72b), enchanter's nightshade (*Circaea lutetiana*), woodruff (*Galium odoratum*), wood avens (*Geum urbanum*) and stinking iris (*Iris foetidissima*) – with

FIG 73. Mosses of base-rich woods. (a) *Plagiomnium undulatum*; (b) *Ctenidium molluscum*; (c) *Eurhynchium striatum*; (d) *Rhytidiadelphus triquetrus*; (e) *Neckera crispa*; (f) *Thamnobryum alopecurum*.

meadowsweet (*Filipendula ulmaria*), bugle (*Ajuga reptans*) and tufted hair-grass (*Deschampsia cespitosa*) in damper places. Ferns include hart's-tongue (*Phyllitis scolopendrium*), hard shield-fern (*Polystichum aculeatum*, mainly N and W) and soft shield-fern (*P. setiferum*, mainly S). Three unspectacular but characteristic plants are wood-sedge (*Carex sylvatica*), false brome (*Brachypodium sylvaticum*) and wood dock (*Rumex sanguineus*). The mosses are generally more calcicole (Fig. 73). The oxlip (Fig. 72d) woods of eastern England belong here.

This woodland has its origins in the wildwood of the more calcareous soils, on the chalk and the various older limestones in Britain, and partakes of the same kind of variation as the last, and has a similar history of use and management; coppice-with-standards examples are often encountered. Essentially the same kind of woodland, but lacking field maple and dog's mercury (and some other species common in southern England), is scattered over the Carboniferous limestone of Ireland, where it has been called (following Braun-Blanquet & Tüxen) the 'Corylo-Fraxinetum' (Fig. 74). In Ireland many examples are dominated by hazel alone, notably the hazel scrubs of the Burren

FIG 74. Ash–hazel wood (Corylo-Fraxinetum) with birch and rowan on a rocky limestone slope, Poulavallan, Burren, Co. Clare, July 1959. The field-layer includes wild strawberry, barren strawberry, wood avens, ivy, yellow pimpernel, wood-sorrel, hart's-tongue, hard shield-fern, bush vetch and common dog-violet. Bryophytes include *Eurhynchium striatum, Loeskeobryum brevirostre, Plagiochila asplenioides, Plagiomnium undulatum, Rhytidiadelphus triquetrus* and *Thuidium tamariscinum.*

in Co. Clare, many of which are species-rich woods in all but their lack of 'trees'.
There is little in the field-layer to differentiate them from ash-dominated sites.

Beechwoods on fertile and calcareous soils

Beech is native with us only in southern England and adjacent parts of Wales,
where it dominates woodland communities parallel to those just considered
dominated by oak. Extensive beechwoods on the deep, acid brown-earth
soils developed on superficial deposits over the chalk of the Chilterns, and
comparable soils from the North Downs to the New Forest, and the Cotswolds
to the Carboniferous limestone of South Wales. Beech casts a deeper shade than
oak, and as one consequence of this these 'plateau beechwoods' (*Fagus sylvatica–
Rubus fruticosus* woodland[W14], Fig. 75) show significant difference from their
oak-dominated counterparts. They lack a shrub-layer (there is very little hazel),
bracken and honeysuckle are much less common and do less well, and bluebell
is infrequent. These beechwoods are not floristically exciting places, but they are
magnificent in their own right, and historically they have provided the timber
on which the Chiltern furniture industry was based: Tansley (1939) included a
photograph of 'chair bodgers' at work in the woods. The canopy is usually almost

FIG 75. Plateau beech wood after the great storm of 15/16 October 1987. Box Hill, Surrey,
November 1987. Patches of low bramble on the floor of the wood; sparse shrub-layer of holly
and box. Occasional catastrophes – storm, drought, fire and flood – are a part of life for plant
communities; the only certainty is that it will happen again – but when?

pure beech, though it may include a proportion of sessile oak, and often there is a sparse or denser subsidiary tier of holly. Other trees that occur occasionally are whitebeam, wild cherry, yew and sycamore. The only constant constituent in the field-layer is bramble, which may cover half the woodland floor, or be only a wispy cover, depending on shade and root-competition with the trees, and grazing. The only other species that is frequent and attains substantial cover is bracken. Otherwise there is only a sparse ground flora, of which the main ingredients are wood-sorrel, occasional yellow archangel and woodruff, a few grasses – most notably wood millet (*Milium effusum*), wood melick (*Melica uniflora*) and creeping soft-grass – and the mosses *Mnium hornum* and *Polytrichastrum formosum*.

The beechwoods of the calcareous rendzina soils of the chalk escarpments are much better known, thanks to Gilbert White's description of the beech 'hanger' in *The Natural History of Selborne*, and to the fact that they are home to some odd and attractive rare plants. The beeches on these thin, poor calcareous soils are not as well grown as the trees on the deeper and richer brown-earth soils we have just considered, but they still cast a deep shade, and the ground beneath them is comparably bare. Beech is usually dominant almost to the exclusion of other trees in this *Fagus sylvatica–Mercurialis perennis* woodland[W12], with just the occasional ash or sycamore, and less often oak. There is often no shrub-layer, or there may be a scatter of hazel, hawthorn, field maple, elder, spindle, wayfaring-tree, dogwood and holly. The character of the field-layer mainly reflects the character of the soil. On the deeper soils, dog's mercury generally forms a light cover, with a sparser version of the ground flora of ash–oak-dominated woodland on similar calcareous soils, often with a good deal of bramble. On thinner and chalkier soils there is more bare ground. Here, sanicle rivals or exceeds dog's mercury in a very characteristic field-layer, with tall slender plants of wall lettuce (*Mycelis muralis*) in the deep shade, wood melick and wood meadow-grass (*Poa nemoralis*); black bryony (*Tamus communis*) and hairy brome (*Bromopsis ramosa*) are also frequent. A very characteristic plant here is the white helleborine orchid (Fig. 76). This community provides a habitat for some notable rarities, including long-leaved helleborine (*Cephalanthera longifolia*), red helleborine (*C. rubra*), narrow-lipped helleborine (*Epipactis leptochila*), bird's-nest orchid (*Neottia nidus-avis*), yellow bird's-nest (*Monotropa hypopitys*), wood barley (*Hordelymus europaeus*) and green hound's-tongue (*Cynoglossum germanicum*). On thin dry soils on southerly slopes, yew can become more constant and form a shrub-layer under the beech; on Box Hill in Surrey, box (*Buxus sempervirens*) can behave similarly. This creates deep shade and, with the dryness of the soil, the ground flora becomes very sparse.

FIG 76. White helleborine (*Cephalanthera damasonium*), a very characteristic plant of escarpment beechwoods on the chalk. Note the otherwise rather sparse field layer. Hambledon Hill, Dorset, June 1970.

Beech has been very widely planted in both Britain and Ireland, and regenerates freely far outside the range it had reached naturally. With most of the other species already there, it has carried what are often to all intents and purposes the last two communities with it to its new-found territory.

Yew woods on the chalk, and elsewhere

Yew locally forms woods on thin soils over the southern English chalk, most typically on the steep slopes of the sides and heads of dry valleys, but always with a preference for a south- to west-facing aspect[W13]. Yew-woods are rather frequent along the North Downs (where the main escarpment is south-facing), at the western end of the South Downs in West Sussex (Fig. 77) and Hampshire, and at the southwestern end of the Chilterns; there is a nice outlying example on the south side of Hambledon Hill in Dorset. Farther north and west, there are patches of yew wood in the Cotswolds and on Carboniferous limestone in the Wye Valley and northern England, and there is a notable Irish yew wood on the Carboniferous limestone of the Muckross peninsula at Killarney.

On the chalk of southern England, yew forms a dense canopy, alone or with a scatter of whitebeams and the occasional ash. Typically the shade is so dense that

FIG 77. Kingley Vale, West Sussex, June 1985. General view of the lower part of the yew wood; escarpment beechwood beyond. Inside the yew wood the shade is so dense that practically nothing grows under the trees. Hawthorn in flower in the foreground.

almost nothing grows beneath the canopy beyond a few tree and shrub seedlings. Whitebeam is an opportunist invader of gaps in the yew canopy, as is elder, the most frequent shrub. Where the canopy is a little more open, a scatter of shrubs and a thin field-layer can become established. A sparse growth of spindle, wild privet and dogwood accompany somewhat more frequent elder. The field-layer consists mostly of a patchy cover of dog's mercury. The slightly 'weedy' tone set by the elder is echoed by occasional patches of stinging nettles; an uncommon ruderal that occurs occasionally in this situation is deadly nightshade (*Atropa belladonna*). On Carboniferous limestone in the Wye Valley and the north of England, yew is seldom as uncompromisingly dominant as it is on the chalk, and the field-layer of yew wood is an impoverished version of that of the neighbouring ashwoods.

The Killarney yew wood is on limestone pavement (Fig. 78). The higher and drier areas, with only skeletal soils, are dominated by yew with little else apart from scattered holly and occasional patches of hazel. Wherever there is some accumulation of soil, taller and more mixed woodland has been able to develop, with ash as a substantial component of the canopy, and an open shrub-layer of hazel, with some spindle. Bramble, hawthorn, ivy, honeysuckle and

FIG 78. Yew wood on Carboniferous limestone, Muckross peninsula, Killarney, Co. Kerry. Here yew is co-dominant with ash, there is a sparse field layer, and the limestone rocks are covered with bryophytes, mostly *Thamnobryum alopecurum*.

holly are scattered throughout. Because the ground is so rocky, herbs needing any substantial depth of soil are sparse: false brome is common; other species include sanicle, hart's-tongue and soft shield-fern. The limestone rocks are thickly covered with bryophytes including *Thamnobryum alopecurum*, *Eurhynchium striatum*, *Thuidium tamariscinum*, *Rhytidiadelphus triquetrus*, *Loeskeobryum brevirostre*, *Neckera complanata*, *Ctenidium molluscum*, *Tortella tortuosa* and *Marchesinia mackaii*.

Box woods

In a few places in southern England box forms woods on steep dry south-facing slopes in a similar way to yew. The largest and best known of these is at Box Hill in Surrey, where the River Mole has cut a steep cliff in the chalk of the North Downs (Fig. 79). The evergreen box casts a deep shade, and this together with the steepness of the slope and the dryness of the soil means that little can grow beneath the canopy. There was formerly a box wood at Boxley near Maidstone in Kent, recorded by John Ray in 1695. Box is also plentiful and locally forms dense scrub at Ellesborough Warren near Wendover in the Chilterns. The box

FIG 79. Box Hill, Surrey, July 1958. The box-wood occupies the steep southwest-facing chalk slope cut by a meander of the River Mole. The high-crowned trees are beech; the trees with grey-green foliage are whitebeams.

wood at Boxwell Court, west of Tetbury in the Cotswolds, is known to have been in existence since at least the thirteenth century. It covers 5.3 ha on a steep southwest-facing limestone slope. Compartments have been regularly coppiced. Box wood is prized for its hardness, light colour and close uniform grain, and the wood brought in a large income to the Huntley family, and no doubt to the Abbots of Gloucester before them. An area cut over during the First World War had regenerated by the 1970s, but still with traveller's-joy (*Clematis vitalba*) among the box trees. On a second area, cutting in 1971 was followed by profuse growth of rosebay willowherb (*Chamerion angustifolium*), nettles and elder – and the odd plant of henbane (*Hyoscyamus niger*) – amongst the regenerating box, all destined to be shaded out as the trees grew up.

Box has been widely planted over the centuries, and does well in our climate, so it is impossible to say with certainty where it is native. 'Box' place-names on the chalk and Jurassic limestones – as at Box in Wiltshire and Gloucestershire, Boxford in Berkshire and Suffolk, Boxted in Suffolk and Boxgrove in Sussex – suggest long historical associations with this tree, which may have been more widely indigenous. Other 'Box' and 'Bux' place-names seem to have other origins (Eckwall 1960).

UPLAND WOODS ON BROWN-EARTH SOILS
AND LIMESTONE

In upland country the woods gradually change in character. The oaks (here most often *Quercus petraea*) are less well grown, and birches (*Betula pendula* and *B. pubescens*) become more frequent and may be dominant. Hazel, hawthorn and honeysuckle are less common, and hornbeam and small-leaved lime are virtually absent from the upland woods. Bracken remains prominent, but brambles are much less so. The field-layer is more grassy, perhaps partly reflecting the fact that these upland woods are more often open to grazing. Sweet vernal-grass (*Anthoxanthum odoratum*), common bent (*Agrostis capillaris*), wavy hair-grass (*Deschampsia flexuosa*) and creeping soft-grass (*Holcus mollis*) are constant or nearly so, and a group of large mosses are common, including *Rhytidiadelphus squarrosus* and *R. triquetrus, Pseudoscleropodium purum, Thuidium tamariscinum, Hylocomium splendens, Pleurozium schreberi* and *Dicranum majus*. The more frequent herbs on the woodland floor include wood-sorrel, common dog-violet, heath bedstraw (*Galium saxatile*), tormentil (*Potentilla erecta*), bluebell, germander speedwell (*Veronica chamaedrys*), wood anemone, primrose, hairy wood-rush (*Luzula pilosa*) and pignut; chickweed-wintergreen (*Trientalis europaea*) is particularly characteristic of stands of this community in the eastern Highlands. Ferns (other than bracken) are usually not prominent; they include scaly male-fern (*Dryopteris affinis*), broad buckler-fern and hard fern (*Blechnum spicant*)[W11].

This upland *Quercus petraea–Betula pubescens–Oxalis acetosella* woodland (Fig. 80) is very widespread in the north and west of Britain (some woods on high 'plateau gravels' in the southeast arguably belong here), and could be matched in Ireland. It is a prime example of a woodland in which the tree-canopy can change with little effect on the character of the vegetation on the ground, so it embraces woods dominated by both the oak species, and many woods dominated by birch (usually *Betula pubescens*, but sometimes *B. pendula*), and even hazel scrub in Skye and elsewhere in western Scotland (Birks 1973).

The northern and upland ashwoods on limestone also differ from their lowland counterparts, the ash–field-maple–dog's-mercury woods, but (perhaps because suitable habitats for them are more fragmented) they do not present as coherent a picture as the oak–birch–wood-sorrel woods of more acid soils. Overall, they lack field maple, the calcicolous shrubs of the south, and some other lowland species. Conversely, rowan and downy birch are commoner and bird cherry occurs more widely. In the field-layer, wood-sorrel, common dog-violet, male-fern and the mosses *Thuidium tamariscinum, Plagiomnium undulatum* and *Eurhynchium striatum* are much more conspicuous than they are in the

FIG 80. Upland oak (*Quercus robur*) wood, near Shaugh Prior, Dartmoor, Devon. The grassy field-layer is dominated by sweet vernal-grass, with some creeping soft-grass, bluebells, bracken, great woodrush and abundant wood-sorrel.

lowland woods. Northern species occurring in some of the damper of these upland ashwoods in northern England and Scotland include wood crane's-bill (*Geranium sylvaticum*), globeflower (*Trollius europaeus*), marsh hawk's-beard (*Crepis paludosa*), melancholy thistle (*Cirsium heterophyllum*) and various lady's-mantles (*Alchemilla* spp.)[W9].

This ash–rowan–dog's-mercury woodland has a northern and western distribution on limestone (or other base-rich rocks) from upland Wales and northern England to the north of Scotland. The woods on the steep sides of the high Yorkshire Dales (closely matched by some Irish woods) are in general this community, by contrast with the Peak District woods (e.g. in Dovedale, and along the Via Gellia), which generally have a decidedly lowland character. Some of these Yorkshire Dales woods are on limestone pavement. Colt Park Wood near Ribblehead is an extreme example (Fig. 81). It occupies a strip of deeply dissected limestone pavement above a cliff a few metres high in the stepped limestone landscape. The flattish blocks of limestone forming the surface ('clints') are separated by deep fissures ('grikes'), many wide enough to hold a sheep (occasionally sheep are trapped in the grikes and perish miserably if not retrieved in time). The clints bear a patchy thin soil, but many plants, including the trees, are

FIG 81. Colt Park Wood, Ribblehead, North Yorkshire, July 1966. Farther back from the cliff edge the vegetation often hides the clints and grikes of the Carboniferous limestone rock.

rooted in the soil at the bottom of the grikes, or in crevices in the limestone rock. Colt Park Wood has a light tree-canopy, mainly of ash, with hazel, hawthorn, bird cherry, blackthorn and wych elm. The ground flora is rich in species to match the diversity of habitat; among the more notable plants are alpine cinquefoil (*Potentilla crantzii*), northern bedstraw (*Galium boreale*) and angular Solomon's-seal (*Polygonatum odoratum*) rooted on the clints, and baneberry (Fig. 82a), globeflower, northern (*Crepis mollis*) and marsh hawk's-beards, wood crane's-bill and melancholy thistle in the grikes. The nearby limestone ravine of Ling Gill lacks the clints and grikes, but has a generally similar flora.

Limestone pavements

Limestone pavements are complex mosaics of diverse habitats, and have correspondingly diverse vegetation (Fig. 83). The grikes supply many of the ingredients of a woodland environment: shelter from drying winds, and shade, which becomes deeper with the depth of the grike. Many bare limestone pavements have a woodland flora dominated by dog's mercury at the bottom of the grikes, with ferns – hard shield-fern (Fig. 82c), male-fern, hart's-tongue and wall-rue, maidenhair and green spleenworts (*Asplenium ruta-muraria, A. trichomanes, A. viride*, Fig. 82d) – and other plants rooted deep in the grikes, or in crevices on their

FIG 82. Plants of limestone pavement. (a) Baneberry (*Actaea spicata*); (b) bloody crane's-bill (*Geranium sanguineum*); (c) hard shield-fern (*Polystichum aculeatum*); (d) green spleenwort (*Asplenium viride*).

sides. Where there is a soil cover on the clints, it can carry a grassland or heathy flora depending on its thickness. Samples from limestone pavements in the Burren (where dog's mercury is very rare) yielded a long list of grassland species; woodland was represented by false brome, ivy, tutsan (*Hypericum androsaemum*), wood-sorrel and sanicle, scrub fringes by bloody crane's-bill (Fig. 82b) – and

FIG 83. Scar Close, Ingleborough, North Yorkshire, July 1964. Limestone pavement, with and without a soil cover; scattered ash trees rooted in the grikes, which have since grown up into patches of woodland.

seedlings or saplings of all the woody ingredients of typical Burren limestone scrub or woodland. The grikes provide a congenial habitat for the establishment of tree and shrub seedlings, but if the pavement is grazed these are browsed level with the clints and cannot grow to maturity. If grazing is relaxed that can quickly alter. Scar Close on the west side of Ingleborough is ungrazed, and has changed within the last half-century from an almost bare expanse of limestone pavement with scattered trees to a no-less-interesting site with abundant trees and some extensive patches of open ash–rowan–dog's-mercury woodland.

Scottish upland ashwoods

There are some woods in Scotland that represent the northern fringe of ashwood in Britain. John Birks described ashwoods from Durness limestone in Skye, which are clearly the same kind of vegetation as the upland ashwoods of the Yorkshire Dales (false brome seems to increase progressively northwards). However, the distinctive tall herbs of the grike flora at Colt Park Wood are missing, and appear in another community, the '*Betula pubescens–Cirsium heterophyllum* association' at low altitudes on inaccessible ledges in damp wooded ravines on basic rocks, dominated by birch but in which ash also occurs (Birks 1973). Maybe there is a lost woodland type here, of which we see a ghost in the

FIG 84. Rassal ashwood, above Loch Kishorn, Wester Ross, June 1981. Open to grazing, with a grassy field-layer and bracken; mosses covering the limestone rocks.

traditional flowery hay meadows of the north of England, intergrading and locally forming mosaics with upland ashwood. We may visualise it dominated by downy birch with ash, willows and bird cherry, with a tall-herb field-layer, something like the little 'Globeflower Wood' near Malham Tarn. There are such woods in Scandinavia, and we have the ingredients for them in Britain.

Rassal ashwood, which lies on an outcrop of Durness limestone sloping gently westwards at about 300 m above Loch Kishorn, is the northernmost substantial ashwood in Britain. It was open to grazing when it was declared a National Nature Reserve in 1956; the limestone rocks were thickly covered with mosses, and the less rocky parts were grassy with a patchy cover of bracken. After the wood became a nature reserve some areas were fenced; the two photographs taken in 1981 (Figs 84, 85) show the remarkable contrast in vegetation between grazed and ungrazed areas. The ash trees, some of them large, form a dense canopy in places but in others are more open. Hazel is abundant, with occasional downy birch, goat willow and rowan, and some hawthorn and blackthorn scrub. The grazed, grassy parts are dominated by sheep's fescue (*Festuca ovina*) and common bent with crested dog's-tail (*Cynosurus cristatus*) and tufted hair-grass. Only a few woodland plants were apparent in the crevices between the rocks (e.g. false brome, sanicle, primrose) but surprisingly luxuriant woodland vegetation

FIG 85. Rassal ashwood.
Ungrazed exclosure,
with luxuriant field-layer
including male-fern,
creeping buttercup and
meadowsweet. June 1981.

(including wood avens, meadowsweet and creeping buttercup, *Ranunculus repens*)
grew up when grazing was excluded.

WET WOODS AND SALLOW CARRS

Sallow carr

Patches and tracts of country dominated by sallows are common in both
countries. In Britain, Rodwell (1991a) divides the wet sallow-dominated
vegetation widely scattered over lowland England and Wales, usually on wet
mineral soils ('*Salix cinerea–Galium palustre* woodland')[W1], from the fen carr of
East Anglia and the Shropshire–Cheshire plain, usually on peat ('*Salix cinerea–
Betula pubescens–Phragmites australis* woodland')[W2]. Both are exceedingly variable,

and not well characterised. There is all-too-little good descriptive information about sallow scrub in Britain. It is often difficult and uncomfortable to work in, and not a superficially attractive habitat. The *Sphagnum* sub-community of the East Anglian fen carr would probably be better united with the birchwoods on peat (Chapter 7), leaving the *Alnus–Filipendula* sub-community (mainly of eutrophic peats) grouped with the widespread *Salix cinerea–Galium palustre* woodland and the Irish sallow sites. In Ireland sallow scrub grows mainly on alluvial soils, and seems to be better characterised, with constant sallow, near-constant alder, marsh bedstraw (*Galium palustre*), creeping bent (*Agrostis stolonifera*), and the mosses *Kindbergia praelonga* and *Calliergonella cuspidata*. There is a long list of frequent species, including ash, purple-loosestrife (*Lythrum salicaria*), creeping buttercup, yellow iris (*Iris pseudacorus*), marsh bedstraw and water mint (*Mentha aquatica*) (Kelly & Iremonger 1997).

The abundant supply of nutrients brought in by silting makes riverside habitats fertile places, which favours plants with ruderal tendencies such as nettles, elder, reed canary-grass (*Phalaris arundinacea*) and great willowherb (*Epilobium hirsutum*). Wet woods along streams and riversides and around eutrophic lakes can be dominated interchangeably by sallow, alder, various tree willows and osiers, for which 'Alnus glutinosa–Urtica dioica woodland'[W6] and 'Salicetum albo-fragilis' are equally misnomers, because neither alder nor tree-willows are necessarily present, and large areas of eutrophic vegetation of this kind are dominated by sallow. This is the nearest we have in Britain and Ireland to the floodplain woodland that lines many central-European rivers. Frequent species include crack-willow, sallow, alder, elder, cleavers (*Galium aparine*), bittersweet (*Solanum dulcamara*), bramble and hemlock water-dropwort (*Oenanthe crocata*). Meadowsweet, greater and lesser pond-sedges (*Carex riparia* and *C. acutiformis*), yellow iris, common valerian (*Valeriana officinalis*), cuckooflower (*Cardamine pratensis*), marsh-marigold (*Caltha palustris*), wild angelica (*Angelica sylvestris*), gypsywort (*Lycopus europaeus*) and opposite-leaved golden-saxifrage (*Chrysosplenium oppositifolium*) lend interest and colour to this otherwise rather forbidding community.

In northern England and Scotland a distinctive fen carr occurs, often dominated by bay willow (*Salix pentandra* – a beautiful species), sometimes other northern *Salix* species, and grey (common) sallow[W3]. This is a well-characterised community with a long list of constant or near-constant species, including cuckooflower, marsh bedstraw, marsh-marigold, meadowsweet, bottle sedge (*Carex rostrata*), water avens (*Geum rivale*) and frequent bogbean (*Menyanthes trifoliata*), marsh hawk's-beard, common valerian, water mint and marsh cinquefoil (*Potentilla palustris*), and bryophytes including *Calliergonella cuspidata*,

Rhizomnium punctatum and *Climacium dendroides*. Bay willow occurs in Ireland, especially in the north, but never seems to become prominent in carr.

Alder woods

Grey sallow commonly forms extensive stands in largely stagnant sites. By contrast, alder usually grows in places where there is some water flow, either close to rivers or streams (alder is a common riverbank tree), at spring-lines, or in valleys with water flow through the subsoil.

A very characteristic type of wet woodland is dominated by alder, with variable amounts of silver birch, sallow and ash, with greater tussock-sedge (*Carex paniculata*) dominant in the understorey[W5]. This is a variable but striking and rather species-rich wet woodland with few constants but a long list of species that are frequent wherever it occurs. These include marsh bedstraw, bittersweet, cuckooflower, marsh thistle (*Cirsium palustre*), water mint, stinging nettle, yellow iris and marsh valerian (*Valeriana dioica*); this is a habitat of the beautiful moss *Hookeria lucens*. Alder buckthorn (*Frangula alnus*, a food-plant of the brimstone butterfly) is common in some stands, and royal fern (*Osmunda regalis*) occurs occasionally. This community is widely scattered over England

FIG 86. Alder–great tussock-sedge (*Alnus glutinosa–Carex paniculata*) woodland at the spring-line between permeable Cretaceous sands and gravels capping the hill, and the Permian marls beneath. Pen Hill, Sidbury, Devon, June 1977.

and Wales, in habitats ranging from some of the Norfolk Broads and the banks of Broadland rivers, to spring-lines on the East Devon hills (Fig. 86) and seepage tracks through New Forest and Dorset valley bogs. There are a few sites in Scotland, but it is apparently rare in Ireland.

Ireland still has a surviving fragment of the Geragh, near Macroom in Co. Cork, a unique labyrinth of braided river channels and islets clothed in tangled woodland, which was largely destroyed by a hydroelectric scheme in the 1950s. Trees include sessile oak and ash up to 15–18 m high, with alder, birch, hazel, hawthorn, grey sallow, goat willow, holly, bird cherry, spindle and guelder-rose. The ground flora includes many plants of base-rich woodland; patches of ramsons are conspicuous, and in the wetter places the widespread woodland herbs give way to patches of e.g. marsh-marigold, opposite-leaved golden-saxifrage, water mint and water ragwort (*Senecio aquaticus*). Many of the island margins are conspicuously populated with hemlock water-dropwort and royal fern (White 1985).

In hilly parts of both Britain and Ireland, flushes or wet gullies in otherwise well-drained oakwoods on mildly acid brown-earth soils are often occupied by patches of a distinctive wet woodland in which alder is often dominant, usually accompanied by some ash, hazel and occasionally birch[W7]. Meadowsweet is usually conspicuous in the field-layer, with creeping buttercup and other plants of wet places such as opposite-leaved golden-saxifrage, wild angelica, marsh thistle, tufted hair-grass, common valerian and soft-rush (*Juncus effusus*); two species particularly characteristic of the habitat are yellow pimpernel (*Lysimachia nemorum*) and remote sedge (*Carex remota*); other sedges that find this a congenial habitat are pendulous sedge (*Carex pendula*, a stately plant) and smooth-stalked sedge (*Carex laevigata*). Bryophytes are common: among the more frequent species are the mosses *Kindbergia praelonga, Brachythecium rutabulum, Atrichum undulatum, Plagiomnium undulatum* and *Rhizomnium punctatum*, and the liverworts *Pellia epiphylla* and *Lophocolea cuspidata*. Striking but rarer species that occur in this habitat are the moss *Hookeria lucens*, which has broad flat leaves and cells so big they can be seen with the naked eye, and the beautifully fringed leafy liverwort *Trichocolea tomentella*.

Most of the common woodland flowers of the oakwood are still present here, and this kind of woodland is often rather species-rich. However, its composition is variable. In the more eutrophic examples (a minority), elder is rather frequent, nettles and cleavers are common, and opposite-leaved golden-saxifrage covers a substantial fraction of the ground. In the wetter centre of the flush meadowsweet and its wet-ground associates are most prominent; occasionally (in both islands) great horsetail (*Equisetum telmateia*) can become dominant in this situation, perhaps always where the percolating water is highly calcareous. Tufted hair-grass dominates the field-layer on ground that is damp rather than wet.

CHAPTER 7

Broad-Leaved Woods on Acid Soils, and Scottish Pinewoods

DECIDUOUS WOODS ON POOR ACID SOILS

Beech, oak and birch in the south and east of England

THE OAK AND BEECH WOODS of the last chapter have their counterparts on leached, acid soils. Soils of this kind are widespread in the west and north, but in the southeast their more localised occurrences stand out as something different. Burnham Beeches, Epping Forest and the beechwoods on leached sands and gravels in the New Forest are examples (Fig. 87). These can still be magnificent woods, but generally the trees are less well grown than in woods on the better brown-earth soils. The trees generally cast a deep shade, and the only shrub of any consequence is holly; the sparse field-layer consists mainly of scattered bracken and wavy hair-grass (*Deschampsia flexuosa*). Calcifuge mosses are often rather prominent, typically including *Mnium hornum*, *Dicranella heteromalla*, *Polytrichastrum formosum* and *Leucobryum glaucum*[W15]. Where the beech casts a particularly deep shade very few plants can survive; even holly and wavy hair-grass may be shaded out, and the ground flora reduced to a few mosses. Usually there is a patchy and thin cover of bracken, occasional flowering plants (e.g. hairy wood-rush (*Luzula pilosa*), bluebell, common cow-wheat (*Melampyrum pratense*), creeping soft-grass (*Holcus mollis*)) and bryophytes including such mosses as *Dicranum scoparium* and *Isopterygium elegans* and the liverworts *Diplophyllum albicans* and *Lepidozia reptans*.

Essentially the same community develops in many places where beech has been planted and has made itself at home on acid soils outside its native range

FIG 87. Heathy beechwood on acid soil; an old wood-pasture. Matley Wood, New Forest, Hampshire, May 1994.

in Britain, for instance on Haldon, near Exeter, and elsewhere in Devon. This kind of beechwood intergrades with dry oakwood on similar soils; examples with mixed dominance occur, and beech can sow into woods that were formerly dominated by oak alone.

A century ago Tansley and others wrote of 'oak–birch heath' in southeast England, and on the cycle rides of my schooldays in the 1940s it was immediately recognisable on Harrow Weald and Stanmore Commons. The same sort of vegetation, dominated by various mixtures of the two oaks and the two tree birches, is still widespread over the lowlands of England on dry acid soils. This is an unexciting kind of woodland; many of us must have examples close to home, to which we probably never give a second thought. Some woods of this kind are of long standing, but many are the result of relatively recent colonisation of formerly open heathland with the decline of traditional usage. The only constant constituents are bracken, wavy hair-grass and trees – either birch (usually *Betula pendula*) or one of the oaks. The most frequent other trees are downy birch, holly and rowan. Heather, bilberry (*Vaccinium myrtillus*), broad buckler-fern (*Dryopteris dilatata*) and brambles sometimes occur, but otherwise the flora consists of a rather random selection of species, of generally 'heathy' nature, characterised as much by what is absent as what is present. Bluebells, wood-sorrel (*Oxalis acetosella*), tormentil (*Potentilla erecta*), honeysuckle (*Lonicera periclymenum*), foxglove (*Digitalis purpurea*) and gorse (*Ulex europaeus*) all occur but none of them

FIG 88. Bryophyte-rich western oak and birch woods. (a) Sessile-oak–birch wood, with field-layer of bilberry and great wood-rush, and abundant mosses, Bridford Wood, Devon, May 1972. (b) Coed-y-Rhygen, an oak–birch[–holly] wood with field-layer of bilberry and mosses covering the boulders, near Trawsfynydd, north Wales, April 1994. (c) Birch wood near Shieldaig, Loch Torridon, with bracken and bilberry in the field layer, and mosses carpeting much of the floor of the wood, June 1981: a well-known bryophyte-rich locality.

is frequent[W16]. In general this is a species-poor community. There is a variant from the (English) Midlands northwards with bilberry and mosses somewhat more prominent, and in southwest England bristle bent (*Agrostis curtisii*) can substitute for wavy hair-grass in a local version of this community.

The moss-rich western woods

The common upland oak (or birch) woodland of the west and north is an altogether different proposition. It is widely distributed, from Dartmoor to Caithness and from Kerry to Donegal, much richer in species (many of them bryophytes), and much more coherent and better characterised (Fig. 88). The commonest dominant is sessile oak but pedunculate oak is dominant on the Dartmoor granite and in some other places. The oaks are usually accompanied by lesser amounts of downy birch, rowan and holly – and birch is not uncommonly the dominant tree, especially in Scotland. Hazel sometimes forms a patchy shrub-layer, usually indicating more base-rich soil. In the field-layer bilberry is constant or nearly so, and heath bedstraw (*Galium saxatile*), heather (*Calluna*), bracken, wood-sorrel and wavy hair-grass are frequent; great wood-rush (*Luzula sylvatica*) is often conspicuous where grazing allows. The community is characterised by a long list of constant or near-constant bryophytes (Fig. 89), many of them large and conspicuous enough to catch the eye of even a casual observer. These include *Dicranum majus, Rhytidiadelphus loreus, Hylocomium splendens, Isothecium myosuroides, Polytrichastrum formosum, Pleurozium schreberi* and *Plagiothecium undulatum*; only a little less frequent are *Dicranum scoparium, Mnium hornum* and *Thuidium tamariscinum*[W17]. The 'Blechno-Quercetum' of Braun-Blanquet & Tüxen (1952) from Ireland is essentially the same community. Much of this woodland in western Britain was traditionally coppiced as a source of charcoal and of bark for tanning, so rather uniform multi-stemmed stands of almost pure oak are common; other trees were probably often weeded out. By no means all the western oakwoods were treated like this, and birch (and hazel and holly) are commoner in stands managed as high forest – as some woods in southwest England, Wales, Cumbria and at Killarney evidently were.

Naturally, such a widely distributed community varies in composition from place to place. Much of the variation is related to the gradient of oceanicity from the west coast eastwards, and the epiphyte flora is impoverished if there is much air pollution. In sheltered situations near the west coast, in the valleys of southwest England, west Wales, Cumbria, western Scotland and western Ireland, bryophytes occur in profusion, and such species as *Scapania gracilis, Plagiochila spinulosa, Sphagnum quinquefarium* and *Bazzania trilobata* (and many more) become a regular part of the flora, together with hard fern (*Blechnum spicant*) and filmy

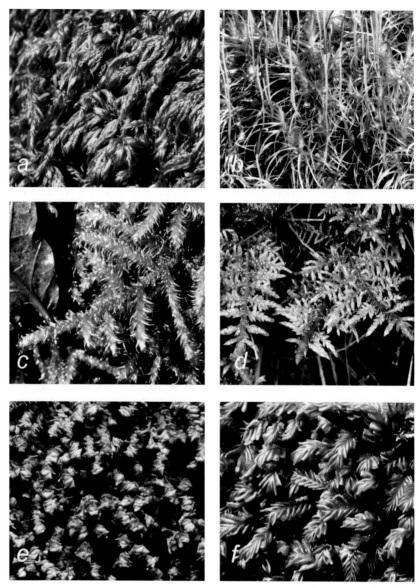

FIG 89. Some typical bryophytes of mossy western oakwoods. (a) *Isothecium myosuroides*, a common species of tree-bases and rocks; (b) *Dicranum majus*, a big upright-growing moss on the ground; (c) *Rhytidiadelphus loreus*, often carpeting rocks on the floor of the wood; (d) *Hylocomium splendens*, another large moss of the floor of the wood; (e) *Scapania gracilis*, a leafy liverwort of rock surfaces and tree trunks in humid places; (f) *Plagiochila spinulosa*, a leafy liverwort of sheltered humid woods.

ferns (*Hymenophyllum* spp.), and oceanic lichens such as *Hypotrachyna* (*Parmelia*) *laevigata, Sphaerophorus globosus* and the beard lichens, *Usnea* spp (Fig. 90). Paul Richards (1938) described the bryophyte communities of the Killarney woods, including cyclical micro-successions on the trees and rocks; parallel features can be seen in the Dartmoor and Welsh woods. In some places (notably in western Scotland and Ireland), shaggy patches of the big green (grey when dry) lichen *Lobaria pulmonaria* ('tree lungwort') cover the branches of the oaks. Several related species occur more rarely; all are sensitive to atmospheric pollution, probably because they are intolerant of low pH and have a significant requirement for calcium (Bates 1992). Eastwards, or in more exposed situations, the less tolerant bryophytes progressively drop out, until we are left with only the bare list of constant species of the community. Where the wood is open to grazing the

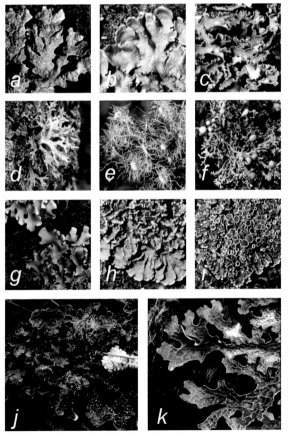

FIG 90. Lichens in western woods. (a) *Parmelia sulcata,* (b) *Parmelia caperata,* (c) *Platismatia glauca* and (d) *Evernia prunastri* are common on trees in all parts of Britain and Ireland. (e) *Usnea florida* is common in Wales and southwest England, and (f) *Sphaerophorus globosus* is common in upland woods. (g) *Hypotrachyna laevigata,* (h) *Degelia plumbea,* (i) *Pannaria rubiginosa* and (j) *Pseudo-cyphellaria norvegica* are lichens of oceanic, western districts. (k) *Lobaria pulmonaria* is one of several big lichens of bases and large branches of old trees in unpolluted areas, now most frequent in hilly western districts.

FIG 91. (a) Wistman's Wood, Dartmoor, from the northwest, June 1971. (b) Wistman's Wood, photographed *c.* 1920, from Christy & Worth (1922, Fig. 10). (c) Wistman's Wood: photographed from the same viewpoint, April 2003. Most of the big trees in the earlier photograph are still recognisable 80 years later, but the epiphyte burden of the trees has decreased, the canopy has risen, and the space under the boughs of the old trees is clearer, allowing a view through to the mid-twentieth-century lateral extension to the wood, of much more 'normal' growth.

field-layer is often more grassy, with sweet vernal-grass (*Anthoxanthum odoratum*), common bent (*Agrostis capillaris*), heath bedstraw and wood-sorrel, all of which are common in the upland oak and birch woods of the previous chapter. In the eastern Highlands, there is a further shift to a variant with prominent heather, bilberry, the moss *Rhytidiadelphus triquetrus*, and occasional chickweed-wintergreen (*Trientalis europaea*), dwarf cornel (*Cornus suecica*) and the orchid *Goodyera repens*. The proportion of birch increases northwards in Scotland, and birch often assumes dominance. Usually the northern subspecies of the downy birch (ssp. *tortuosa*) is involved, but silver birch dominates some woods in the eastern Highlands. These woods are a remarkable community. The richer more western examples are our own temperate rain forest, and the habitat of many of our (and Europe's) rarest bryophytes. We should value them accordingly.

A few woods of this kind have acquired celebrity status. Wistman's Wood, at about 400 m on Dartmoor, has long been noted for the sprawling and tortuous growth-habit of its old oaks (Fig. 91). These are probably 'pioneer' trees that became established among granite boulders when the hillside was less vegetated and more exposed than now, in a more severe climate, and under (sporadically) heavy grazing. Growth since the classic photographs were taken a century ago

FIG 92. Black Tor Copse, a pedunculate oakwood at about 400 m on the Dartmoor granite. The moss carpet on the granite boulders is dominated by *Rhytidiadelphus loreus*, the tree-bases are covered with *Isothecium myosuroides*, their smaller branches with *Hypnum andoi*, and the upper twigs bear *Ulota* spp. and *Frullania dilatata*.

has been much more 'normal': the canopy of the old trees has risen greatly, the younger trees are much more upright in growth, and the wood has expanded a good deal in area (Proctor *et al.* 1980, Mountford *et al.* 2001). The puzzle is not so much why the old oaks are so tortuous, as why they are so big and have lived so long. Estimates based on ring counts and twentieth-century growth rates put their age at around 300 years. There are pioneer trees at Black Tor Copse at similar altitude on northwest Dartmoor which can match them in form (Fig. 92), but few trees there live for much over a century, and none are anything like as big. Birkrigg and Keskadale oaks in Cumbria are at a similar altitude to the Dartmoor woods, but unlike them are *Quercus petraea* woods. Tansley saw all these woods as 'under extreme conditions' or 'at the highest altitudes', and in the context of our pastoral upland landscape that may be so. But they are still far below the potential altitudinal forest limit, which probably lies at around 650 m (Chapter 17).

The Killarney woods have long held an almost hallowed place in the awareness of British and Irish plant geographers and ecologists (Fig. 93). The strawberry-tree (*Arbutus unedo*) is a west-Mediterranean tree, which reaches its northern limit in Ireland, where it grows around the lakes at Killarney. In the nineteenth century the area also became known as a hotspot for rare liverworts and mosses, and filmy ferns (Hymenophyllaceae) including the Killarney fern (*Trichomanes speciosum*) whose population was ruthlessly plundered to decorate 'Wardian cases' in Victorian drawing rooms. After visiting Killarney in 1911, the great Swiss phytogeographer Eduard Rübel (1912) concluded:

> *The extreme oceanic character of the climate of Ireland, which brings the alpines down to the seashore, mixing them with southern plants, unites in the Killarney*

FIG 93. Bryophyte-rich sessile-oak–holly wood, Derrycunihy Wood, Killarney, June 2010.

woods elements which in a warmer or less oceanic country form different formations in different altitudinal belts between 200 and 1400 m. The Atlantic laurineous wood, the Canary heath, the insular Macchia …, the Corsican beech- holly-wood, all show great affinities to the Irish Quercetum sessiliflorae laurineum or Quercetum sessiliflorae aquifoliosum as we might call it.

Perhaps the Killarney woods have been somewhat de-mystified since then (Kelly 1981). There are swathes of oakwood at Killarney without *Arbutus*, and we now see holly as a common component of western woods in Britain too – and some western Welsh and Scottish woods can run them pretty close as bryophyte habitat – but the woods around Killarney remain a special and fascinating place.

Wet birchwoods on peat

The wet woodlands of the last chapter are mostly dominated by sallow or alder, and mostly grow on essentially mineral or alluvial soils. One locally common type of wet birchwood breaks this mould, and grows on acid peat, often cut- over raised-bog peat (Chapter 15). It is dominated by downy birch, often with some grey sallow, and occasionally other willows or alder. The most constant (and usually dominant) species in the field-layer is purple moor-grass (*Molinia caerulea*), usually with a carpet of *Sphagnum fallax* covering most of the ground not occupied by anything else[W4]. Broad and narrow buckler-ferns (*Dryopteris dilatata* and *D. carthusiana*) are often rather conspicuous, and scattered tormentil, soft-rush (*Juncus effusus*) and creeping soft-grass scattered among the *Molinia* are the most frequent of a long list of thinly distributed associates. Deep cushions of *Polytrichum commune* are often a striking feature of the moss-layer. This is usually an easily recognisable community, but it is somewhat variable. Stands on rather dry peat often have a sprinkling of acid oakwood species, such as bramble, honeysuckle and the moss *Mnium hornum*. Some areas vary towards acid fen meadow, with more rushes and a greater variety of grasses, and such species as marsh violet (*Viola palustris*), marsh thistle (*Cirsium palustre*) and the moss *Calliergonella cuspidata*. In wetter areas *Sphagnum* is more luxuriant, such species as *S. palustre* and *S. fimbriatum* join *S. fallax*, and wet-loving vascular plants such as bogbean (*Menyanthes trifoliata*), bottle sedge (*Carex rostrata*) and water horsetail (*Equisetum fluviatile*) grow up through the carpet of *Sphagnum* (Fig. 175).

This *Betula pubescens–Molinia caerulea* woodland is a paradox. It is in one sense a pivotal community. The development of many ombrotrophic bogs must have passed through a stage like this. It is clearly 'odd man out' among the other wet woods. It shows floristic relationships with bogs, poor-fens, fen meadows and acid oakwoods, but it is none of these. It 'belongs' nowhere, yet it is common.

SCOTTISH NATIVE PINEWOODS AND SOME UPLAND SHRUB COMMUNITIES

The Boreal forest zone

In the Scottish Highlands we have the westernmost European outpost of the great Boreal coniferous forest belt that encircles the northern hemisphere. Scots pine (*Pinus sylvestris*), the world's second most widely distributed conifer, is our only representative of the conifers that dominate the Boreal forest. It is joined in Scandinavia, Finland and northern Russia by Norway spruce (*Picea abies*). This is another widely distributed tree, which, as ssp. *obovata* (*Picea obovata*), reaches to the Pacific in Kamchatka. The Siberian taiga has a few other conifers, including a larch (*Larix gmelinii*), Siberian 'cedar' (*Pinus sibirica*) and a fir (*Abies sibirica*), and yezo spruce (*Picea jezoensis*) comes in near the Pacific coast. In North America the dominant trees of the Boreal forest are much more varied. Sitka spruce (*Picea sitchensis*), Douglas fir (*Pseudotsuga menziesii*), western hemlock (*Tsuga heterophylla*) and lodgepole pine (*Pinus contorta*) are common in the Pacific Northwest, the black and white spruces (*Picea mariana* and *P. glauca*) and tamarack (*Larix laricina*) extend right across North America, and red spruce (*Picea rubens*) and eastern hemlock (*Tsuga canadensis*) are confined to the eastern provinces. The relative paucity of the European conifer flora has been attributed to the east–west orientation of the major barriers to plant migration in Europe. Many of the American conifer genera were in Europe before the glaciations, and successive interglacials show a pattern of progressive impoverishment. In North America the barriers to migration run north–south, and two rich floras have survived, one on either side of the Rockies, with common ground in the far north.

Many of the Boreal-forest understorey plants share the wide distributions of the trees, and some of the mosses surpass them, but in general they do not share the circumpolar ranges common among mountain and Arctic tundra plants.

The Scottish native pinewoods: the Caledonian forest

Scots pine is as essential a part of the scene in the eastern Highlands as oak trees in the Weald or in the valleys of southwest England. Much of Deeside from above Braemar to Ballater is wooded, with stands of dense pine forest in Ballochbuie Forest and elsewhere. There are open rocky, grassy and heathery areas too, but pine is never far away. Strathspey is a broader valley, and the pine forests correspondingly more expansive; Abernethy Forest, Rothiemurchus Forest and the forest running up Glen Feshie clothe much of the lower northwestern fringes of the Cairngorms, reaching up at one point to probably the only natural tree-line in Britain. The flatter landscape in Strathspey has more farmland, and more roadside birches and birchwoods (mostly *Betula pendula*), and Abernethy Forest includes a number of raised bogs, looking quite Scandinavian in their pine-forest setting. There is a centuries-old tradition of forestry management for timber in these woods (Fig. 94a), but there are (less tidy) unmanaged areas too (Fig. 94b); both have similar ground floras.

One of the best known and largest tracts of native pine forest elsewhere in the Highlands is the Black Wood of Rannoch, on the south shore of Loch Rannoch some 70 km southwest of the Cairngorms. West of this, patches and fragments of native Scots pine are scattered in an irregular arc through the west Highlands from near Tyndrum to just north of Ullapool, with substantial stands in Glen Affric and Glen Cannich, and around Loch Torridon and Loch Maree (Fig. 95). Scots pine of various provenances has been widely planted in the Highlands, and these plantations may develop a ground flora virtually identical with native pinewood.

FIG 94.
Speyside pinewoods.
(a) Managed native
Scots pine, near
Loch an Eilean,
Rothiemurchus Forest,
June 1981.
(b) Pine forest at about
550 m; Coire Buidhe,
Cairngorms, June 1981.

FIG 95. Native Scots pine, Loch Maree, Wester Ross, June 1981.

FIG 96. Pine-forest plants: (a) creeping lady's tresses (*Goodyera repens*); (b) the moss *Ptilium crista-castrensis*.

In the core area of native Scots pine forest north and south of the Cairngorms, the tree cover is often dense, up to 70% or so, and the pines cast a substantial degree of shade – but far lighter than that cast by beech in the south. Pine is usually the only tree, and there is no shrub-layer apart from the occasional juniper. The ground flora (unlike that of most of the deciduous woodlands) shares almost all of its ingredients with the surrounding heather–bilberry moor, which has two consequences. On the one hand it, is easy to visualise the heather moorland as derived from former forest; on the other, the forest has no sharp boundary but tails off into moorland, perhaps with an irregular intervening zone of juniper scrub. Heather (*Calluna vulgaris*) is almost always important in the ground flora of the pine forest, along with a group of large mosses, including *Hylocomium splendens*, *Pleurozium schreberi*, *Dicranum scoparium* and *Rhytidiadelphus triquetrus*, which build up a mattress intertwined with the stems of the heather. Wavy hair-grass can usually be found growing amongst the mosses. At low altitude in dense shade or after disturbance, there may be few species to add to this list, except for the frequent occurrence of the creeping lady's-tresses orchid (Fig. 96a). This rather pretty species often colonises conifer plantations. More commonly, and over a range of altitudes, the heather is joined by bilberry (*Vaccinium myrtillus*), cowberry (*V. vitis-idaea*), crowberry (*Empetrum nigrum*) and common cow-wheat, and often by the mosses *Dicranum fuscescens*, *Rhytidiadelphus loreus* and *Ptilium crista-castrensis* – this last a particularly beautiful plant (Fig. 96b). A variant of this field-layer, with constant hairy wood-rush and frequent heath bedstraw, wood-sorrel and hard fern is perhaps associated with slightly richer soils (or grazing?), and is widespread not only around the Cairngorms but in the more scattered fragments of native Scots pine forest in the west Highlands as well.

The western pinewoods are often much more open, with a canopy cover of no more than 25% – indeed, what constitutes a 'wood' becomes a rather arbitrary matter! There is space for other trees, and a scatter of rowan and downy birch amongst the pines is common. These open western pinewoods make fine scenery, and their openness makes it easy to appreciate the form of the trees. Their ground flora is more sheltered but otherwise little different from the moss-rich heather–bilberry moorland of their surroundings. In the humid northwest Highlands small-leaved *Sphagnum* species often become prominent, especially *S. capillifolium* and *S. quinquefarium*, with a suite of mosses and liverworts, which approaches the characteristic bryophyte-rich heather–bilberry–*Sphagnum capillifolium* heath of shaded slopes on our western mountains (Chapter 17).

An orchid that occurs thinly throughout the pine forest (and a test for sharp eyes!) is lesser twayblade (*Listera cordata*), with a pair of opposite leaves the size and colour of small bilberry leaves, and a small inflorescence of brown flowers. This little orchid can be found in mossy heather–bilberry moorland from Exmoor to the northern Highlands. As long as it has a degree of shade and a modicum of moisture it seems indifferent to whether there is a tree (or juniper) canopy or not. Lesser twayblade is one of a number of species that are characteristic of coniferous forests in Scandinavia, and grow in similar or related habitats with us, but generally much more rarely. They include twinflower (*Linnaea borealis*), common wintergreen (*Pyrola minor*), intermediate wintergreen (*Pyrola media*), serrated wintergreen (*Orthilia secunda*), one-flowered wintergreen (*Moneses uniflora*), *Goodyera* and coralroot (*Corallorhiza trifida*).

Juniper scrub

Juniper (*Juniperus communis*) is an interesting plant. It is the most widely distributed conifer in the world, widely native in temperate Eurasia and North America. With us it forms scrub on the chalk (where it has declined greatly over the past century), and on Carboniferous limestone in the north of England (Fig. 97) and in the Burren. It forms tracts of scrub on acid soils in the Lake District and the Highlands, and (as ssp. *nana*) it grows in exposed places on the most sterile and unpromising mountain rocks. We shall postpone the sites on chalk to the next chapter and the mountain occurrences to Chapter 17, and

FIG 97. Juniper scrub, Moughton, Ingleborough, July 1966.

consider here only juniper 'woodland'. This is of curiously local occurrence, perhaps because of its particular vulnerabilities – to burning, grazing, succession to taller woodland, and changes of land use, particularly afforestation. The main concentration of juniper scrub now is in the eastern Highlands, often in areas that also support pine forest, but there are interesting scattered outliers in northern England, which it is always nice to encounter unawares!

Well-developed juniper scrub generally has a rather uniform and well-characterised flora, with bilberry, the mosses *Hylocomium splendens* and *Thuidium tamariscinum*, wood-sorrel, heath bedstraw, hairy wood-rush, velvet bent (*Agrostis canina*) and common bent[W19]. On more acid soils calcifuge mosses such as *Dicranum scoparium*, *Rhytidiadelphus loreus* and *Pleurozium schreberi* become more prominent, and the bryophyte mat is often colonised by wavy hair-grass, cowberry and heather. More base-rich soils are marked by more common dog-violet (*Viola riviniana*), wood anemone (*Anemone nemorosa*) and harebell (*Campanula rotundifolia* – 'bluebell' in Scotland) around the junipers, and such mosses as *Rhytidiadelphus triquetrus* and *Plagiomnium undulatum*. Tormentil and (in Scotland) chickweed-wintergreen are frequent in both variants, and lesser twayblade and wintergreens (*Pyrola* spp.) occur occasionally.

Arctic-alpine willow scrub

Knee-high (to waist-high) scrub of arctic and subarctic willows is commonplace in northern Scandinavia (Fig. 24), as it must have been here in the Late-glacial. Nowadays only fragmentary relics remain, on steep rocky mountainsides and mountain cliff ledges inaccessible to sheep; our largest stand is barely half a hectare, and most are far smaller than that. It is arguably out of place in this chapter because most willows of this kind do not form 'woods' even in Scandinavia, and they generally grow on soils that are moist and at least reasonably base-rich. Our commonest and most widely distributed species is downy willow (*Salix lapponum*), a straggling low shrub with grey-green hairy leaves. Next most widespread is whortle-leaved willow (*S. myrsinites*), usually lower-growing, with bright green leaves shaped like bilberry leaves but larger and glossier. Mountain willow (*S. arbuscula*) is restricted to higher altitudes than the last two, and is particularly associated with the Ben Lawers range. The rarest, woolly willow (*Salix lanata*) shares this restriction to high altitudes and mainly occurs in the Clova–Caenlochan area. McVean & Ratcliffe (1962) and Huntley (1979) give lists from 19 sites, ranging in altitude from 630 to 914 m, but the ranges of the individual species extend substantially outside these limits[W20]. Apart from the willows, the vegetation does not differ significantly from that general on mountain cliff ledges[U16][U17] (Chapter 17).

THE FLORA OF CONIFER PLANTATIONS

As pointed out earlier, the recurrent glaciations left western Europe with an impoverished conifer flora, and in particular without a dominant conifer well adapted to our equable and humid climate. The Pacific Northwest of North America is rich in conifers well adapted to a climate very like our own, notably Douglas fir and Sitka spruce. These do well with us (both regenerate from seed), and vast tracts of upland Britain and Ireland have been planted with Sitka spruce – reputedly now the commonest tree in Britain, and probably in Ireland too. Douglas fir has been widely planted too, generally on better soils than the spruce. Sites too poor or too exposed for Sitka spruce to do well may be planted with lodgepole pine, a tree widespread at moderate altitudes in the northwestern Rockies. Dry sites are often planted with pine, either Continental provenances of Scots pine, or Corsican pine (*Pinus nigra* ssp. *laricio*), which does well on sand-dunes.

Ground floras in conifer plantations generally resemble an impoverished version of the ground flora that would be expected under a deciduous wood on the same soil. They are impoverished for several reasons. A dense even-aged stand of *any* tree, conifer or broadleaf, casts a deep shade and sets up intense root competition. A woodland ground flora does not assemble instantly when trees are planted. There will probably be a weed flora of non-woodland plants to be shaded out, and it will take time for the ingredients of a woodland field-layer to reach the site and become established. Many of our familiar woodland plants are notoriously slow to colonise new woodland of any kind (Chapter 5).

Conifer plantations, including their ground floras, can come up with interesting surprises. It is now many years since *Goodyera repens* turned up in Scots pine plantations in Norfolk, where it is now well established. Whether the plant came from wind-borne seed or whether fragments of rhizome came with the young pines is unknown. In recent years the tiny but very distinctive liverwort *Colura calyptrifolia* has been appearing on spruce trunks in Welsh conifer plantations, and more recently the even rarer moss *Daltonia splachnoides* has been similarly colonising Sitka spruce and sallows in plantations in southwest Ireland – an unexpected habitat for such rare bryophytes. Amongst commoner species, there is probably more *Plagiothecium curvifolium* growing on the ground in conifer plantations than in any other habitat in Britain, and a few years ago I was sent remarkable specimens of common woodland mosses, which had grown hanging from branches in the shade and shelter of a western Irish Sitka spruce plantation. The BSBI Local Change project revealed an unexpected increase in lesser twayblade in conifer plantations. Is this apparent increase real, or is it a wake-up call to look harder at a habitat that many of us dismiss as 'dull'?

CHAPTER 8

Wood Margins, Scrub and Hedgerows

W OODS HAVE EDGES, glades and rides, and gaps left by windthrow or
the death of old trees, and these places often have distinctive floras.
Cultivated fields that are abandoned or grassland that ceases to
be grazed is colonised by hawthorn, gorse or other shrubs, and 'tumbles down to
woodland' within a few decades. Traditionally we divide fields or mark boundaries
with hedges, and these again often have rich and varied floras, with much in
common with wood margins. Many species (and vegetation types) are characteristic
of these edge (or temporary) situations, and rarely or never occur in the extensive
plant communities with a degree of permanence – of which you can sample a
'uniform stand' with a quadrat and hence are beloved of traditional plant ecologists!

WOOD MARGINS

Once you are fairly into them, woods are rather uniform places. The neighbouring
patch is similarly shady to the one you are standing in, and it doesn't much matter
whether the leaf-fall, or nutrients leached from the canopy, or the seed rain,
comes from this patch or the next. The birds and insects will be almost entirely
woodland species, born and bred in the wood. The same, *mutatis mutandis*, is true
of a grassland or heath. None of these things are true at the edge of the wood.

German plant ecologists have two useful concepts for wood margins, a zone
of distinctive woody vegetation, which they call the *Mantel* (jacket, overcoat),
and outside that a zone of herbaceous vegetation, the *Saum* (hem) (Fig. 98). The
dominant trees within the wood are mostly pollinated by wind, and generally

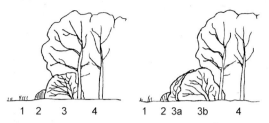

FIG 98. 'Forest-border ecotone. 1. Open area (fields, pastures, heathland, etc.). 2. *Saum* of perennial herbs. 3. Shrub *Mantel*, (a) *Vormantel*, (b) *Hauptmantel*. 4. Forest.' From Wilmanns & Brun-Hool (1982).

their seeds do not rely primarily on animals for dispersal – though oak, beech and hazel undoubtedly benefit from the activities of seed hoarders like squirrels and dormice, and yew has bird-dispersed fruits. The trees and shrubs of the *Mantel* are generally insect-pollinated, and they often produce fruits that are eaten and the seeds dispersed by birds such as thrushes that range widely over the landscape. Typical *Mantel* shrubs are hawthorn (*Crataegus monogyna*), blackthorn (*Prunus spinosa*), dogwood (*Cornus sanguinea*), wayfaring-tree (*Viburnum lantana*), buckthorn (*Rhamnus cathartica*) and brambles (*Rubus fruticosus* agg.); many of them are thorny or spiny. The *Mantel* is most prominent at the edge of woods dominated by light-demanding trees, such as oak or ash. The most shade-tolerant trees (and in general those that cast the deepest shade), like beech or yew, tend to form a *Mantel* from their own lower branches.

The *Saum*, of herbaceous species, is distinct from both the *Mantel* and the neighbouring vegetation, which may be farmland, grassland or heath. It benefits from the shelter and influx of nutrient from leaf-fall of the woody vegetation on the one side, and light, pollinators and possibilities of dispersal on the other, and often shows 'ruderal' tendencies (Chapter 10). Common 'wood-margin' species include red campion (*Silene dioica*), greater stitchwort (*Stellaria holostea*), garlic mustard (*Alliaria petiolata*), ground-ivy (*Glechoma hederacea*), tufted vetch (*Vicia cracca*), rough chervil (*Chaerophyllum temulum*), nipplewort (*Lapsana communis*), germander speedwell (*Veronica chamaedrys*), agrimony (*Agrimonia eupatoria*) and many others (Fig. 99). These herbaceous wood-margin and hedgerow communities are placed in the order Glechometalia by phytosociologists on the Continent. In warm dry situations, particularly on chalk and limestone, a different range of species occurs, including some favourite aromatic herbs. Wild marjoram (*Origanum vulgare*), wild basil (*Clinopodium vulgare*) and common calamint (*Clinopodium ascendens*) grow in vegetation of this kind fringing scrub and woodland, and on hedgebanks (order Origanetalia). Wood sage (*Teucrium scorodonia*) has a rather similar range of habitats, but extends farther north and onto more acid soils.

These two zones may be narrow or broad, compressed within a metre or less, or expanded to many times that.

FIG 99. Typical species of woodland fringes and hedgerows: (a) garlic mustard (*Alliaria petiolata*), also known as jack-by-the-hedge; (b) rough chervil (*Chaerophyllum temulum*); (c) ground-ivy (*Glechoma hederacea*); (d) sweet violet (*Viola odorata*); (e) nipplewort (*Lapsana communis*); (f) germander speedwell (*Veronica chamaedrys*).

GAPS IN WOODS

Gaps in woods come in different sizes and forms. Where a tree has died standing, the crown will probably have thinned gradually, allowing suppressed saplings or seedlings on the floor of the wood to fill the gap almost imperceptibly. If one or a group of trees is windthrown it will almost invariably create a sudden gap, letting light in to the floor of the wood over a significant area. What happens then will depend very much on the trees and other species in the neighbourhood of the gap. Over the first year or two the most immediate change will be increased vigour of the ground vegetation, and several species are often conspicuous in the early stages of colonisation of gaps, however caused. One of these, particularly in woods on acid soils, is foxglove (*Digitalis purpurea*). Foxglove is biennial, forming a large non-flowering leaf rosette in the first year and producing its spectacular spike of bumblebee-pollinated flowers in the second year. It sets a profusion of tiny seeds. In the lowlands, and on any but the poorest soils, by the second year brambles will be noticeable, and (unless there is significant grazing by deer or other animals) within five years the whole of the floor of the gap will probably be a tangle of their prickly arching stems.

What about the trees in regeneration gaps? This depends very much on the available seed parents. If there are already tree seedlings or saplings on the floor of the wood, light will stimulate them into rapid growth. Otherwise ash, with its wind-dispersed 'keys', is quick to colonise any patch of open ground. Ten years or so after a gap becomes vacant it is common to see it occupied by a dense growth of ash saplings, tallest in the middle of the gap where there is most light. As the ash trees grow, the weaker saplings are suppressed, and the group opens up into a patch of ashwood in which, in due course, more shade-tolerant trees such as oak or beech can become established. On acid soils, the first to colonise the gap will probably be birch or sallow, both with wind-borne seeds, and both relatively light-demanding. Shrubs such as hawthorn are seldom important in the closure of small gaps.

We have already considered the special case where routine cutting of coppice is followed by spectacular displays of the spring woodland flowers – but the coppice-stools are poised to grow up quickly and re-establish woodland conditions. Large gaps left by windthrow or felling of the dominant canopy trees have more opportunity to develop distinctive communities and to show clearly the stages in succession back to closed woodland. Wild strawberries (*Fragaria vesca* – immensely better flavoured than cultivated varieties!) often become abundant in the early years of a gap opening and, especially if there has been some burning, rosebay willowherb (*Chamerion angustifolium*) can lay on a display

Apomixis

Brambles are *apomictic*, but not wholly so. That is to say, they can form embryos and produce viable seed without fertilisation, and the offspring are genetic duplicates (in effect, clones) of the parent plant – but brambles also produce a proportion of embryos following normal fertilisation. These are variable in the normal way, and can produce new genotypes. So the bramble hedges its bets! It can adapt by natural selection, but when a genotype arises particularly well adapted to the local situation, it can be duplicated quickly. As the geneticist Jens Clausen wrote in 1954 [somewhat abridged]:

> The apomicts discovered the effectiveness of mass production long before Mr Henry Ford applied it to the automobile. The engineering of apomictic replicas from little seeds packed with genes is incomparably more ingenious than the use of the assembly line, but there are striking similarities between the two which may help us to understand the apomictic process. Once a model of an apomict has been developed, innumerable replicas of it can be produced. If it is a successful one, the replicated model becomes very common until the requirements change and it no more is fitted to its task. Then new models evolved through more or less radical interchanges and alterations get their chance, and swarms of replicas of them are turned out. Not all the old models disappear, but old and new travel together, although the new may outrun the old.

In business terms the brambles have come up with a business model to cope with the 'predictably unpredictable', and they have not made a bad job of it. Some other genera have exploited comparable situations in the same way. On a longer timescale, the apomictic whitebeams (*Sorbus* spp.) and their relatives have exploited the chancy opportunities for colonisation in rocky woods and cliff-edges, treading a hazardous path beween being shaded out by dominant woodland trees and the manifold insecurities of life on bare rock outcrops. If the apomictic whitebeams are not as manifestly successful as the brambles, there are few major areas of limestone crags in Britain and Ireland wholly without them (Rich *et al.* 2010). One might conjecture that the apomictic lady's-mantles (*Alchemilla* spp.) and hawkweeds (*Hieracium* spp.) originated in similarly fragmented habitats in gaps in the forests in the mountains of northern Europe. The apomictic dandelions (*Taraxacum*) have been conspicuously successful in exploiting the constant supply of opportunities that gardens (and natural open habitats) provide.

of pink rivalling foxglove[OV27]. Deadly nightshade (*Atropa belladonna*) occupies a similar ecological niche in gaps in beechwoods on the Continent, but it is rarer and more capricious with us.

SCRUB

The word itself is abrupt and abrasive, and the picture it calls up is hardly more inviting. Yet scrub – shrub-dominated vegetation – is varied, dynamic and, taken as a whole, far from dull. Probably the commonest kind of scrub is that which appears in fields or grassland that are no longer grazed or mown, but the nature of the scrub that develops depends on the soil and situation.

Scrub and succession to woodland

Hawthorn and brambles are by far the commonest woody plants invading neglected farmland, and remain virtually constant throughout the development of scrub. Blackthorn and dog-roses (*Rosa canina* agg.) are also typical early colonists, and where the soil has been disturbed or enriched elder may gain a foothold. In the early stages of succession the bushes form a patchwork with the grassland, but as the leafy canopies of the dominant bushes meet, the grassland species are mostly shaded out. The scrub at this stage can be pretty forbidding. The hawthorn and its spiny and prickly associates form a near-impenetrable thicket. In summer little light penetrates to the largely bare ground. The only other plants that are anything like constantly present are ivy, which, being evergreen, can take advantage of the winter season to maintain at least a thin (and sometimes total) cover over the ground, and a scatter of stinging nettles and cleavers (*Galium aparine*). Apart from that, and occasional ash and sycamore saplings, this hawthorn–ivy scrub[W21] is generally species-poor. There is a long list of grassland, woodland-fringe (*Saum*) and woodland species that *can* occur, but seldom in any quantity or with any constancy. There are two exceptions to this generalisation, both on calcareous soils, and probably both related to the proximity of seed parents. Quite often, the species list recalls a sparse skeletal version of the ground flora of a chalk or limestone wood, with patches of dog's mercury. In freely drained sites where the scrub has colonised open ground on chalk or limestone dog's mercury and most other woodland species are missing, but false brome (*Brachypodium sylvaticum*) can be prominent.

As the scrub ages and the canopy of the hawthorn rises and thins, more woodland species can establish, and the scrub becomes in effect a hawthorn wood. But, by this time, some of the ash and sycamore saplings will have grown into substantial trees. From this stage it is only a matter of time (perhaps half a century) before the site becomes a mature ashwood, maybe with an admixture of sycamore, and (if a jay has dropped a few acorns) maybe a young oak or two, but with a field-layer largely derived from woodland-fringe species. In the 1960s, when I first knew Hooken Cliff (Fig. 100) at Branscombe in East Devon there was a

FIG 100. Hooken Cliff, Branscombe, Devon, May 1970. Chalk grassland in foreground, a patchwork of grassland, hawthorn scrub and ash trees in the middle distance.

FIG 101. Old hawthorn–ivy scrub, Axmouth–Lyme Regis Undercliffs NNR, May 1973. This scrub is around 6 m high, and must be many decades old. It contains some well-grown ash trees; the field-layer is principally ivy, but stinking iris (*Iris foetidissima*) and hart's-tongue fern (*Phyllitis scolopendrium*) are also conspicuous.

patchwork of hawthorn scrub and grassland. My two boys enjoyed the dense scrub, which had small boy-sized tunnels underneath that you could explore, coming up in unexpected places. The patches of grassland have long since gone, and much of the former 'impenetrable' scrub has opened up ready to be colonised by seed from the ash trees, which must have established decades before I knew the site. All stages of succession from hawthorn scrub to ashwood can be seen at the Axmouth–Lyme Regis Undercliffs NNR, a few kilometres east along the coast (Fig. 101).

Chalk scrub

My generation of Cambridge botany students was taught 'the five calcicolous shrubs' by the late Mr Humphrey Gilbert-Carter: dogwood (*Cornus sanguinea*), wayfaring-tree (*Viburnum lantana*), wild privet (*Ligustrum vulgare*), spindle (*Euonymus*

FIG 102. Calcicolous scrub species: (a) dogwood (*Cornus sanguinea*); (b) wild privet (*Ligustrum vulgare*); (c) wayfaring-tree (*Viburnum lantana*); (d) traveller's-joy (*Clematis vitalba*).

europaeus) and buckthorn (*Rhamnus cathartica* – he would *never* have allowed us to use English names!). The first three of these (Fig. 102) characterise a distinctive variant of hawthorn–ivy scrub on chalk and limestone in the south and east of England (Fig. 103). Traveller's-joy or 'old man's beard' (*Clematis vitalba*, one of our few woody climbers) is often conspicuous scrambling over chalk scrub (Fig. 102d); another common scrambler is black bryony (*Tamus communis*), particularly noticeable in autumn from its glossy vivid orange-red berries. Brambles and wild roses are prominent in chalk scrub (sweet-briar, *Rosa rubiginosa*, is often common), and ivy is still dominant in the field-layer. False brome, wild marjoram and wood sage (*Teucrium scorodonia*) are frequent around the fringes of the scrub.

Sixty years ago it would have been impossible to write about chalk scrub without mentioning rabbits. Rabbit warrens were common on the downs. Disturbance and enrichment of the soil round the warrens provided niches for elder bushes to colonise and, once fairly established, their corky bark made them relatively immune to rabbit attack. Juniper (*Juniperus communis*) is still quite widespread on the chalk of southern England, but it was formerly much commoner than it is now, probably largely because it needs very short turf or open ground for the seedlings to establish, and young plants are vulnerable

FIG 103. Hawthorn colonising chalk grassland, Malacombe Bottom, near Tollard Royal, Dorset, May 1970. Hawthorn just past flowering, whitebeam showing the silvery grey undersides of the newly expanded leaves, ash leaves not yet fully expanded. The ridges in the foreground are *terracettes*, formed on the slopes by the trampling of grazing livestock.

to heavy grazing. Sheep and rabbits will eat it, but it is low on their list of preferences. When rabbits were abundant, juniper served as a nurse species for yew and other woodland trees, which were able to establish shielded by its relative unpalatability. Much of the yew wood in Juniper Bottom near Box Hill in Surrey grew up in this way; the dead and dry juniper bushes beneath the yews long kept their characteristic 'cedar-pencil' smell. Establishment of hawthorn and the other chalk shrubs required some relaxation of rabbit grazing, but most of them (except juniper) can sow themselves into even quite rank grassland.

Blackthorn scrub

All habitual makers of sloe gin will surely have their favourite patch of blackthorn scrub! It is not easy to pinpoint habitat differences between scrub dominated by hawthorn and by blackthorn. Hawthorn usually occupies well-drained sites on a wide range of soils from mildly acid to highly calcareous. Blackthorn favours soils that are at least reasonably base-rich and generally rather moist, and it withstands wind-exposure better than hawthorn. Hawthorn (and chalk) scrub is commonly seral – on its way from colonising grassland (or whatever) to becoming woodland. Blackthorn scrub often gives the impression of greater permanence, along wood margins, or on windswept sea-cliffs. Relative stability of habitat may be a factor in the wide distribution of that rather secretive and sedentary butterfly the brown hairstreak, whose larvae feed on blackthorn. Brambles, roses and ivy are less ubiquitous in blackthorn than in hawthorn scrub, but the field-layer frequently includes bracken, red campion, common dog-violet (*Viola riviniana*), germander speedwell, cleavers and scattered grasses, and sometimes a thin scatter of woodland species such as dog's mercury, wood-sorrel, herb-robert (*Geranium robertianum*) and primrose.

Blackthorn scrub on sea-cliffs (Chapter 20) often has a seasoning of species from the neighbouring maritime grasslands, particularly grasses such as cock's-foot (*Dactylis glomerata*), red fescue (*Festuca rubra*) and Yorkshire-fog (*Holcus lanatus*). On exposed coasts, blackthorn, ivy and bracken can interact in bizarre ways. On Alderney I have seen low blackthorn overtopped by bracken growing up through it to unfurl a leafy canopy tens of centimetres above the branches and leaves of the scrub. On the south coast of Guernsey, the erect flowering phase of ivy can form a significant fraction of the canopy of the wind-pruned blackthorn scrub on the cliffs.

Gorse (*Ulex europaeus*) scrub

Sir J. E. Smith (1759–1828), founder of the Linnean Society, wrote in Rees's *New Cyclopaedia* that 'Linnaeus was so enchanted with the gorse in full flower on Putney

FIG 104. Common gorse (*Ulex europaeus*) scrub in full flower in early May, East Budleigh Common, Devon. The lower-growing western gorse (*U. gallii*), abundant in the rest of the heath, will flower later in the summer.

Heath that he flung himself on his knees before it.' The story may be apocryphal, but I have experienced almost equally emotional reactions from Continental visitors at the mass of golden-yellow on the roadsides between Exeter and the Lizard when gorse is at its peak (Fig. 104). Common gorse has a different ecological niche from our other two species, which are plants of heathlands on poor, acid podzolic soils, and have a shorter flowering season in late summer (Chapter 16); if common gorse occurs on these heaths it is generally a sign of disturbance or local enrichment of the soil. It is a plant of moderately acid brown-earth soils, occasional in acid oakwoods, but probably essentially a wood-margin species.

Common gorse is frequently the dominant of patches of scrub in suitable country, generally in company with a low cover of brambles, with various acid-grassland species, including common bent (*Agrostis capillaris*), Yorkshire-fog, sweet vernal-grass (*Anthoxanthum odoratum*), heath bedstraw (*Galium saxatile*), tormentil (*Potentilla erecta*), cat's-ear (*Hypochaeris radicata*), common dog-violet and sheep's sorrel (*Rumex acetosella*); bracken is often present, but seldom dense or particularly well grown[W23]. Roads or tracks across heaths, or through acid woodland, are often lined with gorse scrub, and the cuttings of new trunk roads in southwest England have provided wide stretches of suitable habitat for it. Gorse scrub is widespread on exposed sea-cliffs, where it is probably a permanent element of the vegetation pattern determined by interaction of topography and wind (Chapter 20).

Bramble (and other) fringes

Brambles often grow as a fringe covering the base of a hedge – a *Vormantel* in terms of Wilmanns & Brun-Hool's diagram in Figure 98. The bramble–Yorkshire-fog (*Rubus fruticosus–Holcus lanatus*) underscrub[W24] described in Rodwell (1991a) is in effect a bramble *Vormantel* plus a *Saum* typical of a contact with false oat-grass (*Arrhenatherum*) grassland[W24b] and some other communities[W24a]. Obviously, the details of such transitions will vary with what is on either side; this community provides a couple of fairly typical examples from among many possibilities.

Wilmanns & Brun-Hool (1982) described two *Vormantel* communities from Ireland, a field rose (*Rosa arvensis*) community in Co. Offaly (which is also in central Europe, and must surely occur elsewhere in Ireland and in Britain), and a wild madder (*Rubia peregrina*) community from Killarney. Wild madder has its main distribution in the Mediterranean, but its northern limit follows the winter isotherms up the Atlantic coast to Ireland and southwest Britain (Fig. 6), where it is common in calcareous districts scrambling through scrub and hedges.

Bracken underscrub and other lowland bracken communities

Bracken finds its optimum with us on well-drained acid soils, and is often a troublesome invader of bent–fescue pastures in the hills[U20]. It also dominant on some rather better soils in the lowlands. Here, in a 'bracken–bramble underscrub' it is usually accompanied by a thin understorey of bramble[W25]. On deeper and moister soils it is usually accompanied by bluebells with, for example, creeping soft-grass (*Holcus mollis*), cock's-foot, ground-ivy, herb-robert, male-fern (*Dryopteris filix-mas*), pignut (*Conopodium majus*), common dog-violet, red campion and occasional patches of stinging nettles and cleavers. This variant sometimes occurs inland, usually in association with woodland, or maybe marking sites of former woodland or wood-pasture, but is common on the sea-cliff-slopes of southwest England and Wales (Fig. 105), and no doubt of Ireland too. It is very striking in spring when the bluebells are in flower but the bracken is not yet unfurled. It has probably become more widespread on the coast of southwest England in the last half-century, with the decline of grazing on coastal cliff-slopes. On drier, thinner and more acid soils the community usually lacks bluebell and many of its accompanying herbs, bramble is often more prominent, wood sage (*Teucrium scorodonia*) is common, and Yorkshire-fog and foxglove are frequent (Fig. 106).

ABOVE: **FIG 105.**
Bracken–bramble underscrub, bluebell sub-community[W25a]. North side of Start Point, Devon, May 1992: bracken just unfurling, bluebells in full flower colouring the whole of the north side of the headland a misty blue.

LEFT: **FIG 106.**
Bracken–bramble underscrub: wood-sage sub-community, here with an abundance of foxgloves as the bracken is just unfurling. Later in the summer the cliff-slopes bear a uniform green covering of bracken. Prawle Point, Devon, May 1992.

HEDGES AND HEDGEROWS

A 'hedge' means different things to different people, and in different contexts. Medieval farmers knew all about 'dead hedges', barriers of stakes interwoven with brushwood and thorns, but these were probably always temporary expedients to meet an immediate need, or while a live hedge grew up – dead hedges were prodigal of labour and materials, whereas live hedges were self-renewing. Hedges in Devon and Cornwall are commonly on banks, sometimes with a facing of stones on the sides, and a Cornish 'hedge' is the bank itself regardless of whether anything grows along the top. Much the same is true of many exposed western districts of Britain and Ireland, though the words used to describe it may differ. There are pine hedges on Breckland, and beech hedges on Exmoor. But we are concerned with ordinary live field and roadside hedges here. In the mid twentieth century it was estimated that there were some 750,000 km of hedges in Britain; the total for Ireland would have been perhaps a third of that. Square fields 4 ha in area imply 10 km of hedge per square kilometre. That may be not unrepresentative of traditional farming country (in some areas density is certainly greater), but of course over wide areas of Britain and Ireland hedges are (and have always been) few and far between.

Origins of hedges

Modern hedges are planted, but there are several other ways in which a hedge can originate. Hedges can be relics of former woodland, left defining the boundary between the cleared fields on either side. Hedges can arise by growth of self-sown tree and shrub seedlings along a fence-line or bank. Planted hedges may use whatever saplings are available locally (in which case they will probably be a mixture of species from the outset), or they may be planted with one or two species, usually in recent times nursery-grown.

Some hedges that we know to be ancient from documentary records are likely to be woodland relics, and so too are some other hedges, judging from the evidence of their constituent trees and other characteristics. Hedges that have arisen as 'secondary woodland' along some kind of linear feature (and may be comparably old) are probably the hardest to be certain about. By their nature, their origin is unlikely to have been recorded. The frequency and rapidity with which boundaries become colonised by woody plants in the modern landscape – the Fleam Dyke in Cambridgeshire following myxomatosis, abandoned railway lines, and wire or wooden fences – show that this sort of process could have operated in the past, and should not be discounted lightly. Many hedges were planted from medieval times onwards in districts where open fields never dominated the landscape, and the enclosure of open fields around villages began early.

All planted hedges are not the same. We have to consider the purposes for which the hedge was planted (which will influence the species chosen), and the source of material available for planting. Marking boundaries and providing a barrier against straying of livestock are the two most obvious, but in past centuries people expected much more from their hedges. Hedgerow trees were an important source of timber (a surveyor's report on a seventeenth-century mid-Devon house I once lived in spoke of 'hedgerow scantlings' for many of the smaller timbers), and hazel, ash and elm in hedges were potential livestock feed. During the period of Parliamentary enclosure many kilometres of hedge were needed in a hurry, and hawthorn 'quicks' could be mass-produced from seed in nurseries, and could provide a stockproof barrier after a few years' growth.

The age of hedges: 'Hooper's rule'

In the 1970s Max Hooper and his colleagues at Monk's Wood Research Station, Ernest Pollard and Norman Moore, published a relationship between the number of trees and shrubs in a hedge and its age, which has caught popular imagination (Pollard *et al.* 1974). From a study of 227 hedges in Devon, Lincolnshire, Cambridgeshire, Huntingdonshire and Northamptonshire for which there was documentary evidence of planting date, they proposed the equation $A = 110S + 30$, where A is the age of the hedge in years, and S is the average number of woody species in a 30-yard (27.4 m) length of hedge (Fig. 107).

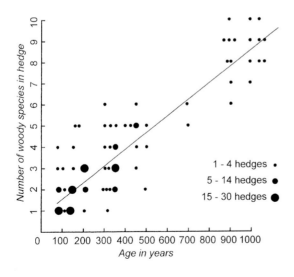

FIG 107. 'Hooper's rule'. Number of woody species plotted against documented age for 227 hedges in the south and east of England (from Pollard *et al.* 1974).

Taken uncritically (which the original authors specifically warned against), this would mean that the number of woody species in the hedge is roughly equal to its age in centuries, but a glance at the graph shows how misleading this simple assumption can be!

There can be no doubt that 'Hooper's rule' enshrines a real truth that, other things being equal, old hedges are richer in woody species than young hedges. Hooper and his colleagues envisaged two reasons why this should be. The oldest hedges might be relics of former woodland. Alternatively (or in addition) old hedges might be species-rich because they had had the most time for colonisation. This second suggestion obviously applies to *all* hedges, regardless of their origin. A third possibility has already been touched on: that hedges may have been planted with a desired mix of species, or that what was planted may have been dictated by the mix of saplings that were to hand, whereas enclosure hedges were usually single-species plantings of hawthorn.

The evidence for some hedges being woodland relics is persuasive. Abundance of trees normally confined to woodland, such as small-leaved lime or wild service-tree, is good evidence of a woodland origin. Hazel, dogwood, spindle and field maple are other indicators. Hedges with several of these trees and shrubs generally have a good number of the herbaceous plants regarded as indicators of 'ancient woodland', such as dog's mercury, primrose, bluebell, wood anemone and yellow archangel (Table 4), but the distribution of some of these seems to have more to do with soil than with the age of the hedge. In the mild southwest a number of woodland species are common in open situations that they would not tolerate farther east. Herbaceous 'indicators of ancient woodland' should be used with even more caution than woody species as evidence of the age of hedges – particularly in richly wooded neighbourhoods.

Old hedges can show uncharacteristically low numbers of woody species for several reasons. One is competition, particularly in hedges dominated by suckering elms, which can exclude virtually all other woody plants. Another is that a number of woodland trees and shrubs tend to be gregarious, so that successive 30-yard lengths of hedge are dominated by one or two species, even though the hedge as a whole is quite diverse.

'Hooper's rule' has its place as a useful rule of thumb, provided it is used critically. It will usually differentiate enclosure hedges (1750 onwards), with only one or two woody species, from those of earlier times, even though those oldest hedges amongst themselves often show almost no relation between the number of woody species and the date when they were first recorded.

Hedges and hedge management

A line of shrubs is not of itself a stockproof barrier, and hedges were traditionally 'plashed' or 'laid'. In winter, after cutting back and removing dead stems and unwanted growth (and weeding out undesirable species), the main stems (pleachers) were cut roughly three-quarters through, and bent over to overlap their neighbours so that, when growth resumed in spring, they would form a continuous impenetrable wall of vegetation. There were many regional (and probably individual) styles of plashing or laying. In the classic Midlands style, vertical poles of ash or hazel (stabbers) were driven in at intervals of some 70 cm and the pleachers woven between them, and the hedge was finished at the top with a band of pliable hazel stakes woven in and out of the stabbers to keep these and the pleachers in place. In Wales and southwest England, some of the niceties of the Midland style were not observed and hedge-laying was aimed at producing a denser and often wider hedge. In a multi-species hedge there must be an element of catch-as-catch-can about laying hedges, and in a neglected hedge many of the woody stems will have grown too thick for laying. In Devon, it is common to see ash laid horizontally for several metres, often in both directions from the base, creating a living post-and-rail fence, and I have seen the same in Co. Sligo. It is also common to see gaps in hedges patched with whatever stop-gap came to hand, often sheep-netting, or an old bedstead. Hedge-laying championships give a very partial view of what happens in practice in real countryside! Traditional hedgerow management is labour-intensive, and most hedges, particularly along roadsides, are now cut with tractor-mounted flail cutters. The immediate results with a flail cutter are not pretty, but cutting this way is far better than no cutting at all.

Hedges are a maintenance liability, they take up a significant amount of land that could be growing crops, and small fields are awkward and time-wasting for large-scale farm machinery. For these reasons many hedges have been removed, especially from arable country in the south and east of England. Norman Moore and his colleagues (1967) compared sample areas in aerial photographs taken in 1946 and 1962–63, and found losses ranging from 0.12 km per square kilometre in Cornwall to 2.13 km per square kilometre in Huntingdonshire. They saw evidence of destruction of hedges in every one of 23 English counties, in 11 out of 12 Scottish counties and in all of the six Irish counties they passed through in the course of their journeys. The length of hedge on three manors in Huntingdonshire, covering 16 km², reached a peak of about 122 km in 1850 following enclosure, and was only a little less in 1946. By 1965 it had fallen to 32 km, almost the same as in the early fourteenth century at the height of medieval open-field farming. The process has continued, but in most places the 1950s and 1960s probably saw the greatest losses.

The structure and flora of hedges

Hedges are complex habitats and, potentially at least, correspondingly rich in species. The hedgerow trees represent a woodland core, most important in the older hedges of 'ancient countryside', but often lacking, especially in hedges planted in the last century or two. Typically, most of the woody species of the hedge are thorny shrubs or small trees, which might equally be found along the edge of a wood; these may be fringed by brambles and roses, and other scrambling plants. But a very large component of the hedge flora, including many of the species we think of as most characteristic of hedgerows, is made up of essentially woodland-margin plants – species of neither woodland nor grassland, but more or less tied to the linear habitat where the two meet. At its foot the hedge grades into *Arrhenatherum* grassland, meadow or pasture, or the ruderal vegetation of arable fields or roadside. The proportions of these components vary enormously from hedge to hedge; some hedges in a district can be very rich in species (especially in 'ancient countryside'), whilst others are disappointingly poor. Hedges are a conservation resource we do not take seriously enough.

The hedges of Farley Farm near Chudleigh in South Devon may serve as an example (Michelmore & Proctor 1994). Farley is a small farm (28 ha), typical of the area, mostly on the south-facing slopes of a small coombe on Carboniferous shales, about 100 m above sea level. Counts of woody species in 30-yard samples of hedge yielded a mean of 5.9 species (s.d. 1.6) from 81 samples, so from Hooper's rule the hedges are evidently old, with no significant variation from one part of the farm to another. Of Farley's 12 fields, 10 were under grass and two were arable at the time the survey was made. For recording, the hedges were divided into segments by intersections and gates, and the two sides of the hedge were recorded separately. This gave 70 stretches of hedge, in which the species were recorded on a subjective scale from 1 (very rare) to 7 (dominant); the product of these two figures gave a 'hedge abundance index' (HAI) from 1 to 490 for each species. Table 5 lists the species with the top 25 HAI scores. The lists from the 70 hedge segments were also analysed statistically. They fell into eight main groups, three (A–C) representing the common dry hedgerow flora, including such species as shining crane's-bill, garlic mustard, rough chervil and nipplewort. Group D was the largest, embracing the most ubiquitous species in all the hedges. The more abundant of these are listed in Table 6, and (tentatively) assigned to zones in the hedge. Group E are less widespread but are another central group, including such characteristic hedgerow plants as agrimony (*Agrimonia eupatoria*), meadow vetchling (*Lathyrus pratensis*), honeysuckle (*Lonicera periclymenum*), wild madder, wood sage and tufted vetch. Group F adds some

TABLE 5. Farley Farm, Devon. The 25 most abundant 'characteristic hedge species', defined arbitrarily as 'those whose total score in all other habitats was not more than double their score in hedges (HAI, see text), and whose hedge habitat score is not exceeded by that in any other single habitat'. This excludes brambles and some woodland plants (notably pedunculate oak). In the habitat-formula column, g = grassland, a = arable, h = hedge, e = edge (of woodland), w = woodland; '-' indicates absence; lowercase, uppercase, and uppercase bold indicate minimal, minor and major presence respectively.

HAI	Species	Occurrence out of 70	Habitat formula
328	Dog-rose (*Rosa canina*)	69	g-**HEW**
306	Blackthorn (*Prunus spinosa*)	68	--HeW
296	Hedge bedstraw (*Galium mollugo*)	70	gAH-W
281	Hawthorn (*Crataegus monogyna*)	67	g-**HEW**
251	Rough chervil (*Chaerophyllum temulum*)	64	-AH-w
191	Creeping soft-grass (*Holcus mollis*)	44	---HEW
157	Hedge woundwort (*Stachys sylvatica*)	57	--H-W
152	Field maple (*Acer campestre*)	44	--HEW
150	Spindle (*Euonymus europaeus*)	54	--HeW
145	Stone parsley (*Sison amomum*)	44	--H--
140	Barren brome (*Anisantha sterilis*)	41	g-H--
133	Nipplewort (*Lapsana communis*)	51	-aHew
112	Foxglove (*Digitalis purpurea*)	42	--HeW
93	Garlic mustard (*Alliaria petiolata*)	35	--H-w
89	Shining crane's-bill (*Geranium lucidum*)	28	--h--
84	Greater chickweed (*Stellaria neglecta*)	30	--H-w
79	Prickly sow-thistle (*Sonchus asper*)	42	gAH--
66	Wild plum, bullace (*Prunus domestica*)	21	g-h-w
43	American willowherb (*Epilobium ciliatum*)	16	--h-w
40	Wall pennywort (*Umbilicus rupestris*)	15	--h-w
36	Turkey oak (*Quercus cerris*)	19	--h--
33	Dogwood (*Cornus sanguinea*)	16	g-h-w
28	Wild madder (*Rubia peregrina*)	17	--H-w
26	Hoary willowherb (*Epilobium parviflorum*)	9	g-h-w
25	Black spleenwort (*Asplenium adiantum-nigrum*)	15	--h--

species that avoid the driest sites, or favour some degree of shade, such as bugle (*Ajuga reptans*), hazel, bluebell, dog's mercury, primrose and fleabane (*Pulicaria dysenterica*). Groups G and H are almost all species of moist or shady place, such as wild angelica (*Angelica sylvestris*), marsh thistle (*Cirsium palustre*), meadowsweet (*Filipendula ulmaria*), greater bird's-foot trefoil (*Lotus pedunculatus*), hart's-tongue fern (*Phyllitis scolopendrium*), red campion, lady-fern (*Athyrium filix-femina*) and

TABLE 6. A different view of the Farley hedges. The species with 50% or more of possible occurrences in the largest species-group (D) to emerge from a multivariate statistical analysis. This group embraces most of the ubiquitous hedgerow species. In this table they have been roughly divided between the main components of the hedgerow: grassland species G (including some ruderals), herbaceous woodland-fringe species H (*Saum*), woodland-edge shrubs and their associates S (*Mantel*), species of woodland W.

Species	G	H(S)	S(M)	W
Yarrow (*Achillea millefolium*)	***			
Common daisy (*Bellis perennis*)	***			
Meadow foxtail (*Alopecurus pratensis*)	***			
Common mouse-ear (*Cerastium fontanum*)	***			
Field thistle (*Cirsium arvense*)	***			
Spear thistle (*Cirsium vulgare*)	***			
Crested dog's-tail (*Cynosurus cristatus*)	***			
Cock's-foot (*Dactylis glomerata*)	***			
Yorkshire-fog (*Holcus lanatus*)	***			
Perennial rye-grass (*Lolium perenne*)	***			
Ribwort plantain (*Plantago lanceolata*)	***			
Rough meadow-grass (*Poa trivialis*)	***			
Meadow buttercup (*Ranunculus acris*)	***			
Creeping buttercup (*Ranunculus repens*)	***			
Creeping cinquefoil (*Potentilla reptans*)	**	*		
Broad-leaved dock (*Rumex obtusifolius*)	**	*		
Dandelion (*Taraxacum officinale* agg.)	**	*		
Hogweed (*Heracleum sphondylium*)	*	**		
Prickly sow-thistle (*Sonchus asper*)	*	**		
Herb-robert (*Geranium robertianum*)		***		
Ground-ivy (*Glechoma hederacea*)		***		
Hedge woundwort (*Stachys sylvatica*)		***		
Greater stitchwort (*Stellaria holostea*)		***		
Germander speedwell (*Veronica chamaedrys*)		***		

opposite-leaved golden-saxifrage (*Chrysosplenium oppositifolium*), concentrated in the damp valley bottom.

The main variation at Farley is between the hedges associated with well-drained arable fields up on the spurs, and those associated with permanent pasture in the moist valley. A subsidiary trend can be detected from a more ruderal to a more stable, late-successional flora, but there is clearly a great deal of chance (or unexplained) variation. The variation at Farley is paralleled elsewhere in Devon, and many mid-Devon lanes are lined by hedges dominated by blackthorn, hawthorn

Species	G	H(S)	S(M)	W
Stone parsley (*Sison amomum*)		***		
Wood avens (*Geum urbanum*)		**	*	
Hedge bedstraw (*Galium mollugo*)		*	**	
Hawthorn (*Crataegus monogyna*)			***	
Spindle (*Euonymus europaeus*)			***	
Blackthorn (*Prunus spinosa*)			***	
Field rose (*Rosa arvensis*)			***	
Dog-rose (*Rosa canina*)			***	
Black bryony (*Tamus communis*)			***	
English elm (*Ulmus procera* – sucker growth)			***	
Field maple (*Acer campestre*)			**	*
Holly (*Ilex aquifolium*)			**	*
Bramble (*Rubus fruticosus* agg.)			**	*
Lords-and-ladies (*Arum maculatum*)			*	**
Ash (*Fraxinus excelsior*)			*	**
Ivy (*Hedera helix*)			*	**
Creeping soft-grass (*Holcus mollis*)			*	**
Lesser celandine (*Ranunculus ficaria*)			*	**
Wood dock (*Rumex sanguineus*)				***
Pedunculate oak (*Quercus robur*)				***
False brome (*Brachypodium sylvaticum*)		*	*	*
Foxglove (*Digitalis purpurea*)		*	*	*
Cleavers (*Galium aparine*)		*	*	*
Stinging nettle (*Urtica dioica*)		*	*	*

and hazel, and bright in spring with red campion, bluebells, greater stitchwort and germander speedwell (Fig. 108). Hedges on wetter and poorer soils often contain an abundance of sallow, rare at Farley. On the Devonian limestone around Torbay or on chalky soils in east Devon different species again are prominent – dog's mercury, yellow archangel and ramsons (*Allium ursinum*) amongst others. And hedges in other parts of the country pose their own questions.

Declan Doogue and Daniel Kelly (2006) analysed the woody species in hedge samples widely spread over eastern Ireland (Table 7). They found a primary

TABLE 7. Eastern Irish hedges. Frequency of trees, shrubs and woody climbers in hedgerows across Leinster from 90 aggregate sites (10 pooled samples per 10 km grid square) and from 865 individual 30 m hedge samples. Species in less than 10% of the aggregate sites omitted. From Doogue & Kelly (2006).

Species	Aggregate sites (%)	30 m samples
Hawthorn (*Crataegus monogyna*)	100	83
Bramble (*Rubus* 'non-*ulmifolius*')	99	78
Ivy (*Hedera helix*)	96	68
Ash (*Fraxinus excelsior*)	93	60
Blackthorn (*Prunus spinosa*)	91	45
Dog-rose (*Rosa canina*)	90	51
Elder (*Sambucus nigra*)	79	31
Bramble (*Rubus ulmifolius*)	70	38
Sycamore (*Acer pseudoplatanus*)	68	16
Gorse, furze (*Ulex europaeus*)	68	26
Honeysuckle (*Lonicera periclymenum*)	66	25
Grey sallow (*Salix cinerea*)	63	18
Wild privet (*Ligustrum vulgare*)	57	23
Hazel (*Corylus avellana*)	48	16
Holly (*Ilex aquifolium*)	47	20
Field rose (*Rosa arvensis*)	34	12
Raspberry (*Rubus idaeus*)	34	6
Beech (*Fagus sylvatica*)	33	5
Sherard's downy-rose (*Rosa sherardii*)	30	6
Spindle (*Euonymus europaeus*)	28	9
Wild plum (*Prunus domestica*)	26	4
Guelder-rose (*Viburnum opulus*)	20	5
Eared sallow (*Salix aurita*)	19	5
Snowberry (*Symphoricarpos albus*)	19	3
Alder (*Alnus glutinosa*)	18	6
Bittersweet (*Solanum dulcamara*)	14	3
Downy birch (*Betula pubescens*)	14	4
Wild cherry (*Prunus avium*)	13	3
Pedunculate oak (*Quercus robur*)	12	2
Goat willow (*Salix caprea*)	11	2
Rowan (*Sorbus aucuparia*)	11	3
Wych elm (*Ulmus glabra*)	11	3
Small-leaved elm (*Ulmus minor*)	11	3
Apple (*Malus domestica*)	11	2

FIG 108. (a) Roadside near Morchard Bishop, Devon, May 2000; hedge of hazel, oak, ash, blackthorn etc. (b) Detail of hedge bank; red campion, creeping buttercup, greater stitchwort and germander speedwell in flower.

gradient from base-rich soils, supporting hedges rich in spindle, hazel, wild privet and guelder-rose, to acid soils, with rowan, gorse and downy birch. There was a clear secondary dry–wet gradient, with sycamore, wych elm, ash, elder and honeysuckle at the dry end, contrasting with alder, sallows, buckthorn and birch, which favoured wetter habitats. They found roadside hedges to be richer in woody species than hedges dividing fields (average 7.43 as against 6.47 species per 30 m sample); the difference is statistically significant, but not dramatic. The answer to the question posed in the authors' title is clear: the assemblages of woody species in the hedges of eastern Ireland are mainly the products of environment – but with detectable historical influences.

CHAPTER 9

Meadows and Pastures

'ENGLAND'S GREEN AND PLEASANT LAND' and 'the Emerald Isle' both take for granted the green fields of the traditional mixed farming landscape that dominated the scene in both countries, and to a large extent still does. However, what has happened in this landscape in the last 50 years is an awful cautionary tale for conservationists with a fixation on 'rare' or 'threatened' species.

Traditional mixed farms relied upon a balance of arable crops and livestock. The balance varied depending on soil conditions and climate. On well-drained soils in the south and east the emphasis was on arable farming, especially cereals. On the less well-drained soils and farther west, the emphasis swung towards livestock, until hill farms relied almost entirely on sheep and hardy cattle. But even in predominantly arable areas some livestock were kept. Horses were needed for ploughing and tillage operations, and for transport. Cows and sheep, pigs and chickens were kept at least for domestic needs, and a market existed for their produce everywhere. Conversely, in predominantly livestock-farming areas, some cereals and roots were grown as stock feed, and vegetables for the needs of the family and the local market. In every area, some substantial proportion of the farm was permanent grassland, which would have been used as pasture, or more or less regularly shut up for hay in spring and early summer. Then, following mowing and haymaking, stock were grazed on the 'aftermath' (when we use that word we often forget its farming origins). The permanent grass would have received farmyard manure from time to time and, especially in the rainy west, dressings of lime were important to counter leaching and acidification of the soil. Lime burning and lime kilns were very widespread.

THE MAIN TYPES OF TRADITIONAL 'PERMANENT GRASSLAND'

What grasslands has traditional farming bequeathed to us? On fertile, well-drained base-rich brown-earth soils throughout much of Britain and Ireland, the likelihood is that any grassland that has not been ploughed in the past century will be dominated by a mixture of red fescue (*Festuca rubra*), crested dog's-tail (*Cynosurus cristatus*) and some combination of cock's-foot (*Dactylis glomerata*), Yorkshire-fog (*Holcus lanatus*), variable amounts of perennial rye-grass (*Lolium perenne*), common bent (*Agrostis capillaris*) and sweet vernal-grass (*Anthoxanthum odoratum*) – the last providing the traditional smell of new-mown hay. An abundance of traditional meadow flowers grow with these grasses, including common knapweed (*Centaurea nigra*), oxeye daisy (*Leucanthemum vulgare*), red and white clovers (*Trifolium pratense* and *T. repens*), bird's-foot trefoil (*Lotus corniculatus*), ribwort plantain (*Plantago lanceolata*), daisy (*Bellis perennis*), meadow vetchling (*Lathyrus pratensis*), yarrow (*Achillea millefolium*), common sorrel (*Rumex acetosa*), self-heal (*Prunella vulgaris*), bulbous and meadow buttercups (*Ranunculus bulbosus* and *R. acris*) and many more. This 'Centaureo-Cynosuretum' grassland^{MG5} averages over 20 species in a 4 m² sample (Fig. 109). It is variable, partly because it is so widespread. In chalk and limestone districts lady's bedstraw (*Galium verum*), yellow oat-grass (*Trisetum flavescens*), rough hawkbit (*Leontodon hispidus*) and cowslips (*Primula veris*) are locally common, and some of the characteristic grasses of limestone grassland (Chapter 11) begin to appear, such as crested hair-grass (*Koeleria macrantha*) and quaking-grass (*Briza media*). In sites at the base-poor end of its range, species favouring rather acid soils are more prominent, including common bent, tormentil (*Potentilla erecta*), field wood-rush (*Luzula campestris*), heath-grass (*Danthonia decumbens*), devil's-bit scabious (*Succisa pratensis*) and betony (*Stachys officinalis*). Corky-fruited water-dropwort (*Oenanthe pimpinelloides*), the only species of the genus to reach into dry grassland, is common in this community in an area extending from Gloucester to Exeter to Portsmouth. In Britain, the Centaureo-Cynosuretum is the commonest habitat of meadow saxifrage (*Saxifraga granulata*), and in both islands probably more cowslips and green-winged orchids (*Orchis morio*) grow (or grew) in this kind of grassland than in any other habitat. Perhaps the only really rare plant with its headquarters in this community is the downy-fruited sedge (*Carex filiformis*) – hardly a species to make headlines.

This is the kind of grassland of which good examples are now most often described (and cherished) as 'old meadow'. Fifty or sixty years ago few people would have given a thought to its conservation, because it was a commonplace

FIG 109. Centaureo-Cynosuretum: an unusually species-rich example at Hardington Moor NNR, Somerset, with cowslip, green-winged orchid, daisy and bulbous buttercup in flower and dandelions in fruit[MG5], May 2000.

FIG 110. Upland hay meadow, Ravenstonedale, Cumbria[MG3]. (a) General view: wood crane's-bill in foreground; meadow buttercup and white umbellifers are the most abundant flowers. (b) Close-up: wood crane's-bill, meadow buttercup, pignut, meadowsweet, ribwort plantain, great burnet and sweet vernal-grass can be seen in the picture. June 2005.

part of the rural scene – and would surely *always* be there. How wrong they were! First the fashion (and agricultural advice) was to plough and reseed old grassland. Then there was a change of thinking to improving old grasslands by repeated application of nitrogenous fertiliser without ploughing. The result was much the same. The productivity of the grassland increased, but its floristic richness declined as the broad-leaved herbs were out-competed by the grasses. A less obvious further factor in the declining species-richness of grasslands is increased nitrogen deposition in rain (Stevens *et al.* 2010, 2011).

Two other, more localised, traditional grassland types have long been seen as 'special'. When I first visited the Pennine dales 60 years ago, it was impossible not to be struck by the spectacular wealth of flowers in the hay meadows. There were not only the familiar meadow flowers of the lowlands, but abundant wood crane's-bill (*Geranium sylvaticum*), great burnet (*Sanguisorba officinalis*) and lady's-mantle (*Alchemilla glabra*). These meadows[MG3] (Fig. 110) are the product of a particular pattern of hill farming in the Pennines and Lake District (and to a lesser extent in southern Scotland north to Tayside). They form part of the 'in-by' land of the farm. Traditionally, they were grazed in winter, mainly by sheep. In late April or early May the stock were moved to the 'out-by' summer grazing of the unenclosed hill pastures, and the meadows were given a light dressing of farmyard manure and shut up for hay. Mowing generally took place in late July to early August depending on the weather ('Make hay while the sun shines'), but in unfavourable seasons it might be delayed as late as September. The aftermath was then grazed until winter closed in. The meadows received no other treatment apart from occasional liming.

The grasses are not visually as dominant as in most other grasslands but nevertheless make up a substantial fraction of the herbage. The commonest species are sweet vernal-grass, cock's-foot, rough meadow-grass (*Poa trivialis*), red fescue, common bent and Yorkshire-fog, with crested dog's-tail, soft brome (*Bromus hordeaceus*), yellow oat-grass and perennial rye-grass playing a subsidiary role. The most constant of the long list of associated species, apart from wood crane's-bill, lady's-mantle and great burnet mentioned already, include common sorrel, pignut (*Conopodium majus*), ribwort plantain, meadow buttercup, common mouse-ear (*Cerastium fontanum*) and white clover. Yellow-rattle (*Rhinanthus minor*), bulbous buttercup, rough hawkbit, field wood-rush and red clover are other frequent associates. A clutch of rare lady's-mantles – *Alchemilla monticola, A. acutiloba* and *A. subcrenata* – grow in meadows in (and around) upper Teesdale (Bradshaw 1962). In damper parts of the meadows, melancholy thistle (*Cirsium heterophyllum*), globeflower (*Trollius europaeus*), common bistort (*Persicaria bistorta*), marsh hawk's-beard (*Crepis paludosa*) and water avens (*Geum rivale*) show the beginnings of a gradual transition to poorly drained meadows with marsh-marigold (*Caltha palustris*). On roadsides or in undisturbed corners, these meadow species are joined by giant bellflower (*Campanula latifolia*) and sweet cicely (*Myrrhis odorata*).

The traditional alluvial hay meadows bordering some of our lowland rivers that drain from chalk or limestone are a far cry from the northern upland hay meadows, but they were shaped by a comparable pattern of agricultural use, tailored to a different setting. These meadows are flooded in winter; and the lime-rich water and silt input of the annual flood was crucial in maintaining their fertility. Typically, the meadows were shut up for hay in early spring, and mown in July. Then the aftermath was grazed through late summer and autumn, mainly by cattle. Management certainly varied from place to place, and over recent decades many former alluvial hay meadows have been drained and reseeded, so that the area of traditionally managed alluvial meadows is now a fraction of what it was a couple of centuries ago. Some of the richest remaining examples are where common rights have kept traditional practices in being (Fig. 111), or where the meadows have been valued for non-agricultural reasons, as in some of the meadows round Oxford.

The turf of these meadows is typically dominated by red fescue and crested dog's-tail, usually with a substantial component of meadow foxtail (*Alopecurus pratensis*) and perennial rye-grass, and often cock's-foot and yellow oat-grass. A

FIG 111. Lowland alluvial meadow, North Meadow NNR, Wiltshire[MG4]: (a) in spring, with dandelions and fritillary in flower, 5 May 1977; (b) a month later the grass has grown and oxeye daisies and abundant meadow buttercups are in flower, 1 June 1984.

profusion of broad-leaved herbs accompany these grasses, including great burnet, ribwort plantain, meadow buttercup, common sorrel, dandelion, red clover, meadow vetchling, pepper saxifrage (*Silaum silaus*), common knapweed, bird's-foot trefoil, creeping buttercup (*Ranunculus repens*) and much else[MG4]. The most notable species native to this habitat is fritillary (*Fritillaria meleagris*), the 'chequered daffodil', often grown in gardens but rare in the wild. W. H. Hudson (1919) vividly described an encounter with this species in his *Book of a Naturalist*. Some of the hay meadows in drier parts of the Shannon callows (Heery 1991, 2003) may be an Irish equivalent of these English flood-meadows, but conditions are rather different and several of their most characteristic species are lacking in Ireland.

GRASSLANDS PRODUCED BY MORE INTENSIVE FARMING

The commonest kind of permanent farm grassland in lowland Britain and much of Ireland is much less species-rich than the communities described above. It is generally dominated by perennial rye-grass, crested dog's-tail, red fescue and Yorkshire-fog, with white clover, common mouse-ear, meadow buttercup,

FIG 112. Upland Lolio-Cynosuretum above Llanrwst, Clwyd[MG6]. The fields in the foreground are badly infested with creeping thistle and soft-rush (*Juncus effusus*). The mountains of Snowdonia in the distance. August 1975.

ribwort plantain, yarrow and creeping thistle (*Cirsium arvense*). A wide range of other species occurs more sparsely. This 'ordinary' farm grassland (Lolio-Cynosuretum[MG6]) varies depending on base status, fertility and moisture content of the soil, and on the management it receives (Fig. 112). In some instances, sweet vernal-grass becomes constant, sometimes in substantial quantity with common bent, and frequent common sorrel, cock's-foot, smooth meadow-grass (*Poa pratensis*) and a rather richer associated flora. Fields regularly mown for hay often contain a good deal of meadow foxtail. Tufted hair-grass (*Deschampsia cespitosa*) sometimes occurs in ill-drained areas, and riverside Lolio-Cynosuretum grassland is often dotted with clumps of yellow iris (*Iris pseudacorus*).

Short-term leys sown with improved forms of perennial rye-grass, alternating with arable, have been a common part of farming in Britain and Ireland since at least the end of the eighteenth century, especially in Scotland and Northern Ireland. Seed spread from these old leys may be responsible for the prevalence of rye-grass in most semi-natural farm grasslands at the present day (Beddows 1967). In recent decades there has been a swing away from permanent grass towards short-term leys (principally for silage) or reseeded pasture sown with modern high-yielding rye-grass cultivars. Rye-grasses (*Lolium* spp.) outdo all our other common grasses in productivity. For short-term (1–2-year) silage leys Italian ryegrass (*L. multiflorum*) is generally preferred. For rather longer (3–4-year) silage leys hybrid rye-grass may be used. For more permanent general-purpose grassland the balance of advantage swings to perennial rye-grass, generally with white clover[MG7a], with or without an admixture of timothy (*Phleum pratense*). Landscape and playing-field mixes are usually made up of perennial rye-grass and red fescue or smooth meadow-grass; a standard bank-and-verge mix contains perennial rye-grass, red fescue, smooth meadow-grass, common bent and white clover.

Short-term leys are intended to be species-poor, and usually are. Longer-term sown grasslands pick up at least a modest associated flora, from root and stem fragments in the ploughland, the seed-bank, and seeding-in from the surroundings. If well managed they may remain productive for many years and ultimately become indistinguishable from established Lolio-Cynosuretum. However, there are stages along the way, and these vary depending on the context. Sometimes timothy becomes prominent along with rough meadow-grass and other grasses[MG7b]. In seasonally flooded river valleys and on other moist soils meadow foxtail often makes up a large proportion of the herbage, with or without meadow fescue (*Festuca pratensis*). These rather species-poor meadows may be derived from former water-meadows and flood-meadows, reseeded and managed for hay[MG7c,d]. More commonly, old rye-grass–white clover leys are increasingly invaded by ribwort plantain, greater plantain (*Plantago major*) and dandelions, leading to a sward[MG7e], which intergrades with the trampled track and gateway communities of the next chapter. Some seed mixtures contain smooth meadow-grass, and this species too is a common constituent of the grass in parks and on road verges[MG7f]. Arguably, all these intensive agricultural grasslands could be embraced within a broadly defined 'Lolio-Cynosuretum'. But both *Lolium* and *Cynosurus* are usually lacking in a grassland that is quite commonly encountered, dominated by common bent and red fescue.

ROAD VERGES AND SOME OTHER UNGRAZED GRASSLANDS

False oat-grass (*Arrhenatherum elatius*) is a conspicuous and distinctive grass that is common throughout Britain and Ireland, but remarkably unobtrusive in the literature. It is mentioned several times by Tansley (1939) as an occasional component in plant succession from abandoned farmland to woodland and in open woods on limestone, and as a local and sporadic dominant in chalk grassland ('perhaps only in disturbed soil'). However, Tansley observed that 'a more characteristic habitat for this grass is the highly characteristic roadside "verges" in chalk districts, where it is frequently dominant for long stretches, often in company with common knapweed and wild parsnip (*Pastinaca sativa*). This vegetation is generally cut over in the late summer or early autumn, but is not grazed except quite casually.' The road verges near Bibury in Gloucestershire, recorded annually over many decades by the late Professor Arthur Willis, are this community (Dunnett *et al*. 1998). They were mown once a year in autumn, and showed year-to-year variations with the weather, but little long-term change.

Roadside verges are indeed the commonest habitat for false oat-grass, and not only in chalk and limestone country. It is a tall elegant grass, generally dominant where it occurs, usually accompanied by cock's-foot, Yorkshire-fog and tall herbs including hogweed (*Heracleum sphondylium*), cow parsley (*Anthriscus sylvestris*), rough chervil (*Chaerophyllum temulum*), creeping thistle, ribwort plantain, common knapweed and white dead-nettle (*Lamium album*)[MG1] (Fig. 113). On nutrient-rich verges stinging nettles are common, sometimes with patches of mugwort (*Artemisia vulgaris*) or comfrey (*Symphytum officinale*). Comfrey favours moist soils, and sometimes grows with meadowsweet (*Filipendula ulmaria*) and wild angelica (*Angelica sylvestris*) in a variant of this community on damper stretches of roadside. In chalk districts (as Tansley noted), wild parsnip is a

FIG 113. Ungrazed *Arrhenatherum* road verges, Fiddleford, Dorset, with cow parsley in flower[MG1], May 1977.

regular associate of *Arrhenatherum*, and yarrow, lady's bedstraw, ragwort (*Senecio jacobaea*), field scabious (*Knautia arvensis*), wild basil (*Clinopodium vulgare*), agrimony (*Agrimonia eupatoria*), greater knapweed (*Centaurea scabiosa*) and wild marjoram (*Origanum vulgare*) are frequent. In Ireland, O'Sullivan (1982) similarly noted that 'Oatgrass meadows do not occur any more in normal farm situations … Instead, roadsides, cemeteries and railway banks must be looked to for examples … Cemeteries probably offer the best examples since most have been mown once annually for many generations.'

In Britain, *Arrhenatherum* communities also occur in fields once ploughed but later abandoned (especially on chalk and limestone), in ungrazed chalk grassland, along wood margins, on railway banks and motorway verges, and in churchyards. One very characteristic community is confined (with a few exceptions) to cliffs and screes, often facing north, on Carboniferous limestone in the north of England. Most of these sites are more or less inaccessible to grazing; some are lightly shaded by fragmentary woodland. *Arrhenatherum* is associated with a luxuriant and rich herbaceous vegetation including hogweed, meadowsweet, tufted hair-grass, stinging nettle, common valerian (*Valeriana officinalis*), water avens, male-fern (*Dryopteris filix-mas*), wild angelica (*Angelica sylvestris*), great burnet, lady's-mantle, common sorrel and common figwort (*Scrophularia nodosa*).

FIG 114. Jacob's-ladder (*Polemonium caeruleum*) at Malham Cove, Yorkshire, in ungrazed *Arrhenatherum* grassland in open ashwood on limestone scree[MG2], July 1966.

Many smaller woodland and meadow species grow alongside these. The north-of-England native occurrences of Jacob's-ladder (*Polemonium caeruleum*) are mostly in this kind of vegetation[MG2] (Fig. 114).

SOME GRASSLANDS OF ILL-DRAINED SOILS

The meadows and pastures considered so far are on relatively freely draining soils. They may be flooded in winter, but through the growing season the water-table is sufficiently far below the surface for the soil to be well-enough aerated not to impede root growth or nutrient uptake. If the soil is intermittently or persistently wet during the growing season the consequences are apparent in the vegetation.

This is most obvious in wet meadows near rivers (or on clay soils in high-rainfall districts), where the water-table is persistently high during spring and summer. In some river-valleys in southern England this situation was exploited in the period of 'high farming' from the late seventeenth to early nineteenth centuries to create 'water meadows', with systems of leats which allowed controlled flooding and irrigation of the sward. In late winter river water was warmer than the land, so the water brought both warmth and nutrients to the grass, stimulating an 'early bite' for the cattle. Being labour-intensive to operate and maintain, these systems gradually fell into disuse during the second half of the nineteenth and early twentieth centuries, but the remains of their ditch systems are still often apparent in aerial photographs. Floristically, all these wet pastures combine many of the common grassland species, already encountered, with a contingent of characteristic wet-ground plants. The dominant grasses include crested dog's-tail, red fescue, Yorkshire-fog, rough meadow-grass, sweet vernal-grass and meadow fescue, growing alongside marsh-marigold, meadow buttercup, creeping buttercup, white clover, common sorrel, autumn hawkbit (*Leontodon autumnalis*) and ribwort plantain. Apart from the marsh-marigold, frequent indicators of wetter conditions include creeping bent (*Agrostis stolonifera*), carnation sedge (*Carex panicea*), meadowsweet, cuckooflower (*Cardamine pratensis*), ragged-robin (*Lychnis flos-cuculi*), brown sedge (*Carex disticha*), common spike-rush (*Eleocharis palustris*), jointed rush (*Juncus articulatus*) and the moss *Calliergonella cuspidata*[MG8]. Wet grasslands of this kind (Fig. 115) are still scattered widely but unevenly over England and Ireland (where they occur widely in the 'callows' of the Shannon floodplain) but are sparser in Wales and Scotland.

I was brought up with tufted hair-grass, though it was years before I learnt its name. We lived on the London Clay, and there was a vacant builder's plot next-door-but-one to us, and I well remember the big tussocks of ribbed rough leaves,

FIG 115. Wet upland pasture on boulder-clay near Malham, Yorkshire, with marsh-marigold in flower[MG8], June 1983.

and the graceful silvery inflorescences, taller than I was at the time. *Deschampsia cespitosa* is a versatile grass, which grows in a variety of situations, including wet woods and grasslands that are at least seasonally moist and reasonably nutrient-rich, from the lowlands to high on the mountains. In lowland pastures it is characteristic of heavy clay soils prone to waterlogging; it also occurs in zonations at the margins of pools and fens. It has little agricultural value, tends to be avoided by stock and is difficult to eliminate once it gets a hold in pasture. In grassland, it is usually accompanied by Yorkshire-fog, and its only other (nearly) regular associates are rough meadow-grass, cock's-foot, red fescue, creeping bent and creeping buttercup, though such common pasture species as common sorrel, ribwort plantain, meadow foxtail and creeping thistle are still present[MG9]. In grazed examples a number of familiar pasture species are frequent or occasional – such things as meadow buttercup, rushes, dandelion, clovers, marsh thistle (*Cirsium palustre*) and the ragworts. Where there is no grazing, most of the smaller pasture plants are shaded out and false oat-grass becomes a substantial component, with abundant cock's-foot and frequent common knapweed, but the community remains poor in species. More detail about tufted hair-grass communities is given by Davy (1980).

Damp pastures with clumps of rushes are a common sight throughout the lowlands of Britain and Ireland, and to a substantial altitude on the hills. Soft-rush (*Juncus effusus*) tends to predominate in the north and west, hard rush (*J. inflexus*) in the south and east, and on lime-rich soils. Views on rush infestation depend on the point of view of the farmer. If maintaining the productivity of the grazing for cattle is his priority, he will probably regard rushes as unmitigated weeds. If he is keeping sheep on an exposed upland farm he may feel that 'a clump of rushes is worth sixpence' for the shelter they give to young lambs. Once rushes have become established they are hard to eradicate. They produce abundant readily dispersed seed, and the seedlings quickly develop deep fibrous roots. The grassy matrix of this 'Holco-Juncetum'[MG10] is in effect a run-down damp Lolio-Cynosuretum in which the better grasses – perennial rye-grass and crested dog's-tail – though still present, have largely given way to Yorkshire-fog, creeping bent and rough meadow-grass (Fig. 112). It is a matter of choice whether rush-infested pasture is regarded as yet another variant of a broadly defined Lolio-Cynosuretum, or as a community in its own right, part way to a fen meadow (Chapter 14).

No such doubts beset the diverse but distinctive grasslands in which creeping bent and silverweed (*Potentilla anserina*) are among the most conspicuous components (Fig. 116). These are mainly plant-communities of freely draining but intermittently flooded habitats (or habitats in which the water-table

FIG 116. Intermittently flooded grassland with creeping bent, silverweed, creeping buttercup, marsh bedstraw, meadowsweet. Newtown turlough, Co. Galway, June 2010. The surface of the soil is cracked after a long dry spell.

fluctuates widely), generally calcareous, such as the goose green round the village pond (*anserina* is Latin for 'of geese') or flood-prone hollows and ground along frequently flooded streams and rivers. In Britain, several probably rather disparate communities of this kind have been recognised[MG11]. The most frequent has much in common with ordinary Lolio-Cynosuretum agricultural grassland, with abundant perennial rye-grass and Yorkshire-fog, and associated species including creeping buttercup, dandelion and broad-leaved dock[MG11a]. Another silverweed-rich community is associated with upper saltmarshes, with frequent spear-leaved orache (*Atriplex prostrata*), sea mayweed (*Tripleurospermum maritimum*), knotgrass (*Polygonum aviculare*) and parsley water-dropwort (*Oenanthe lachenalii*)[MG11b]. A third is associated with maritime shingle and dune slacks, with a sprinkling of shingle-beach and sand-dune species[MG11c]. Silverweed is often prominent in dune-slack communities, but traditionally plant ecologists have relegated these to another context (Chapter 19)!

Silverweed-rich flooding communities inland in Britain have probably been rather neglected. However, in Ireland abundant examples are provided by the turloughs, in which complete zonations can be seen (up to 5 m or so in vertical extent) from permanently dry pastures – Centaureo-Cynosuretum or Lolio-

Cynosuretum – to open water or weedy communities of wet mud (Chapter 12).

Tall fescue (*Festuca arundinacea*) is a striking grass, which grows in various habitats, including scrub and wood margins. It is particularly common near the sea, and a very characteristic situation for it is the banks of tidal rivers and the upper fringes of saltmarshes, where it is liable to occasional flooding by brackish water. It is usually accompanied by silverweed and many of the common species of agricultural grassland, including creeping bent, red fescue, Yorkshire-fog, perennial rye-grass, white and red clovers, ribwort plantain, bird's-foot trefoil and often by a sprinkling of saltmarsh and brackish-water species, such as parsley water-dropwort, sea-milkwort (*Glaux maritima*), saltmarsh rush (*Juncus gerardii*), false fox-sedge (*Carex otrubae*) and distant sedge (*C. distans*)[MG12]. This community is commonly ungrazed. It occurs in suitable places all along the south and west coasts of Britain at least as far north as Arran, and in south and west Ireland.

In hollows in riverside pastures and meadows where water stands late into the spring and after flooding (or low in the turlough zonation) a community develops dominated by creeping bent and marsh foxtail (*Alopecurus geniculatus*), usually with creeping buttercup and scattered tall docks (*Rumex crispus, R. obtusifolius*) (Fig. 117). Other frequent or occasional associates are floating sweet-grass (*Glyceria fluitans*), water forget-me-not (*Myosotis scorpioides*), water

FIG 117. Meadows beside the River Stour at Fiddleford, Dorset, 25 May 1977. Freely drained meadow on the left with abundant meadow foxtail and meadow buttercup[MG6]; flood-meadow on the right dominated by creeping bent and marsh foxtail[MG13]. Since ploughed and reseeded.

mint (*Mentha aquatica*), marsh bedstraw (*Galium palustre*), marsh-marigold and water-pepper (*Persicaria hydropiper*)[MG13]. Silverweed occurs, but is not nearly as prominent as it is in grasslands that are better drained between floodings. A similar community occurs in Irish turloughs.

WHAT FACTORS DIFFERENTIATE THESE GRASSLANDS?

Almost all the grasslands described in this chapter, from the most species-rich old meadows to short-term rye-grass leys, reflect in some measure the influence of man, and usually of his grazing animals. They must reflect too the physical factors of location, soil and climate. The effects of some different agricultural treatments are demonstrated in the Park Grass Experiment at Rothamsted, Hertfordshire. The experiment was established in 1856 (–1875), in a level, species-rich old hay meadow, chosen for its uniform herbage. The traditional management had been to take one hay crop a year, and graze the aftermath with sheep. The land had received occasional farmyard manure. The experiment was divided into plots, which received various fertiliser treatments every year, two plots being left unmanured as controls. Sheep grazing ceased in 1875, to be replaced by a second hay cut. Most of the plots were split into two around a century ago, one half receiving lime in addition to the fertiliser treatments, which continued as before. Some of the differences between the fertiliser treatments are impressive, both in species composition and in sward height and production of hay (Silvertown *et al.* 2006).

Unsurprisingly, the plots that received no fertiliser additions, or only a light dressing of nitrogen, leached somewhat, but otherwise changed little and remained as Centaureo-Cynosuretum (MG5). Unlimed plots that received the higher nitrogen treatments as ammonium sulphate progressively acidified, ending up with only a few species, which most nearly matched gorse–bramble scrub (W23). Liming corrected this, the corresponding plots coming to match most closely the species-poor perennial rye-grass–meadow foxtail grassland (MG7D). Plots that received a PKMg mineral fertiliser tended to change in the direction of false oat-grass grassland, with a deep sward and such tall species as hogweed and meadowsweet. An interesting feature of the Park Grass Experiment is the rarity of perennial rye-grass, and with that, few plots matching 'ordinary' Lolio-Cynosuretum farm grassland (MG6). This may be because the experiment is exclusively a hay regime, with no grazing, and partly because no reseeding (or seeding-in of rye-grass) has taken place for the 150-years-plus of the experiment – two eventualities unlikely in normal farming practice. One of the aims of 'good

management' is to maintain the proportion of productive rye-grass, which tends to decline in many pasture situations (Dodd *et al.* 1994).

Obviously, in the real world grazing (or lack of it) is important. Grassy road verges are no less a product of the cultural landscape than the fields they border, and before the days of motor vehicles they were often grazed, at least casually. Nowadays they are usually cut once a year, and not grazed at all. Traditional meadow regimes generally involved cutting in mid to late summer, and grazing the aftermath. This provided suitable conditions for a range of tall perennials to flower and fruit, and included a grazing component, and of course the animals on the farm were instrumental in providing the farmyard manure that went on the hay meadow. Many grasslands were used purely as pasture, and grazing was important in their maintenance as grassland. Hawthorn soon sows into undergrazed grassland, and periods of agricultural recession are marked by a generation of scattered hawthorn bushes, especially in hill-farming districts such as North Wales, Dartmoor and western Ireland.

We return to the water-table and aeration. Agriculturally, the ideal grassland soil is retentive of moisture but freely drained. Not all soils are like that, and a great deal of work is expended on trying to bring them nearer the ideal. For soils on heavy, impermeable clays (grasslands dominated by tufted hair-grass and many rushy pastures are examples) such remedies as there are lie in the hands of the farmer. For grasslands that may potentially be freely drained, but which flood intermittently, or in which the water-table fluctuates near the surface, the alternatives are to adapt to conditions as they are, or deeper drainage. Traditional farming has many examples of adaptation to particular conditions. Drainage has been going on at least since Roman times in Britain, and much earlier than that in the older civilisations of the Middle East and Asia. Engineering the environment is nothing new, but it is costly, and has social and political consequences, often unintended and often controversial.

Meadows and pastures are all green and grass-dominated, so it is easy to see them as the continuum that they are, with no sharp boundaries except the boundaries we impose upon them. It is possible to recognise 'noda', frequently recurring types of grassland, within the continuum, but intergradations exist in every direction. On chalk and limestone, the agricultural grasslands grade into the calcareous grasslands of Chapter 11, on wetter ground they intergrade with the fens and fen meadows of Chapter 14, in hill country on hard acid rocks they grade into the upland grasslands of Chapter 16, and near the sea they intergrade with the upper-saltmarsh swards and the fixed-dune, machair and cliff-top grasslands of Chapters 18–20.

CHAPTER 10

Weeds and Ruderals: Ecological Opportunism

P ROBABLY ALL OF US KEEP 'weeds' in a different mental category from the plants that make up the predominant kinds of long-lived vegetation. Certainly anyone with a garden is keenly aware of the opportunist species that germinate and quickly grow to occupy any area of bare ground unless constant war is waged against them. Some of the characteristic features of weedy species have already been touched on in Chapter 4. However, as any gardener or farmer should know, weeds are very diverse. Most establish readily in suitable habitats, and grow quickly. Weedy plants typically take up nutrients rapidly, and respond strongly to the levels of plant nutrients in the soil, especially phosphorus (P) and nitrogen (N). Weedy species are often spoken of as 'nitrophilous' (and stinging nettles do indeed have exceptionally high levels of the nitrogen-assimilating enzyme nitrate reductase), but weeds are probably more often limited by the availability of P than of N. Another characteristic of weeds of cultivated ground is that they generally have a persistent 'seed-bank', of seeds dormant in the soil. In sheer number of species weeds make a major contribution to the richness of our flora. It is a common experience of botanists 'square-bashing' – recording species from Ordnance Survey grid-squares – that the squares yielding the longest species-lists are those with a good range of weedy habitats.

Weedy species are typically plants of short-lived or intermittent habitats (Fig. 118). These can arise from disturbance by man, as in our arable fields and gardens, or from occasional but unpredictable natural events such as gales, floods or landslips. Weedy species may also take advantage of the more or less constant disturbance (and nutrient enrichment) of such sites as seabird colonies (Chapter 20) and rabbit warrens (Chapters 8, 11). Some intermittent habitats

FIG 118. Ruderals on abandoned farmland, Alderney, July 1969. A spectacular mass of colour, mostly common poppy (*Papaver rhoeas*) and common mallow (*Malva sylvestris*).

are predictably seasonal. Thin soils sun-baked in summer can support rich assemblages of short-lived plants, which germinate in autumn or winter, flower in early spring, and set seed and die before heat and dryness make the habitat untenable. Another group of seasonal habitats are flooded in winter, but become free for plant colonisation when the water level falls in summer. In addition, there are many more or less 'weedy' plants particularly associated with 'edges' – abrupt changes in habitat and vegetation – as on wood margins (Chapter 8), riverbanks and seashores (Chapters 18–20).

WEEDS OF CULTIVATION

Plants that are 'weeds' from a farmer's or gardener's point of view can come from any of these categories – but from some more than others. All our weeds must have originated from 'natural' situations that existed before Neolithic man began stock raising and arable cultivation. Many of our familiar annual weeds probably had their wild origins in the seasonally dry spring-annual communities of the Mediterranean and (especially) the fertile Middle East, where our principal cereals and some other crop plants such as flax originated. Others are likely to have come from sites disturbed by animals (e.g. chickweed) or recurrent landslips

FIG 119. Colt's-foot (*Tussilago farfara*), on a crumbling clay cliff on the Dorset coast. It finds similar open habitats inland around building sites and on clayey waste ground. Near Chideock, April 1974.

(e.g. colt's-foot, Fig. 119), or liable to seasonal flooding, or on seashores (e.g. dock and goosefoot species). Plants that are troublesome weeds in one situation may be little or no problem in another, so that different soils or different crops tend to have different weed assemblages. The short-lived weeds of arable fields or the vegetable garden are generally little trouble in permanent grassland. A newly sown grass field can produce a remarkable diversity of small arable weeds in the first year, which disappear in the second year as the grasses shade them out, with no treatment apart from mowing. Conversely, perennial weeds such as ragwort, docks and thistles are characteristically a problem of permanent grassland. In anything other than a plot constantly cultivated for vegetables or annual bedding, invasive and persistent perennial weeds – such as dandelions, docks, bindweeds, couch-grass, ground-elder and brambles – are the bane of the gardener.

Arable weeds

Weeds of arable crops are much less abundant and conspicuous nowadays than they were before the days of chemical selective weedkillers. Nevertheless, most of them are still there, but at lower density and more localised, so that they need more looking for. No weed occurs everywhere, but probably the most-nearly universal weeds of arable fields throughout Britain and Ireland are chickweed (Fig. 120c), annual meadow-grass (*Poa annua*), groundsel (*Senecio vulgaris*), shepherd's-purse (*Capsella bursa-pastoris*), knotgrass (*Polygonum aviculare*) and fat-hen (Fig. 121b). On fertile loams and clays in the lowlands these plants are often accompanied by the daisy-like scentless mayweed (*Tripleurospermum inodorum*), scarlet pimpernel (Fig. 120b), the tiny field pansy (*Viola arvensis*) and common couch-grass (*Elytrigia repens*)[OV13]. Particularly in patches where there has been some disturbance

FIG 120. Small arable weeds: (a) common field speedwell (*Veronica persica*); (b) scarlet pimpernel (*Anagallis arvensis*); (c) chickweed (*Stellaria media*); (d) sharp-leaved fluellen (*Kickxia elatine*).

or trampling, annual meadow-grass and groundsel may be far-and-away the commonest species[OV10]. These are very widespread field-weed assemblages. On somewhat lighter well-drained soils, particularly among cereal crops in southern and eastern parts of the country, the blue-flowered annual common and grey field-speedwells (*Veronica persica*, Fig. 120a, and *V. polita*) are prominent, often accompanied by black-bindweed (*Fallopia convolvulus*) and sometimes with the red and cut-leaved dead-nettles (*Lamium purpureum*, *L. hybridum*), sun spurge (*Euphorbia helioscopia*) and occasional common (corn) poppies (*Papaver rhoeas*)[OV7]. Locally, in winter-sown crops in southern and eastern England, black-grass (*Alopecurus myosuroides*) can be a prominent member of this community[OV8].

On light, friable, neutral-to-acid, freely draining soils, the picture is rather different. In the south and east, this is a preferred habitat of common poppies (Fig. 121a), usually with field pansy, common field-speedwell, scarlet pimpernel and hop trefoil (*Medicago lupulina*) and sometimes with the prickly poppy (*Papaver argemone*)[OV3]. The common poppy is mostly a plant of lowland Britain and Ireland, becoming rare in the north and west. The corn marigold (*Chrysanthemum segetum*) with its spectacular yellow flowers is more widespread, and can dominate weed communities on a variety of well-drained acid soils from southern England to western Ireland and the Hebrides (Fig. 122), and indeed over most of Atlantic Europe. It is usually accompanied by corn spurrey (*Spergula arvensis*), knotgrass, annual meadow-grass and other common weeds, and on more moisture-retentive soils and in the wetter climate of the north and west by creeping buttercup (*Ranunculus repens*)[OV4]. But weed assemblages on acid soils vary greatly, and in west Cornwall, for example, the little annual toad rush (*Juncus bufonius*) is common and ramping-fumitories (*Fumaria occidentalis*, *F. bastardii* and *F. muralis*, Fig. 123b) are often prominent, with a number of small annuals such as sticky mouse-ear (*Cerastium glomeratum*) and rough-fruited buttercup (*Ranunculus*

FIG 121. Taller arable weeds: (a) common poppy (*Papaver rhoeas*), Alderney, July 1969; (b) fat-hen (*Chenopodium album*) and scentless mayweed (*Tripleurospermum inodorum*), Shroton, Dorset, August 1975.

FIG 122. A weedy barley field, Islay. September 1985. Abundant corn marigold (*Chrysanthemum segetum*), corn-spurrey (*Spergula arvensis*) and redshank (*Persicaria maculosa*).

FIG 123. (a) Field bindweed (*Convolvulus arvensis*), Coleton Fishacre, Devon, August 1990; (b) common ramping-fumitory (*Fumaria muralis*), Exeter, June 1976.

muricatus)[OV6]. In the extreme case of the bulb fields of the Isles of Scilly, near-constant lesser quaking-grass (*Briza minor*) and small-flowered catchfly (*Silene gallica*) are accompanied by some perennial weeds and a rich flora of winter annuals and arable-field bryophytes including various species of the liverwort genera *Riccia* and *Sphaerocarpus*[OV2].

Light, lime-rich soils also have their characteristic weed communities. The commonest and most conspicuous species include a number that are not confined to calcareous soils, such as the common field-speedwell, scarlet pimpernel, knotgrass and black-bindweed. The influence of the lime-rich soil is apparent in the greater frequency of field bindweed (*Convolvulus arvensis*, Fig. 123a), dwarf spurge (*Euphorbia exigua*), Venus's-looking-glass (*Legousia hybrida*), field madder (*Sherardia arvensis*) and three toadflaxes, small toadflax (*Chaenorhinum minus*) and the two fluellens (*Kickxia elatine* (Fig. 120d) and *K. spuria*)[OV15]. Sometimes on lime-rich, freely drained soils in south and east England (most often with cereal crops), a distinctive assemblage occurs in which the night-flowering catchfly (*Silene noctiflora*) and common poppy are prominent. This community has decreased greatly with more intensive farming, but has benefited in recent decades from the creation of conservation headlands around arable fields[OV16].

This sketch of the weed assemblages of arable fields is inevitably incomplete. Soils and climate vary from place to place and even from field to field, and farming practices vary from farm to farm. Add to that apparently random variation, and it is understandable that weed communities – as David Webb (1954) wrote of vegetation in general – hover tantalisingly on the borderline between being classifiable and unclassifiable.

Garden weeds

Gardens share many weeds with arable farmland. I have probably never had a garden wholly without common field-speedwell, annual meadow-grass, scarlet pimpernel, groundsel and knotgrass. But of five gardens in Devon, very different in situation and altitude and spanning a fair range of soils, in not one was shepherd's-purse (*Capsella bursa-pastoris*) more than a chance casual (I needed some one year for teaching a first-year plant-science class, and eventually found it on the local allotments). In these same five gardens, the really troublesome quick-growing weeds were the bitter-cresses (*Cardamine flexuosa* and *C. hirsuta*) and three species of *Epilobium*, the broad-leaved willowherb (*E. montanum*), the short-fruited willowherb (*E. obscurum*) and the great willowherb (*E. hirsutum*). However, gardens are more complex and variable than arable fields, for a number of reasons. Gardeners are not so single-mindedly bent on growing a large stand of a single crop as farmers – and another crop, the same or different, next year. Gardening is on a smaller scale than farming, so edge effects are greater, and gardens usually include diverse elements that favour different weed floras and allow perennial weeds such as dandelions, docks and ground-elder (*Aegopodium podagraria*) to take hold. Vegetable gardens (and allotments) are nearest to arable farmland, but in most gardens shrubs and trees can shelter what is in effect a woodland element in the weed flora. Beds of herbaceous perennials have something of the same effect. Garden plants that self-sow too freely can be as troublesome as any weed. In my last garden in Exeter self-sown *Cotoneaster* and *Buddleja* were major public enemies, second only to willowherbs, dandelions and docks. Some of us have pet weeds that we tolerate or even encourage. In my parents' garden in Hampshire in the late 1940s I was careful not to exterminate weasel's-snout (*Misopates orontium*), as was a lady botanist of my acquaintance in Devon a decade or two later. In a mid-Devon garden we waged constant war on bitter-cresses (young and tender, they are quite good in salads) but cherished the small-flowered buttercup (*Ranunculus parviflorus*). Now, near the east Devon coast, we wage the same war on bitter-cresses, actively encourage self-sown primrose seedlings, but have reservations about bluebells and foxgloves spreading themselves too widely. We pull out a lot of purple toadflax (*Linaria purpurea*) and red valerian (*Centranthus ruber*) but would stop short of eradicating them completely.

RUDERALS

Cultivated ground is not the only habitat for 'weedy' species that human activity has created. Roadsides and paths, railways and canal sides, rubbish dumps and waste ground in villages and towns all have their own often characteristic assemblages of opportunistic species. These include not only annuals but also many biennials and perennials, depending particularly on frequency of disturbance and how long the ground remains tenable to a plant once it is established. Plants characteristic of these situations are often referred to as *ruderals* (from the Latin for rubble or rubbish), and many rarely or never behave as weeds in the sense of interfering with cultivation. However, the distinction between weeds and ruderals is a matter of convenience rather than a rigid one. It is possible for a species to behave as a weed in one situation, a ruderal in another and to be a regular member of a stable community in a third, as does the stinging nettle, but there are relatively few that attain that versatility.

There are some characteristic ruderal communities that span both rural and urban habitats. No country walk would be complete without patches of

FIG 124. Stinging nettle (*Urtica dioica*); female (left) and male (right) plants growing together. Nettles grow in single-sex patches; it is unusual to see the sexes together like this.

that arch-ruderal the stinging nettle (Fig. 124; Wheeler 2007) by hedge-sides and around outbuildings, usually accompanied by cleavers (*Galium aparine*) and often by rough meadow-grass (*Poa trivialis*), hogweed (*Heracleum sphondylium*), creeping thistle (*Cirsium arvense*) and barren brome (*Anisantha sterilis*). In neglected corners of suburban gardens and in urban and industrial waste places, patches of nettles are equally at home[OV24]. More isolated patches of nettles in pastures and neglected farmland usually differ in the greater prominence of thistles[OV25].

Heavily trampled ground in farm gateways, on country and town paths and in recreation areas chacteristically bears open mixtures of pineapple-weed, perennial rye-grass (*Lolium perenne*), knotgrasses (*Polygonum aviculare* and *P. arenastrum*), annual meadow-grass, greater plantain and shepherd's-purse in various proportions[OV18, 21] (Fig. 125a). A diminutive but distinctive community occurs in the cracks between paving stones and cobbles where there is regular trampling or perhaps light motor traffic, with annual meadow-grass, procumbent pearlwort (*Sagina procumbens*) and the little silvery moss *Bryum argenteum* among the most constant and conspicuous species[OV20]. Finally, among these communities we trample underfoot, are those unconsidered patches of grass in towns, playing fields, and around houses, factories and public buildings. Usually

FIG 125. Ruderals: (a) pineapple-weed (*Matricaria discoidea*) with greater plantain (*Plantago major*), a common combination in trampled field gateways; (b) field penny-cress (*Thlaspi arvense*), which appeared in immense abundance on the banks of the road works for the building of the M5 motorway near Exeter in 1976; the following year it was gone.

dominated by the perennial rye-grass with which they were sown, they are commonly colonised by cock's-foot (*Dactylis glomerata*), ribwort plantain (*Plantago lanceolata*), dandelions, often with lesser trefoil (*Trifolium dubium*), annual meadow-grass, 'wall' barley (*Hordeum murinum*), yarrow (*Achillea millefolium*), greater plantain, white clover (*Trifolium repens*), broad-leaved dock (*Rumex obtusifolius*), occasional daisies (*Bellis perennis*) and a long list of other occasional species, mostly common weeds[OV23].

Waste places, roadsides and field edges where trampling is less heavy are occupied by a variable weedy vegetation, in which scentless mayweed, couch-grass and annual meadow-grass are the most constant species, with pineapple-weed, knotweed, fat-hen and spear-leaved orache (*Atriplex prostrata*) often common but less constant. These plants are accompanied by various species from neighbouring arable fields in country waste places, and by such species as the bright yellow-flowered Oxford ragwort (*Senecio squalidus*), the off-white drifts of hoary cress (*Lepidium draba*) and the pink spires of rosebay willowherb (*Chamerion angustifolium*) in urban situations such as railway yards and trading estates in towns[OV19] (Fig. 126). Where there is a degree of disturbance but little trampling a coarse, more-or-less open, weedy vegetation develops. In this, dandelions and annual meadow grass are the only constants, accompanied by a long and varied list of associates, including scentless mayweed, barren brome, Oxford ragwort, spear and creeping thistles, broad-leaved dock, greater plantain, beaked hawk's-beard (*Crepis vesicaria*) and American willowherb (*Epilobium ciliatum*)[OV22]. Many other species are characteristic of this kind of vegetation, ranging from the strikingly beautiful common mallow and field bindweed to frankly dull plants such as hedge mustard (*Sisymbrium officinale*) and Canadian fleabane (*Conyza canadensis*), both with tiny yellowish flowers. Prickly lettuce (*Lactuca serriola*), often growing at the foot of walls, is a 'compass plant', with its leaves characteristically oriented north–south. 'Wall' barley (*Hordeum murinum*) is a misnomer – the name *murinum* has nothing to do with walls but means mouse-like, from the ability of the flower spike (as many children know) to run up your sleeve and emerge at your neck.

FIG 126. 'Railway plants': (a) Oxford ragwort, Marsh Barton trading estate, Exeter, June 1989; (b) hoary cress, railway siding, Exeter, May 1976. Both species are abundant in railway yards and similar situations, especially in the Midlands and southeast England.

ROADSIDES, RIVERBANKS AND OTHER EDGE HABITATS

Plant ecologists doing a vegetation survey look first for areas of 'uniform' vegetation – heath, saltmarsh, oakwood or whatever. When these areas of uniform vegetation have been recognised and described, they can be used to draw a vegetation map. But it is common experience that the recognisable and extensive plant communities are far from embracing all the species in the area. Many species are associated with 'edges' – road- and track-sides, hedgebanks, wood margins, and the banks of pools, ditches, rivers and canals. Such habitats abound in the 'unofficial countryside' of Mabey (1973).

We always tend to think of transitions from land to water in terms of the hydrosere (Chapter 13), but that is often not a true picture. Many river and canal banks are abrupt transitions, in which emergent aquatics like common reed (*Phragmites australis*), marsh plants like yellow iris (*Iris pseudacorus*), marsh woundwort (*Stachys palustris*), gypsywort (*Lycopus europaeus*), hemp agrimony (*Eupatorium cannabinum*) and meadowsweet (*Filipendula ulmaria*), meadow species like tufted vetch (*Vicia cracca*) and meadow vetchling (*Lathyrus pratensis*), and ruderals like great willowherb, hedge bindweed (*Calystegia sepium*), bittersweet (*Solanum dulcamara*) and stinging nettle intermingle to form a rich and colourful community[OV26b]. Stands dominated by great willowherb are common on patches of wet silty ground close to water[OV26a]. The strikingly beautiful introduced Indian balsam (*Impatiens glandulifera*) has made itself at home along many of our rivers, and raises passions both for and against. I find it hard to think ill of a plant that mercifully brings life and colour to so many soot-blackened river- and canal-sides in the industrial north of England. I believe, conservation-wise, it does very little real harm – but I know there are people whose crusading zeal will immediately condemn my tolerance.

Common roadside and hedgerow plants such as white dead-nettle (*Lamium album*), lesser burdock (*Arctium minus*) and garlic mustard (*Alliaria petiolata*) seldom or never grow in the grasslands or woodlands that may lie only a few metres away from them. Although in Britain and Ireland there has been little systematic study of the plant assemblages in these linear habitats, they have been much studied by vegetation ecologists in continental Europe, and many of the groupings they recognise will strike familiar chords to anyone who knows our own countryside well. Ground-ivy (*Glechoma hederacea*), stinging nettles, cow parsley (*Anthriscus sylvestris*), crosswort (*Cruciata laevipes*), rough chervil (*Chaerophyllum temulum*), garlic mustard, herb-robert (*Geranium robertianum*), shining crane's-bill (*G. lucidum*) and wood avens (*Geum urbanum*) give their names to communities listed in Erich Oberdorfer's *Pflanzensoziologische Exkursionflora* (within the order Glechometalia), and they behave in essentially the same way in Germany, the Netherlands and southern England. Another group (order Artemisietalia) embraces a range of communities in which white dead-nettle, black horehound (*Ballota nigra*), mugwort (*Artemisia vulgaris*), hemlock (*Conium maculatum*), burdocks and thistles are often prominent. There are some distinctive associations on well-drained, and generally lime-rich soils (order Onopordetalia), with us mainly on the chalk and Jurassic limestones, including such species as musk thistle (*Carduus nutans*), weld (*Reseda luteola*), woolly thistle (*Cirsium eriophorum*), wild carrot (*Daucus carota*), tansy (*Tanacetum vulgare*) and bristly ox-tongue (*Picris echioides*). Again it should be emphasised that these various combinations of species represent recognisable nodes in a continuous web of variation; there are no sharp boundaries. The plants mentioned in this paragraph are most heavily concentrated in lowland southern England, and few reach Kerry, Connemara, Donegal or northwest Scotland.

Why do so many of these species favour 'edges'? One obvious partial explanation is that many man-made habitats are linear, but there must be more to it than that. It is easy to visualise 'edge' situations providing favourable combinations of conditions that cannot be found in a continuous field or wood. The open side of a hedge-line or woodland fringe lets in the light and pollinators; the closed side provides shelter, combined with an annual influx of nutrients from leaf-fall.

WALLS

Walls have their characteristic plants too (Fig. 127). Wallflowers (*Erysimum cheiri*) do well on the right wall but are very local. Much more widespread is red valerian, an incomer from southwest Europe, which has made itself thoroughly at home with us on walls and other dry places, a very ornamental plant with its red, pink or white flowers. Another incomer is the rather pretty Mexican fleabane (*Erigeron karvinskianus*), which is particularly common on town walls in southwest England. Finally, two plants that have been with us long enough to qualify as natives, pellitory-of-the-wall (*Parietaria judaica*) and ivy-leaved toadflax (*Cymbalaria muralis*), which often accompanies it, are seldom seen away from the habitat embodied in their Latin names[OV41, 42]. They are sometimes accompanied by the wall-rue (*Asplenium ruta-muraria*) and maidenhair spleenwort (*A. trichomanes*), two small ferns common on walls but which have their natural habitat in crevices of lime-rich rock outcrops[OV39]. Walls are good habitat for mosses (Rishbeth 1948), and two distinctive species are virtually ubiquitous even on

FIG 127. Wall plants. (a) Ivy-leaved toadflax, Juniper Hall, Surrey, May 1994. (b) Pellitory-of-the-wall, Sidbury, Devon, June 1967. (c) Mosses on a wall in Exeter; in the foreground, the top cushion (with capsules on long setae) is *Tortula muralis*, the middle one is *Orthotrichum cupulatum*, and the bottom one is *Grimmia pulvinata*. The bright-green patch beyond is *Homalothecium sericeum*. (d) *Grimmia pulvinata* on the parapet of a railway bridge in Exeter, February 1987.

town walls. Both have hair-points to the leaves. *Tortula muralis* forms low hoary cushions or patches, often along the lines of mortar; it has narrow cylindrical capsules standing a centimetre or so above the leaves. *Grimmia pulvinata* is easily recognised by its tight hairy near-hemispherical cushions, with round capsules borne on swan-neck setae that nestle the capsules back amongst the leaves until they are ripe in late spring.

SEASONALLY FLOODED PLACES: 'MUD-PLANTS'

Sir Edward Salisbury (1886–1978), pioneer ecologist and former Director of Kew, listed 45 'mud-species', 'found as pioneers, colonising previously submerged soils', with another five perennials in the same habitat. His list is certainly not exhaustive, and contains a number of true aquatics, which may be stranded on mud when water levels are low. Nevertheless, it highlights an interesting, and diverse, element in our flora and vegetation, and one that has unquestionably declined greatly in the course of the past century. The seasonal flooding of a river floodplain, the annual fluctuations of water level in and around the ponds on a traditional village green, and water standing in winter along tracks or in pools or puddles in heathy ground – all these create very different conditions for plant growth, and support different assemblages of species.

The 'village green' assemblage has declined most. Small fleabane (Fig. 128a) and pennyroyal (Fig. 128b) formerly occurred widely in southern England, but now hang on in only a dozen or two native sites – though pennyroyal has recently recovered much of its former area as an alien introduced with North American seed. Small galingale (*Cyperus fuscus*), never as common as the last two,

FIG 128. 'Village-green plants', small short-lived plants of seasonally flooded sandy and grassy places: (a) small fleabane (*Pulicaria vulgaris*); (b) pennyroyal (*Mentha pulegium*).

has suffered a similar decline. Village greens used to be busy places, providing grazing and watering for farm stock and geese as well as open space for other uses. Nowadays, the farming and other workaday needs that the green formerly served have become incompatible with a tidy, gentrified, increasingly residential village.

The plants of seasonally wet places on heathy ground have fared better, but have shared in the general decline of heathland and our increased dependence on cars and roads rather than shanks's pony and beaten tracks, and they have probably suffered from the declining frequency of heath fires and casual rough grazing. The commonest of these annuals of seasonally wet heathy places is the toad rush. A characteristic plant of such ground that is still wet in summer is water-purslane (*Lythrum portula*), often in company with lesser spearwort (*Ranunculus flammula*), common spike-rush (*Eleocharis palustris*) and other plants such as toad rush, creeping bent (*Agrostis stolonifera*), marsh bedstraw (*Galium palustre*), water forget-me-not (*Myosotis scorpioides*), marsh pennywort (*Hydrocotyle vulgaris*) and water-pepper (*Persicaria hydropiper*)[OV35]. Heathland pools with shelving shores can have varied and fascinating vegetation zonations, including marsh St John's-wort (*Hypericum elodes*), shoreweed (*Littorella uniflora*), bog pondweed (*Potamogeton polygonifolius*) and pillwort (*Pilularia globulifera*), but are exceedingly local. Some of the best sites are in the New Forest and west Cornwall.

Several tiny summer annuals grow in open spots, often on tracks across heaths, which may hold pools of water in winter but are no more than slightly moist in summer (Fig. 129). They include allseed, in effect a miniature flax, chaffweed, a diminutive relative of the scarlet pimpernel, and yellow centaury (*Cicendia filiformis*). *Cicendia* is much rarer than the other two and restricted to the south; it is probably commoner in Kerry than anywhere else in Britain or Ireland. Two rare annual rushes match these three species in miniaturisation; dwarf rush grows on the Lizard heaths and in Anglesey and the Channel Islands but the pigmy rush (*Juncus pygmaeus*) occurs with us only at the Lizard (depauperate plants of the much commoner toad rush are easily mistaken for it). Another plant of comparable situations is the land quillwort, which grows in grazed and windswept clifftop turf, wet in winter but baked dry in summer, along the Lizard cliffs and in the Channel Islands.

Like the small winter and spring annuals of dry places, these small plants of ground that is seasonally wet or inundated are more prominent and number many more species in the Mediterranean region, where they form a distinctive facet of the vegetation in the early part of the year.

FIG 129. Minute annual plants of seasonally damp spots in heathy places: (a) allseed (*Radiola linoides*), Crownhill Down, Dartmoor, August 1986; (b) chaffweed (*Anagallis minima*), Dawlish Warren, Devon, August 1981; (c) dwarf rush (*Juncus capitatus*), Alderney, July 1969; (d) land quillwort (*Isoetes histrix*), Kynance Cliff, Cornwall, May 1965.

SUMMER-EXPOSED MUD IN RIVERS, PONDS AND DITCHES

Despite pressure from farmers and the public at large, and the best efforts of the drainage engineers, many rivers still flood regularly, and many more flood occasionally during unusually wet weather. Flooding during winter when growth is at a standstill has little effect on the vegetation and, as with the Nile flood, the yearly input of silt may be beneficial. Where only winter flooding

FIG 130. Vegetation of seasonally exposed wet mud, with water forget-me-not, silverweed, bur-marigolds etc. Breamore, Hampshire, October 1981.

FIG 131. Riverside summer communities on wet shingle and mud. (a) Marsh yellow-cress (*Rorippa palustris*) on shingle beside the River Exe, Brampford Speke, Devon, June 1973.
(b) Creeping yellow-cress (*R. sylvestris*) on muddy ground by River Stour, Fiddleford, Dorset, August 1969.

occurs the grasslands are of the general kind already described in Chapter 9. Where water lies long into the growing season, or where sporadic flooding occurs frequently during the summer, the dominant creeping bent and marsh foxtail (*Alopecurus geniculatus*) are accompanied by creeping buttercup, water-pepper and other persicarias, curled dock (*Rumex crispus*), creeping yellow-cress (Fig. 131b), silverweed (*Potentilla anserina*), marsh bedstraw, reed canary-grass (*Phalaris arundinacea*) and a sprinkling of other marsh plants and common weeds[OV28]. Similar vegetation to this community and the next occurs in the Irish turloughs (Chapter 12). Very distinctive annual communities can develop on mud and shingle exposed when the water level is at its lowest in summer (Figs 130, 131a), in which the two bur-marigolds (*Bidens cernua, B. tripartita*) are conspicuous and characteristic, along with water-pepper, amphibious bistort (*Persicaria amphibia*), redshank (*P. maculosa*), marsh yellow-cress, marsh foxtail, silverweed, greater plantain and a scatter of other more-or-less aquatic plants and ruderals[OV30]. A somewhat similar niche is occupied by the celery-leaved buttercup (*Ranunculus sceleratus*). This typically grows in seasonally flooded, low-lying, muddy nutrient-enriched pastures beside ponds and slow-flowing rivers, which are grazed, poached and manured by cattle. It often accompanies creeping bent, water forget-me-not, and such species as pink water-speedwell (*Veronica catenata*), water-cress (*Rorippa nasturtium-aquaticum*), creeping buttercup and water-pepper[OV32]. *Ranunculus sceleratus* itself can tolerate high levels of nutrient enrichment, and will even grow on the sludge beds of sewage farms.

CHAPTER 11

Chalk and Limestone Grasslands

T HE CALCAREOUS GRASSLANDS are surely among our best-loved plant communities. Many of us must carry a picture in our minds of their short, springy, flowery turf, always well-drained underfoot, perhaps of the first wild orchids we found, or of blue butterflies, or of a favourite 'bank whereon the wild thyme blows' – though Shakespeare must have been thinking of other things in the rest of that passage from *A Midsummer Night's Dream*! These are semi-natural grasslands in the sense that, apart from grazing, they owe nothing to deliberate agricultural intervention. They have not been cultivated, sown or manured, and all of their species have arrived by natural means. A century ago the chalk downs of the south and east of England were pasture, much of it unploughed since prehistoric times, and grazed mainly by sheep. Some inroads had been made on the extent of the pasture in the course of the nineteenth century, and more again during the two world wars, but at the end of the Second World War a surprising extent of the chalk grassland remained intact. After the war, society and farming were changing. The use of the downs as sheepwalk declined along with the rest of the labour-intensive rural economy of the earlier part of the century, and by 1950 large areas of chalk grassland were kept in being mainly by innumerable wild rabbits. The post-war drive for national self-sufficiency in food production led to the ploughing and conversion to arable of much of the flatter chalk country, and the advent of myxomatosis in 1953 and the ensuing crash in the rabbit population led to rapid invasion of hawthorn scrub over much of the rest. The picture has been similar on the Cotswolds and our other lowland limestones.

The story on the harder and older limestones of the north and west is rather different. Wider open spaces, together with hill-farming subsidies (and planning policies in the National Parks), kept sheep farming economically viable, so the limestone grasslands of the Derbyshire and Yorkshire Dales have changed less in either extent or character than their southern and lowland counterparts. In Ireland the move away from labour-intensive small-scale farming has been more recent, but anyone who has known the limestone country of the Burren over the last half-century cannot fail to be struck by the increase in hazel scrub in recent decades.

THE CALCAREOUS GRASSLAND HABITAT

Soils

Calcareous grasslands do not occur everywhere that chalk or limestone bedrock appears on the geological map. This is because millennia of solution and soil-

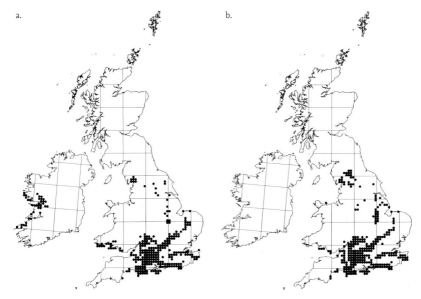

FIG 132. The distribution of (a) squinancywort (*Asperula cynanchica*) and (b) horseshoe vetch (*Hippocrepis comosa*). Both species are restricted to dry lime-rich rendzina soils. The southern part of their distributions reflects that of calcareous grasslands on scarp slopes and steep valley-sides on the chalk and the Jurassic, Permian and Carboniferous limestones. To the north and west their occurrences become more scattered. Both are on the limestone around the head of Morecambe Bay, but curiously both are rare or absent on the north Wales limestone. Horseshoe vetch reaches higher on the Carboniferous limestone in the Pennines, and squinancywort alone reaches Ireland, where it is common on the limestone of Clare and Galway, and in calcareous fixed-dune grasslands.

forming processes have left a layer of weathering residue covering the calcareous bedrock on any more-or-less level surface. Thus much of the chalk of southern England is overlain by a layer of 'clay-with-flints', to which may be added airborne *loess* deposited as dust in the cold periglacial landscape of the last ice-age. The chalk or limestone is only exposed as the parent-material of the soil on the slopes of escarpments and valleys where surface erosion has kept pace with soil-formation. Calcicole species such as squinancywort (Fig. 132a) and horseshoe vetch (Fig. 132b) map the major chalk and limestone escarpments with considerable fidelity. Calcareous grasslands most typically grow on rendzina soils, which have a near-neutral pH, are rich in calcium, but poor in available nitrogen and (especially) phosphate, and in which iron and other metallic trace elements are in relatively short supply. Often, limestone grasslands form part of a *catena*, a sequence of plant communities related to topography and repeated wherever similar topography recurs, with deep soils, sometimes leached and acid, on the plateau at the top, and fertile lime-rich soils as the slope levels off at the bottom (Figs 133, 134).

Chalk and limestone grasslands are well drained, and dry. It is an apparent paradox that many calcareous-grassland species – such as dwarf thistle (Fig. 135b), common rock-rose (Fig. 136a) and salad burnet (Fig. 136b) – are deep rooted,

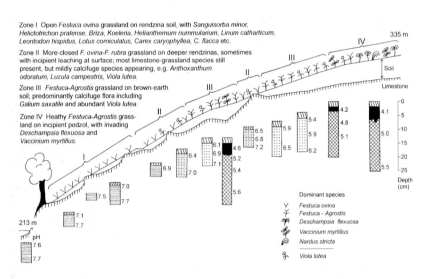

FIG 133. Catena on Carboniferous limestone from Cressbrook Dale to Wardlow Hay Cop, Derbyshire. Horizontal shading in soil profiles, limestone rendzina; diagonal shading, surface root-mat and litter; stipple, brown earth; solid black, raw humus; cross-hatching, orange mineral soil of incipient podzol. After Balme, 1953.

FIG 134. Chalk escarpment near Batcombe, Dorset, June 1967. A catena comparable to that shown in Fig. 133. Rendzina soils with chalk grassland on the steep scarp slope, passing into deeper brown earths on the plateau above, bearing bracken and gorse. Cultivated farmland on the deep fertile soils at the foot of the scarp.

and transpire rapidly even under water stress. This is because in a short turf on a sunny day, with a kilowatt of solar energy beating down on every square metre, the surface and leaves close to it would become insupportably hot without transpirational cooling. The grasses, with their finer upstanding leaves, equilibrate more rapidly with the temperature of the air than the broad-

FIG 135. (a) The chalk cliffs between Lulworth and White Nothe, Dorset, August 1975. Tor grass (*Brachypodium pinnatum*) dominates large areas of chalk grassland along this coast. (b) Dwarf thistle (*Cirsium acaule*) in chalk grassland on Hambledon Hill, Dorset, August 1973.

FIG 136. Some characteristic chalk- and limestone-grassland plants: (a) common rock-rose (*Helianthemum nummularium*), Carreg Cennen, Carmarthenshire, May 1989; (b) salad burnet (*Sanguisorba minor*), Anvil Point, Dorset, May 1980: female plant in flower; (c) wild thyme (*Thymus polytrichus*), Cressbrookdale, Derbyshire, July 1973; (d) squinancywort (*Asperula cynanchica*), Box Hill, Surrey, July 1990.

leaved herbs close to the soil, so they are often shallow rooted and can restrict transpiration when water is short. Chalk, because of its porosity, provides a greater reserve of water than many other limestones.

Slope and aspect

The steepness of the slope and the direction in which it faces have profound effects on the climate of a grassland on a hill in sunny weather (Fig. 8). In our latitude, the sun shines roughly at right angles to a 30° south-facing slope at noon in midsummer. At all other times the sun's rays are more oblique to the surface. On a 30° slope facing north the angle of incidence will be just less than 30°. Consequently, solar radiation at noon will be about half as intense, partly compensated by longer hours of sunshine in the early morning and late evening. An east-facing slope will get its best sunshine in the morning while the air is still cool, and conversely on a west-facing slope sunshine is most intense when the air has already warmed through the day. With a less steep slope these effects are less pronounced. It is easy to see that a south (to west)-facing slope is likely to be most favourable for a southern species. We shall see in Chapter 17 that mountain corries with their arctic-alpine plants occur mostly on the north to east side of mountain summits. These effects are illustrated for a few calcareous grassland species in Figure 137. Horseshoe vetch (*Hippocrepis comosa*) shows a strong preference for southwest-facing slopes, fairy flax (*Linum catharticum*) and glaucous sedge (*Carex flacca*) also prefer slopes (chalky soils), but are less choosy which way they face. Downy oat-grass (*Helictotrichon pubescens*) favours flattish ground (deeper soils), with some preference for a northerly aspect. Moisture loving calcicoles such as hoary plantain (*Plantago media*) and mosses such as *Ctenidium molluscum* tend to favour steepish north-to-east slopes.

Rabbits

Rabbits were the main grazers on many English chalk grasslands until the early 1950s, but are not native to Britain or Ireland. They were introduced to both

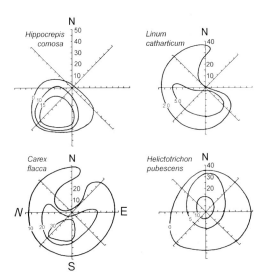

FIG 137. The distribution of some chalk grassland species with slope and aspect in the Blandford–Shaftesbury area of north Dorset. The eight arms represent the points of the compass; slope (°) increasing outwards from the centre. Contours show the areas of the diagram within which the species attain the percentage cover shown. From Perring (1959).

islands from southern Europe by the Normans, and by the fifteenth century had become very abundant. They were valued as a source of both food and fur, but by the first half of the twentieth century they had become a serious pest of agriculture and forestry in England. Various attempts were made to control or eradicate them, but these foundered because (amongst other reasons) many country people valued wild rabbits as a source of food or sport. The situation changed completely with the appearance of myxomatosis in 1953. This, a comparatively mild virus disease of American wild rabbits, had been found to cause over 99% mortality in laboratory populations of European rabbits, and had been used successfully in 1950 to combat rabbit infestations in Australia. In 1952 infected rabbits were released near Paris, from where the disease quickly spread, and in 1953 it appeared in Kent. By the end of that year myxomatosis was present in almost every part of Britain, and in 1955 it was estimated that the rabbit population had been reduced to 10% of its former level. Myxomatosis has remained common, though irregular in its occurrence, and although the numbers of rabbits are now higher than in 1955 they have not regained anything like their former abundance and show no sign of doing so.

CHALK AND LIMESTONE OF LOWLAND ENGLAND

Sheep's fescue and upright brome

The commonest kind of chalk and limestone grassland in lowland England is dominated by sheep's fescue (*Festuca ovina*), which forms a short, somewhat open turf[CG2]. The fescue is almost always accompanied by crested hair-grass (*Koeleria macrantha*), meadow oat-grass (*Helictotrichon pratense*) and quaking-grass (*Briza media*). We students were taught by the late Humphrey Gilbert-Carter that the first thing you noticed when you sat down on chalk grassland in Cambridgeshire was 'a pricking sensation, and a smell of cucumber'; the pricking sensation was dwarf thistle, the smell of cucumber was salad burnet. We might also have noticed the smell of another ubiquitous chalk-grassland plant, wild thyme (Fig. 136c). Other constant or near-constant plants in chalk grassland

OPPOSITE: **FIG 138.** More calcareous grassland plants: (a) pasqueflower (*Pulsatilla vulgaris*), near Royston, Hertfordshire, April 1981; (b) carline thistle (*Carlina vulgaris*), Tenby, Pembrokeshire, August 1955; (c) mouse-ear hawkweed (*Pilosella officinarum*), Braunton, Devon, June 1990; (d) chalk milkwort (*Polygala calcarea*), Hambledon Hill, Dorset, June 1963; (e) clustered bellflower (*Campanula glomerata*), Hod Hill, Dorset, August 1973; (f) yellow-wort (*Blackstonia perfoliata*), Box Hill, Surrey, July 1989.

are glaucous sedge, ribwort plantain (*Plantago lanceolata*), bird's-foot trefoil (*Lotus corniculatus*), rough hawkbit (*Leontodon hispidus*), fairy flax, mouse-ear hawkweed (Fig. 138c) and small scabious (*Scabiosa columbaria*). Other frequent species include squinancywort, horseshoe vetch, spring-sedge (*Carex caryophyllea*), self-heal (*Prunella vulgaris*), hoary plantain, harebell (*Campanula rotundifolia*), common rock-rose and various calcicole mosses, such as *Ctenidium molluscum* (Fig. 73b) and *Homalothecium lutescens*, and the more widespread *Pseudoscleropodium purum*. Dwarf sedge (*Carex humilis*) can be abundant on some south-facing slopes in north Dorset and south Wiltshire. The short species-rich turf provides the habitat for our downland orchids, of which the commonest are pyramidal orchid (*Anacamptis pyramidalis*) and bee orchid (*Ophrys apifera*). This kind of vegetation is common with minor variations in chalk grassland on the North and South Downs, the Chilterns, the Hampshire, Berkshire, Wiltshire and Dorset chalk (Fig. 135), and northwards through Lincolnshire to the Yorkshire Wolds. It occurs on Carboniferous limestone on Mendip, in Gower, along the north Wales coast and in the Derbyshire Dales, amongst other places.

Upright brome (*Bromopsis erecta*) is another common dominant of grasslands on chalk and other limestones of lowland England, including the

FIG 139. Common tall perennials of chalk country: (a) greater knapweed (*Centaurea scabiosa*), Hod Hill, Dorset, August 1973; (b) wild parsnip (*Pastinaca sativa*), Box Hill, Surrey, July 1995.

Cotswolds and the extension of Jurassic limestones northeastwards, and the magnesian limestone of west Yorkshire and County Durham. Upright brome most often dominates sites with little or no grazing[CG3]. Sheep's fescue is usually still present, and most of its characteristic calcicole associated species are still to be found, but with lower constancy and cover, especially of the smaller species. Taller-growing species (such as salad burnet), or those with large robust basal rosettes (such as dwarf thistle) continue to do well, and coarser perennials such as the knapweeds (Fig. 139a), field scabious (*Knautia arvensis*) and wild parsnip (Fig. 139b) are sometimes conspicuous. The shorter downland orchids, such as musk orchid (*Herminium monorchis*) and autumn lady's-tresses (*Spiranthes spiralis*) would be swamped by the taller herbage of the *Bromopsis* grasslands, but taller species such as the pyramidal orchid and the much rarer man orchid (*Aceras anthopophorum*) do as well here as in shorter turf, as does the now rare pasqueflower (Fig. 138a).

Other dominant grasses: tor grass and *Sesleria*

Tor grass (*Brachypodium pinnatum*) is a vigorous grass with much the same distribution as upright brome, which invades poorly grazed or ungrazed grasslands. With its coarse, yellow-green unpalatable foliage it is a much more uncompromising dominant than upright brome. It occurs locally on the North and South Downs, but is commoner towards the western and northern parts of the chalk outcrop (where it is particularly characteristic of the Yorkshire Wolds), and on the Jurassic limestones of the Cotswolds and their extension northwards[CG4]. These grasslands are typically much less species-rich than those dominated by either sheep's fescue or upright brome, but sheep's fescue remains as a near-constant ingredient though with low cover, and the only other near-constant species is glaucous sedge. Frequent or occasional species include bird's-foot trefoil, harebell, salad burnet, fairy flax, quaking-grass, common rock-rose, wild thyme and yellow oat-grass (*Trisetum flavescens*). A long list of other calcicole species (including some rarities) can occur locally or occasionally, but none with any regularity. A much richer community dominated by various mixtures of upright brome and tor grass is particularly characteristic of the Jurassic limestone of the Cotswolds and sites such as the 'hills and holes' of the old quarries at Barnack in Northamptonshire[CG5]. This community is home to a notable list of uncommon species, including pasqueflower and purple milk-vetch (*Astragalus danicus*), as well as some, such as man orchid, chalk milkwort (Fig. 138d) and bastard toadflax (*Thesium humifusum*), with their main distribution on the southern chalk.

We cannot leave English lowland chalk and limestone grasslands without mention of some sites that appear too dry, exposed and deficient in plant

ABOVE: **FIG 140.** Berry Head, Devon, May 1983. Rocky dry grassland with white rock-rose (*Helianthemum apenninum*), horseshoe vetch (*Hippocrepis comosa*) and thrift (*Armeria maritima*).

LEFT: **FIG 141.** Southern plants of dry rocky grassland: (a) honewort (*Trinia glauca*), male plant in flower, Berry Head, Devon, May 1970; (b) small restharrow (*Ononis reclinata*), Berry Head, June 1968.

nutrients to have developed a complete vegetation cover. In the pioneer vegetation on the floor of a derelict chalk quarry in Hampshire Tansley and Adamson in 1926 found that the most abundant species was sheep's fescue, followed by mouse-ear hawkweed, wild thyme, squinancywort, bird's-foot trefoil, hop trefoil (*Medicago lupulina*), rough hawkbit and a number of less constant species. Grasslands described by Alex Watt from sharply drained chalky sands on the Suffolk Breckland produced a similar list, with the addition of crested hair-grass, meadow oat-grass, ribwort plantain, lady's bedstraw (*Galium verum*), purple milk-vetch, a number of winter and spring annuals such as thyme-leaved sandwort (*Arenaria serpyllifolia*), whitlowgrass (*Erophila verna*) and early forget-me-not (*Myosotis ramosissima*) and a long list of bryophytes and lichens. Other dry nutrient-depleted sites, often with a history of disturbance, produce similar species lists – moss-rich, but often with some 'weedy' species such as blue fleabane (*Erigeron acer*) or wild strawberry (*Fragaria vesca*) – in which winter annuals and mosses figure prominently. These sites have enough in common to be considered together[CG7]. They support a number of southern and continental species, uncommon with us (e.g. *Galium parisiense, Medicago minima, Silene conica*), which find a congenial niche in this kind of vegetation in Breckland and elsewhere in dry parts of south and east England.

A better-characterised and more species-rich open community occurs on sunny, dry, usually west- to southwest-facing limestone outcrops close to the south and west coasts of England and Wales. In these places sheep's fescue is usually the dominant grass, but it makes only a very open cover, often equalled or exceeded by salad burnet, wild thyme, mouse-ear hawkweed and bird's-foot trefoil, and usually leaving some bare soil (or limestone) visible amongst the foliage. Carline thistle (Fig. 138b) is a conspicuous and rather constant associate, as is cock's-foot. Ribwort plantain and many of the other common limestone-grassland plants occur more of less frequently[CG1]. This is the habitat of white rock-rose on the headlands around Torbay (Fig. 140) and on Brean Down (Fig. 9) and Purn Hill in Somerset, and of hoary rock-rose (*Helianthemum oelandicum*) in Gower and along the North Wales coast. This community musters a long list of uncommon plants, including small restharrow (Fig. 141b) in Devon and South Wales, honewort (Fig. 141a) in Somerset and Devon, goldilocks aster (*Aster linosyris*) in Devon, Somerset, South Wales and North Wales, and Somerset hair-grass (*Koeleria vallesiana*) on Brean Down and the western end of Mendip, to name only a few of the rarest. The rarer rock-roses seem largely to exclude the common rock-rose from their preferred niche, even though the common species often occupies more run-of-the-mill limestone grasslands nearby.

UPLAND AND WESTERN LIMESTONE GRASSLANDS

With us *Sesleria caerulea* is a northern, western and largely upland plant of Carboniferous limestone country. It has two main areas of distribution. In Britain, its headquarters is the block of limestone country from the Yorkshire and Cumbrian Pennines between Skipton and upper Teesdale, to the lower ground around the head of Morecambe Bay, with a small Scottish outlier on a few mica–schist mountains in the central Highlands. In Ireland its main area in north Clare and southeast Galway includes the Burren; a somewhat smaller area in Sligo and Leitrim includes Ben Bulben and Carrowkeel. This grass is something of an enigma, because it dominates chalk grassland in the north of France along the cliffs of the Seine valley, as at Les Andelys, but occurs nowhere on the English chalk. The nearest it comes to the English chalk is in a few lowland *Sesleria*-dominated sites[CG8] on the magnesian limestone of eastern County Durham, with vegetation otherwise similar to southern-English chalk grassland.

The north of England

Sesleria is ubiquitous in limestone grasslands on steep hill-slopes, screes and rock ledges on the limestones of the Yorkshire Dales (Fig. 142). The typical

FIG 142. The valley of Cowside Beck near its confluence with Littondale, above Arncliffe, Yorkshire, April 2003. *Sesleria* grassland on slopes over Carboniferous limestone. Mountain avens (*Dryas octopetala*, Fig. 146) occurs on northwest-facing crags here, its southernmost English locality.

Sesleria grassland of the mid and northern Pennines[CG9] is a well-characterised community with clear affinities with the lowland limestone grasslands, but a decided flavour of its own. Apart from the dominant *Sesleria*, constant or near-constant species include wild thyme, crested hair-grass, limestone bedstraw (*Galium sterneri* – a northern and upland species with us), harebell, fairy flax, sheep's fescue, quaking-grass, glaucous sedge, spring-sedge and the mosses *Hypnum cupressiforme* and *Dicranum scoparium*. Frequent species include common dog-violet (*Viola riviniana*), meadow oat-grass, bird's-foot trefoil, eyebrights (*Euphrasia* spp.), mouse-ear hawkweed and the striking calcicole mosses *Ctenidium molluscum* and *Tortella tortuosa*. A number of species familiar in lowland calcareous grasslands are sparser here (e.g. salad burnet, ribwort plantain, rough hawkbit) or missing altogether (e.g. small scabious, dwarf thistle). Where seepage of water from the underlying limestone creates somewhat damper conditions, flea sedge (*Carex pulicaris*) and carnation sedge (*C. panicea*) join glaucous sedge as near-constants, bryophytes and lichens become much more prominent, and a sprinkling of species appear that are more at home in the small-sedge rich-fens of Chapter 14, such as grass-of-Parnassus (*Parnassia palustris*), common butterwort (*Pinguicula vulgaris*) and bird's-eye primrose (*Primula farinosa*).

On the 'sugar limestone' of upper Teesdale the *Sesleria–Galium sterneri* grassland plays host to some of the suite of rarities that grow together here. The rare mountain sedge *Kobresia simpliciuscula*, hair sedge (*Carex capillaris*) and spring gentian (*Gentiana verna*) become near-constant in the short *Sesleria*–sheep's fescue–sedge turf on Cronkley and Widdybank fells high in the dale; Teesdale violet (Fig. 143b) is very local (much less common than common dog-violet, which accompanies it) and frequent alpine bistort (*Persicaria vivipara*), mountain everlasting (*Antennaria dioica*) and the lichen *Cetraria islandica* remind us that we are in the uplands. A little to the south and nearer the Pennine escarpment, on the high limestone fells a bryophyte-rich version of this grassland, with frequent mossy saxifrage (*Saxifraga hypnoides*) and alpine scurvygrass (*Cochlearia alpina*), is the southernmost British locality of alpine forget-me-not (*Myosotis alpestris*).

Descending from the high Pennine fells to the lower limestone hills north and east of Morecambe Bay (Fig. 144) we encounter another version of the *Sesleria–Galium sterneri* grassland. This is drier, warmer and sunnier country, reflected in the occurrence of southern species such as squinancywort and small scabious. Horseshoe vetch is common (near its northern limit with us), and this is one of our strongholds of hoary rock-rose, which grows in the most open, stony *Sesleria* grassland (recalling the open *Festuca* grassland of its Welsh localities), while common rock-rose favours the more continuous turf farther back from the cliff edge. There are still reminders that we are in the north of

FIG 143. Notable plants in *Sesleria* grassland in northern England: (a) the lady's-mantle *Alchemilla glaucescens*, a very local species of thin turf over limestone, Pen-y-Ghent, May 1982; (b) Teesdale violet (*Viola rupestris*); (c) rare spring-sedge (*Carex ericetorum*); (d) hoary rock-rose (*Helianthemum oelandicum*); the latter three on 'sugar limestone' in Teesdale, June 1975.

England in the abundance of bryophytes, in species such as bird's-foot sedge (*Carex ornithopoda*) in open *Sesleria* grassland on stabilised scree, and bird's-eye primrose in seepages below the cliffs of Whitbarrow.

The Burren and other Irish limestone grasslands

Irish limestone grasslands have much in common with those of northern England, but some familiar species are missing. Common rock-rose is confined in Ireland to a single locality in Co. Donegal, and meadow oat-grass does not occur at all, despite much apparently suitable ground for both. The most-

ABOVE: **FIG 144.** Underbarrow Scar, Kendal, Cumbria. *Sesleria* grassland, habitat of hoary rock-rose (*Helianthemum oelandicum*) and rare spring-sedge (*Carex ericetorum*). The apomictic whitebeam *Sorbus lancastriensis* grows on the cliffs, and bird's-foot sedge (*Carex ornithopoda*)

on the grassy scree slopes below. The limestone hill of Arnside Knott in the distance.

LEFT: **FIG 145.** The north coast of the Burren looking towards Black Head from Cappanawalla, July 1963. The vegetation is a diverse patchwork of limestone pavement, calcareous grassland with *Dryas* and heath subshrubs and limestone heath.

studied and certainly the richest limestone grasslands in Ireland are those of the Burren, in northern Clare just south of Galway Bay. The Burren hills, formed of almost flat-bedded Carboniferous limestone, rise steeply from sea level to a little over 300 m (Fig. 145). Compared to the limestone of the midland plain or the drumlin-covered country to the south, glacial drift is scanty on the Burren. *Sesleria* is present almost everywhere in the grassland, and commonly dominant. Those who know the Burren will find much that is familiar in the limestone of the Yorkshire Pennines and Cumbria, and vice versa. The big difference is the much more oceanic situation of the Burren, bringing milder winters and cooler summers than in northern England, and a more humid and windier climate.

The patchy vegetation on the slopes of the hills of the western Burren is largely made up of a rich and very characteristic 'grassland' (named 'Asperuleto-Dryadetum' by Braun-Blanquet and Tüxen on their visit to the Ireland in 1949) dominated by varying proportions of *Sesleria* and mountain avens (Fig. 146), with near-constant squinancywort, glaucous sedge, flea sedge, harebell, sheep's fescue, slender St John's-wort (*Hypericum pulchrum*), fairy flax, bird's-foot trefoil, tormentil (*Potentilla erecta*), goldenrod (*Solidago virgaurea*), devil's-bit scabious (*Succisa pratensis*), wild thyme and common dog-violet, with a suite of conspicuous bryophytes including the mosses *Breutelia chrysocoma*, *Ctenidium molluscum*, *Hypnum tectorum*, *Neckera crispa*, *Pseudoscleropodium purum* and *Tortella tortuosa*, and the leafy liverworts *Frullania tamarisci* and *Scapania aspera*. Less constant than these but conspicuous almost everywhere are burnet rose (*Rosa pimpinellifolia*) and bloody crane's-bill (*Geranium sanguineum*). Other frequent species include mountain everlasting, heather, carline thistle, Irish eyebright (*Euphrasia salisburgensis* – an Irish speciality), limestone bedstraw, fragrant orchid (*Gymnadenia conopsea*), sea plantain (*Plantago maritima*) and two common species of limestone grasslands elsewhere, ribwort plantain and crested hair-grass. Hoary rock-rose is very local, but abundant where it occurs.

On the exposed crests and limestone-pavement summits of the hills this community tends to pass into a wind-clipped turf a few centimetres high

FIG 146. Mountain avens (*Dryas octopetala*), Black Head, Co. Clare, July 1970. Strictly a mountain plant over most of Europe (where it is abundant in the limestone Alps), *Dryas* comes down to sea level in the Burren and northwest Scotland.

with constant squinancywort, heather, harebell, carline thistle, glaucous sedge, flea sedge, mountain avens, crowberry (*Empetrum nigrum*), scattered plants of dark-red helleborine (*Epipactis atrorubens*), sheep's fescue, *Sesleria*, devil's-bit scabious and wild thyme, from which the taller and more drought-sensitive species are missing. Where a substantial depth of organic soil has accumulated over the limestone and leaching has acidified the surface layers, *Dryas* may hang on for some time, but the smaller and more demanding calcicoles drop out progressively with the increase of the heathers (*Calluna vulgaris* and *Erica cinerea*), and the calcicole mosses are replaced by common calcifuge species, notably *Hylocomium splendens*, *Hypnum jutlandicum* and *Rhytidiadelphus* spp. Locally, this limestone heath is colonised by bearberry (*Arctostaphylos uva-ursi*), and *Arctostaphylos*-rich heaths are an important component of the vegetation on the high limestone pavements and gently sloping summit plateaux around Black Head and Gleninagh Mountain. In general, the limestone heaths remain surprisingly species-rich, with near-constant *Antennaria*, *Calluna*, harebell, *Carex flacca*, *C. pulicaris*, bell heather, sheep's fescue, slender St John's-wort, bird's-foot trefoil, tormentil, *Sesleria*, devil's-bit scabious and wild thyme.

To the east of the high limestone hills of the western Burren, an almost level expanse of lake-studded almost bare limestone stretches from Corrofin to Galway Bay (Fig. 147). In this area the sparse soils generally contain a greater proportion of glacial drift, and the predominant grassland (the '*Antennaria dioica–Hieracium pilosella* nodum') is rather different. *Dryas* is often absent, but constant or near-constant species include mountain everlasting, kidney vetch (*Anthyllis vulneraria*), squinancywort, harebell, spring-sedge, sheep's fescue, crested hair-grass, oxeye daisy (*Leucanthemum vulgare*), fairy flax, bird's-foot trefoil, mouse-ear hawkweed, ribwort plantain, self-heal, *Sesleria*, devil's-bit scabious and wild thyme. Bryophytes are less conspicuous, but *Hypnum lacunosum* is near-constant, and *Ctenidium molluscum*, *Fissidens adianthoides*, *Frullania tamarisci*, *Scapania aspera*, *Pseudoscleropodium purum* and *Tortella tortuosa* are frequent. Outside the Burren similar grasslands, with or without *Sesleria*, and generally (not always!) poorer in species, occur widely on eskers and other suitable soils on limestone or calcareous drift.

Perhaps the only other limestone grasslands in Ireland to approach those of the Burren in distinctiveness and species-richness are those of the Ben Bulben massif that straddles the Sligo–Leitrim border, and the limestone country to the east and north (Fig. 148). Braun-Blanquet and Tüxen (1952) recorded species lists from mossy *Sesleria* grassland here, which have many species in common with *Sesleria* grasslands both in the north of England and in the Burren. What sets these Ben Bulben grasslands apart is the occurrence of such mountain plants as moss campion (*Silene acaulis*), fringed sandwort (*Arenaria ciliata*), purple

FIG 147. Mullagh More, Co. Clare. A spectacular synclinal hill in the heart of the Burren, July 1966. Hoary rock-rose grows in the fragmentary limestone turf on the upper slopes, and there is a rich and interesting flora in the lakes, fens and turloughs in the flatter surrounding limestone country.

FIG 148. The Ben Bulben massif, Co. Sligo, at the entrance to Glencar. June 2010.

saxifrage (*Saxifraga oppositifolia*), yellow saxifrage (*S. aizoides*) and alpine meadow-rue (*Thalictrum alpinum*) around the outcrops and cliffs, but they intergrade with grasslands that would be unremarkable on limestone elsewhere.

Dryas-rich communities on limestone in northwest Scotland

Cambrian limestone outcrops intermittently from Skye to the north Sutherland coast, and many of these bear a *Dryas–Carex flacca* 'heath'[CG13], which lacks *Sesleria* but has many of its more frequent species in common with the Asperuleto-Dryadetum of the Burren, including *Dryas*, mountain everlasting, heather, glaucous sedge, flea sedge, sheep's fescue, slender St John's-wort, fairy flax, bird's-foot trefoil, sea plantain, tormentil, wild thyme and common dog-violet, and the mosses *Ctenidium molluscum* and *Tortella tortuosa*. Most of the limestone exposures are small and surrounded by acid rocks, so chance may have played an important part in variation between stands. Nevertheless, the average species-list stands up well to comparison with species-rich limestone grasslands farther south. A less common variant on Raasay and the north Sutherland coast lacks tormentil, slender St John's-wort, flea sedge, heather, devil's-bit scabious and the two calcicole mosses, but has creeping willow (*Salix repens*), harebell, lady's bedstraw and the mosses *Pseudoscleropodium purum* and *Rhytidiadelphus triquetrus*.

HEAVY-METAL CONTAMINATION: MINE-SPOIL SITES

The Carboniferous limestone of Britain and Ireland is locally mineralised with seams rich in zinc and lead, which supported an active mining industry in times past, especially in Mendip, Derbyshire and the north Pennines. Lead was usually the metal most sought by the miners, but zinc is generally responsible for most of the toxicity. The distribution of lead-mining in Britain is pretty well mapped by the distribution of alpine penny-cress; much commoner than that species, and almost always accompanying it, is spring sandwort (Fig. 149a). Mine-spoil vegetation on limestone is usually dominated by sheep's fescue and spring sandwort forming a rather open community, usually with harebell, wild thyme and common bent (*Agrostis capillaris*)[OV37]. Bird's-foot trefoil, common sorrel (*Rumex acetosa*) and more locally alpine penny-cress and mountain pansy (*Viola lutea*) are frequent, and sea campion (Fig. 149b) or bladder campion (*Silene vulgaris*) are locally conspicuous; characteristic mosses include *Weissia controversa* var. *densifolia*, *Bryum pallens* and *Dicranella varia*. This open turf intergrades on less-contaminated ground with more normal limestone grassland. Of course metalliferous mining (and smelting) was not confined to limestone, and there are many old mine sites heavily polluted with copper, arsenic and other elements in southwest England and west Wales. The spoil from these has often been slow to recolonise, bearing only a skeletal vegetation of common bent, heather and common sorrel over a brownish

FIG 149. Heavy-metal-tolerant plants: (a) spring sandwort (*Minuartia verna*) and alpine penny-cress (*Thaspi caerulescens*, bottom left) growing in a carpet of the moss *Weissia controversa* var. *densifolia* on old zinc-mine spoil, Pikedaw, Malham, Yorkshire, July 1980; (b) sea campion (*Silene uniflora*) growing on old lead-mine spoil, Charterhouse, Mendip, Somerset, June 1989.

carpet of bryophytes such as *Jungermannia gracillima*, *Cephaloziella* spp., *Pohlia* spp. and *Racomitrium canescens*. Larger pleurocarpous mosses and lichens (e.g. *Rhytidiadelphus squarrosus*, *Cetraria aculeata*, *Cetraria islandica*, *Cladonia* spp., *Peltigera* spp.) often form a mat above the surface, largely escaping the influence of the toxic soil underneath.

WHAT IS THE HISTORY AND ORIGIN OF THE LIMESTONE GRASSLANDS AND THEIR FLORA?

Open limestone grasslands comparable with ours are widely scattered across Europe. Species rare with us, such as pasqueflower, hoary rock-rose, goldilocks aster and others, can be seen in recognisably 'the same' habitat in chalk or limestone grasslands in France, Germany, the Czech Republic or Hungary. But there is general agreement that before deforestation by our Neolithic and Bronze Age forebears the whole of northern Europe, including Britain and Ireland, was substantially forested. The lowland grasslands were (along with the arable fields and hay meadows) a part of the cultural landscape, kept in being by centuries of traditional farming. The chalk and limestone grasslands and their flora, which we have been accustomed to take for granted, had 'never had it so good' as under traditional agriculture – which demanded a great deal of low-paid human labour. With traditional patterns of farming no longer economic, we have to face up to finding alternative ways to maintain elements of the traditional landscape that we value.

Where did the flora of the chalk and limestone grasslands come from? Naturally non-forested habitats, notably mountains above the tree-line, can tell us something about the possibilities. We can also study the (tantalisingly incomplete) evidence provided by pollen analysis of peats and other deposits (Chapter 2). The Late-glacial flora contained not only the expected arctic and mountain plants, but calcicoles such as salad burnet, wild thyme, bird's-foot trefoil, hoary plantain, rock-roses and other plants of open grassland. Thus, a calcicole flora that could have given rise to our limestone grasslands was already established in our part of Europe before the spread of the forests. The question then is, where did the light-loving limestone-grassland species escape being overwhelmed by forest, to provide the nuclei for the limestone grasslands that followed forest clearance by Neolithic farmers?

A number of habitats are conceivable candidates. Dry, exposed crests, ridges and summits are one possibility; riverbanks kept open by erosion are another. We know that many such sites can bear closed forest at the present day, but we reckon without wild populations of browsers and grazers – deer, wild boar, aurochs. We do not know in enough detail how much effect these may have had in maintaining openings in the forest. At least fragmentary open communities could have kept a foothold in the crevices and ledges of inland cliffs of the harder limestones, and more extensively on exposed coasts such as the chalk of the south coast of England from Devon to Kent, and the Carboniferous limestone of Gower, the Great Orme, Humphrey Head, Whitbarrow or the Burren. On the exposed Burren coast a rich calcicole flora existed under open pine until about 500 BC (Feeser & O'Connell 2009). Some of these areas have outstandingly rich floras at the present day. Clearly that reflects ecologically favourable conditions for a rich diversity of species at the present time – but it is a reasonable conjecture that biodiversity 'hotspots' at least partly reflect historical factors. It seems hard to conceive of the British and Irish distribution of *Sesleria* being other than a historical 'accident', and equally hard not to invoke history to explain some of the widely disjunct distributions of species such as hoary rock-rose, mountain avens and shrubby cinquefoil (*Potentilla fruticosa*), or the similarities between the Burren and the Swedish island of Öland, which now enjoy very different climates. By contrast, we may think of the vegetation of the English chalk and Jurassic limestone grasslands as less influenced by its Late-glacial origins, and more by recruitment from the Continent over the 4000 years since Neolithic forest clearance began. But these are speculative thoughts, which future molecular genetic evidence may confirm or disprove.

CHAPTER 12

Lakes and Rivers: Freshwater Aquatic Vegetation

LAKES AND RIVERS SUPPORT RICH and varied plant life and, although they have fresh water in common, many aspects of their ecology are different. Lakes are extensive, and some are deep. The deepest lakes in our islands are in glacially deepened troughs in the mountains. Loch Morar in western Scotland reaches a maximum depth of 310 m, while the largest lake in Britain, Loch Lomond (71 km²), has a maximum depth of 190 m, and the second largest, Loch Ness, reaches 230 m, and its average depth is over 130 m. In surface area, none of the lakes in Britain matches the larger lakes of Ireland. Lough Neagh (388 km²) is the largest, and Lough Corrib (200 km²), Lough Derg (118 km²), Lough Ree and Lough Mask (89 km²) are all larger than any Scottish lake. All these big Irish lakes are relatively shallow.

Geologically, lakes are temporary features of the landscape, because they trap sediment brought in by the rivers entering them, which in the course of time fills the lake basin. Many of our mountain lakes have flat alluvial ground bordering the river at their heads; sometimes former lakes are obvious as flat areas in the bottoms of valleys, as in the Nant Ffrancon in Snowdonia. Most of the lakes in Britain and Ireland owe their existence to the glaciations, either through glacial scour over-deepening valleys (Scottish Highlands, Snowdonia and the English Lake District) or through moraines and other deposits left by the glaciers and ice-sheets damming the surface drainage (many Irish lakes and some in Britain); many lakes in the mountains owe something to both processes. England was less heavily glaciated, so outside the Lake District there are only a few small natural lakes in the Pennines, and in the lowlands the Shropshire and Cheshire meres. Almost all the other lakes in England are man-made, as reservoirs or sometimes as landscape features. The Norfolk Broads are medieval peat diggings (Chapter 13), and have a modern parallel in the flooded gravel pits of the Thames valley. A few natural lakes exist behind shingle-bars on the coast, such as Slapton Ley in Devon and the Loe Pool in Cornwall.

The longest rivers in our islands are the Shannon (386 km), the Severn (354 km) and the Thames (346 km). The longest river in Scotland is the Tay (188 km), closely followed by the Spey and the Clyde; all these are shorter than the Barrow (192 km) in southeast Ireland, but they all drain large tracts of rainy upland country so they are major rivers in the amount of water they discharge. However, we have a multitude of smaller rivers, no less interesting in their vegetation (Fig. 150).

Lakes may be deep, and they provide a large fetch for wind, but flow rates in the water are slow, even in lakes that have substantial rivers entering and leaving

FIG 150. The River Stour below Sturminster Newton, Dorset. A slow-flowing lowland river, draining from the chalk and fertile farming country, with alder and willows, reed canary-grass (*Phalaris arundinacea*) and yellow iris (*Iris pseudacorus*) lining the banks. Emergent aquatics include common reed (*Phragmites australis*), arrowhead (*Sagittaria sagittifolia*), bur-reed (*Sparganium*) and club-rush (*Schoenoplectus lacustris*), with white water-lilies (*Nyphaea alba*) floating in the open water.

them. By contrast all rivers and streams are shallow – never more than a few metres deep – and the rate and character of their flow is a cardinal factor in their ecology and vegetation (Haslam 1978). The large vascular plants, the *macrophytes*, make up the conspicuous part of the vegetation of both lakes and rivers, but in lakes these may be matched or even exceeded in importance by the microscopic *phytoplankton* suspended in the water.

SOME GENERALITIES ABOUT LAKE AND RIVER WATERS

In both lakes and rivers the chemistry of the water (see box on p. 242) has a major influence on the vegetation. The concentration of calcium and other cations varies widely, and with that comes variation in pH. Rainwater always contains dissolved substances, in part derived from sea-spray and in part from blown dust and gases in the atmosphere, and pure unpolluted rainwater on average has a slightly acid pH from dissolved carbon dioxide (CO_2) in the atmosphere. Rain that has percolated through limestone (with CO_2 raised by the respiration of soil organisms) dissolves calcium as bicarbonate, so streams draining from limestone country have high calcium (Ca^{2+}) and bicarbonate (HCO_3^-), and a near-neutral pH (around 7); they are 'hard-water' streams. By contrast, streams draining from non-calcareous rocks (notably the old hard rocks of many upland districts) contain much less Ca^{2+} and HCO_3^-, and their composition is dominated by sodium (Na^+), magnesium (Mg^{2+}), with some Ca^{2+}, balanced by chloride (Cl^-, ultimately from sea-spray) and sulphate (SO_4^{2-}, partly from sea-spray, partly from pollution), and they have a mildly acid pH; they are 'soft-water' streams.

Bicarbonate concentration, CO_2 concentration and pH in natural waters are intimately related. Over a wide range, pH rises with calcium concentration, which, because high Ca^{2+} generally reflects solution of limestone, is balanced by HCO_3^-. Beyond about 20 mg/litre Ca, there is little further increase in pH with increasing Ca, and the pH of calcareous waters is largely determined by the concentration of CO_2. If the Ca concentration is 40 mg/litre, and the water is in equilibrium with atmospheric CO_2, the pH will be about 8.2. Natural waters are rarely in equilibrium with the atmosphere (they come closest to that in the spray of waterfalls), because of the effect of respiration and photosynthesis on CO_2 concentration. Most of the organic matter in river and fen waters comes from neighbouring land habitats, and breakdown processes predominate in the water itself, so CO_2 levels are high – at the same Ca concentration of 40 mg/litre the pH may be no higher than 6.5–7. In a lake, photosynthesis by the phytoplankton can produce wide daily swings of pH. The phytoplankton

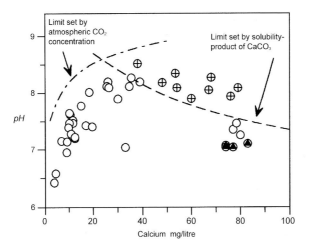

FIG 151. The relation of pH to concentration of calcium in water from Darnbrook and Cowside Becks, Littondale, Yorkshire. All the points fall well below the line of equilibrium with atmospheric CO_2. The broken line is the limit set by the precipitation of calcium carbonate ($CaCO_3$). The black triangles are springhead sites at which water was emerging from deep in the limestone; crosses are sites where tufa deposition is taking place.

(and some other aquatic plants) can deplete CO_2 to far below the concentration in the atmosphere, and raise pH as high as 10. However, we are then well into the region where the carbonate/bicarbonate ratio rules, and the possible HCO_3^- concentration is limited by the insolubility of calcium carbonate, which precipitates out as an incrustation on the plants or as fine calcareous sediment. Paradoxically, beyond a point set by the solubility product of $CaCO_3$, the concentration of Ca can only increase with a concomitant increase of CO_2, and falling pH (Fig. 151).

These mineral solutes are not the whole story, because a number of other elements are essential for the growth of all plants, especially nitrogen (N), phosphorus (P) and potassium (K). Limnologists contrast *eutrophic* waters, rich in plant nutrients, with *oligotrophic* waters in which growth is limited by nutrient supply; *mesotrophic* waters come between these extremes. A fourth category, *dystrophic*, takes in the acid peat pools and lochans in blanket-bog country (Chapter 15) in which peaty dissolved organic matter plays an important role in their water chemistry. Most lowland lakes are both moderately calcareous and moderately eutrophic. Lakes and tarns (lochans) in the mountains are generally non-calcareous (unless calcareous rocks outcrop in their catchments) and oligotrophic.

Water chemistry: ionic balances and carbonate–bicarbonate–carbon dioxide–pH equilibria in natural waters

The mineral nutrients in lake, fen or river waters are in solution as electrically charged *ions*. The positively charged *cations* of the metallic elements such as sodium (Na^+), potassium (K^+), calcium (Ca^{2+}) and magnesium (Mg^{2+}) must be exactly balanced by the negatively charged *anions* such as chloride (Cl^-), sulphate (SO_4^{2-}) and bicarbonate (HCO_3^-), so that the solution is electrically neutral.

The pH of calcareous waters is governed by the relationships:

$$pH = pK_1 + \log([HCO_3^-]/[CO_2]) \qquad (1)$$
$$pH = pK_2 + \log([CO_3^{2-}]/[HCO_3^-]) \qquad (2)$$

where pK_1 is the first dissociation constant of H_2CO_3 (*c.* 6.4) and pK_2 is the second dissociation constant (*c.* 10.4).

If two of the variables in either of these equations are known, the third can be calculated. An ambient atmospheric CO_2 concentration of 360 μlitre/litre is roughly equivalent to 1.6×10^{-5} mol/litre; water dissolves roughly its own volume of gaseous CO_2, so the concentrations of CO_2 in air and water can be assumed to be the same. As most of the Ca in calcareous water is derived from solution of $CaCO_3$, the concentration of HCO_3^- in a lake, river or fen water can be taken as approximately equalling the concentration of Ca. If we assume a 10^{-3} molar concentration of (Ca and) HCO_3^- we can calculate the pH from Equation (1) as

$$pH = 6.4 + \log(1.0 \times 10^{-3}/1.6 \times 10^{-5}) = 8.2$$

This assumes the rare case of a water in equilibrium with atmospheric CO_2. If we measure the calcium concentration as 1.0×10^{-3} molar, and the pH as 6.7, we can calculate the CO_2 concentration in the water as about 5×10^{-4} molar – around 30 times the atmospheric concentration, a result which might be expected in a fen because of microbial respiration in the water and peat.

These two calculations have ignored Equation (2), and a quick calculation will show that at pH 8, with 10^{-3} molar HCO_3^- the concentration of carbonate is negligible. But if we suppose that photosynthesis by the phytoplankton (or a *Chara* bed) has reduced CO_2 in the water to 1.0×10^{-6} molar and we leave HCO_3^- at 10^{-3} molar, the pH would rise to 9.4. From Equation (2), the predicted concentration of carbonate (CO_3^{2-}) would then be 1.0×10^{-4} mol/litre.

However, the molar product $[CO_3^{2-}] \times [Ca^{2+}]$ would be 1.0×10^{-7}, well above the *solubility product* for $CaCO_3$ at 18 °C (1×10^{-8}), and slightly above the solubility product even at 25 °C (6×10^{-8}), so precipitation of solid $CaCO_3$ would be expected. This concentration of Ca^{2+} can only remain in solution at higher CO_2 and lower pH.

FRESHWATER MACROPHYTE VEGETATION

Emergent aquatics

Many waterside plants can grow equally well in permanently wet soil, or rooted at the bottom in shallow water as 'emergent aquatics' – plants rooted under water but with their leaves and flowers above (Fig. 152). The common reed (*Phragmites australis*), one of the few cosmopolitan plants, can dominate tall herb fens on peat in the Norfolk Broads (and elsewhere) or grow in up to half a metre or so of water; its rhizomes often form floating rafts fringing the shore. It can tolerate a wide range of nutrient regimes, moderately acid to calcareous, avoiding only the most acid and nutrient-poor places[S4]. Reed canary-grass (*Phalaris arundinacea*) looks rather like *Phragmites* in build and leaf but is easily recognised by the membranous ligule at the base of the leaf blade, where *Phragmites* has only a line of hairs. *Phalaris* is more nutrient-demanding than *Phragmites*, and often marks places where there is heavy silting, so perhaps does not really qualify as an 'aquatic'[S28]. Another large grass of the banks of eutrophic lakes and rivers, which often forms floating rafts over open water, is reed sweet-grass (*Glyceria maxima*)[S5].

Some big sedges (Cyperaceae) are also characteristic of this habitat. The great fen-sedge (*Cladium mariscus* – just 'sedge' in East Anglia) like *Phragmites* spans the boundary between open water and tall fen on firm peat[S2]. It has long-lived,

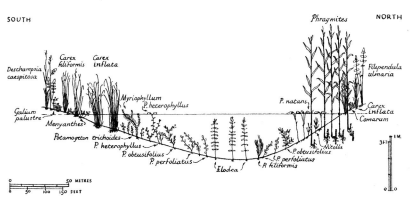

FIG 152. Profile transect across White Moss Loch, just north of the Ochill Hills southwest of Perth. Marsh and fen plants around the lake margin give way first to emergent aquatics (common reed, bogbean, slender sedge, bottle sedge), then to floating-leaved (*Potamogeton natans, P. gramineus*) and submerged aquatics (the pondweeds *Potamogeton trichoides, P. obtusifolius, P. perfoliatus, P. filiformis* and Canadian waterweed *Elodea canadensis*). From Tansley 1939, after Matthews 1914. *Carex filiformis* = *C. lasiocarpa, C. inflata* = *C. rostrata, Comarum* = *Potentilla palustris, Potamogeton heterophyllus* = *P. gramineus*.

FIG 153. Tufted-sedge (*Carex elata*) swamp on the northwest shore of Lough Bunny, Co. Clare, at a time of low water level, September 1975.

coarse saw-toothed leaves, which effectively suppress almost all competing species. Smaller than *Cladium*, but still substantial sedges and important in our vegetation, are the big species of *Carex* (which give their name to the alliance Magnocaricion of Continental vegetation ecologists). They include the tufted-sedge (*Carex elata*) [S1], a local pioneer at the edge of calcareous pools and lakes (Fig. 153), perhaps often associated with fluctuating water level, the greater tussock-sedge (*C. paniculata*), whose massive tussocks fringe the lower courses of the Broadland rivers and line valley-bottoms elsewhere[S3], the greater pond-sedge (*C. riparia*), which often fringes the banks of eutrophic lakes, slow-flowing rivers and canals[S6], and the lesser pond-sedge (*C. acutiformis*), which occurs in similar but perhaps generally more calcareous habitats[S7]; the latter two are commonest in the lowlands of southeast England. Bottle sedge (*C. rostrata*), by contrast, is a plant of the north and west and of oligotrophic waters, fringing many mountain and moorland lakes and tarns, mostly acid, but sometimes calcareous as at Malham Tarn[S9]. In sheltered situations it is often accompanied by such species as marsh cinquefoil (*Potentilla palustris*), bogbean (*Menyanthes trifoliata*) and water horsetail (*Equisetum fluviatile*), in effect a transition to fen[S27]. The common club-rush (*Schoenoplectus lacustris*) grows in lakes, slow-flowing rivers and canals in water usually 0.5–1.5 m deep, and ranging from acid and base-poor to highly calcareous, and from oligotrophic to eutrophic[S8]. The grey club-rush (*S. tabernaemontani*) is much less bound to deep water and is salt-tolerant; it is mainly coastal but occurs widely inland in more or less calcareous and eutrophic sites in the English Midlands[S20].

The bulrushes (*Typha*) are locally conspicuous reedswamp plants. *Typha latifolia* occurs in standing or slow-moving, circum-neutral and more or less eutrophic waters on silty substrates throughout Britain and Ireland; it is often common around lowland lakes, ponds and reservoirs and in sluggish streams and river backwaters[S12]. The seeds are dispersed by wind, and *T. latifolia* is often quick to colonise newly excavated pools. Lesser bulrush (*T. angustifolia*) tends to grow in deeper water and in less eutrophic conditions[S13], but there is a great deal of overlap between the two species.

A shorter but striking plant, often a prominent feature of moderately deep still or slow-flowing waters, is arrowhead (Fig. 154a), with long ribbon-like submerged leaves and arrow-shaped emergent leaves held, like the three-petalled white flowers, well clear of the surface of the water[S16].

The remaining emergent aquatics all make less impact on the landscape than the reedswamp dominants just discussed, but one common and often dominant species, branched bur-reed (Fig. 154b), provides the occasion to mention some of the others. Branched bur-reed commonly fringes eutrophic ponds, ditches, canals and rivers on mineral soil, and like a number of other reedswamp

FIG 154. Emergent aquatics: (a) arrowhead (*Sagittaria sagittifolia*), Fiddleford, Dorset; (b) branched bur-reed (*Sparganium erectum*), Slapton, Devon.

FIG 155. Emergent aquatics: (a) flowering-rush (*Butomus umbellatus*), Othery, Somerset Levels; (b) greater spearwort (*Ranunculus lingua*), Woodford, Co. Galway, July 1966.

plants can grow either in permanently wet soil or in shallow water. Common associated species include water plantain (*Alisma plantago-aquatica*), common spike-rush (*Eleocharis palustris*), water mint (*Mentha aquatica*), water-cress (*Rorippa nasturtium-aquaticum*), fool's-water-cress (*Apium nodiflorum*), water forget-me-not (*Myosotis scorpioides*), gypsywort (*Lycopus europaeus*), yellow iris (*Iris pseudacorus*), meadowsweet (*Filipendula ulmaria*), skullcap (*Scutellaria galericulata*) and hemlock water-dropwort (*Oenanthe crocata*), a western plant with us (and in Europe as a whole) and one of the most poisonous plants in our flora[S14]. Muddier, shelving water-margins are often dominated by floating sweet-grass (*Glyceria fluitans*) with a similar range of associates, which sometimes include marsh foxtail (*Alopecurus geniculatus*) and false fox-sedge (*Carex otrubae*)[S22]. Two rather rare plants of these water-margin communities are the greater spearwort (Fig. 155b), a magnificent spear-leaved buttercup, and the beautiful 'flowering-rush' (Fig. 155a) – nothing to do with the true rushes despite its rush-like flower stems, but more nearly related to the water-plantains.

Floating-leaved aquatics

Many water plants are rooted but have floating leaves, most obviously the water-lilies. The white water-lily (*Nymphaea alba*) can tolerate a very wide range of waters in pools and slow-flowing rivers. Because it has no submerged leaves it is

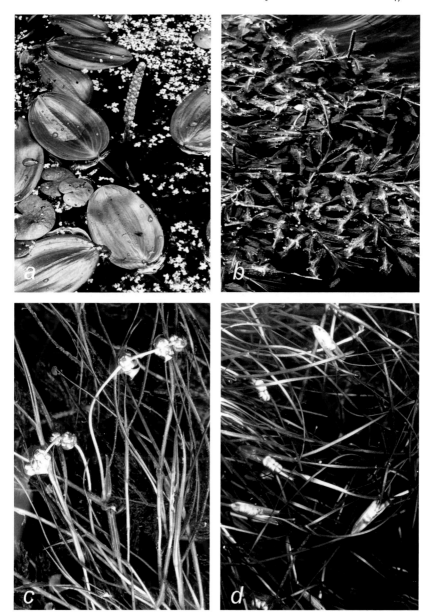

FIG 156. Pondweeds: (a) broad-leaved pondweed (*Potamogeton natans*), Exminster, Devon; (b) shining pondweed (*P. lucens*), Exeter Canal; (c) fennel pondweed (*P. pectinatus*), Fiddleford, Dorset; (d) beaked tasselweed (*Ruppia maritima*), Cuckmere Haven, Sussex.

easily damaged by boat traffic. It occurs throughout Britain and Ireland, but is probably most naturally at home in acid, often peaty, oligotrophic lakes and tarns of the north and west, where it grows with occasional pondweeds (*Potamogeton natans* and *P. polygonifolius*), bulbous rush (*Juncus bulbosus*) and a thin scattering of other aquatic plants[A7]. The yellow water-lily (*Nuphar lutea*, Fig. 158a), by contrast, prefers at least moderately calcareous and eutrophic waters, so is commonest in the southeast and the Midlands of England and the midland plain of Ireland. It occurs widely in slow-flowing rivers, little-used canals (as it has submerged leaves it is less vulnerable to disturbance by boats), dykes, lakes and pools, with a longer and more diverse list of associates than is usual for the white water-lily, of which perhaps the most frequent are the common water-starwort (*Callitriche stagnalis*), the duckweeds *Lemna minor* and *L. trisulca*, Canadian waterweed (*Elodea canadensis*) and amphibious bistort (*Persicaria amphibia*)[A8].

Amphibious bistort itself is common and sometimes dominant in the shallows of still and slow-flowing, sometimes fluctuating, waters (it is frequent in some of the Irish turloughs). It occurs widely in Britain and Ireland, mostly in the lowlands in moderately base-rich and moderately eutrophic water. It can grow in wet ground but flowers much more freely as an aquatic plant, with the flower spikes held stiffly above the surface[A10]. A number of pondweeds (*Potamogeton*) can have floating leaves. Much the most widespread is the broad-leaved pondweed (Fig. 156a), which occurs throughout Britain and Ireland in still or slowly flowing waters (and even occasionally in a fully submerged form in fast-flowing streams). It has extraordinarily wide ecological tolerance, and can grow in acid or calcareous, oligotrophic or eutrophic waters, and from near-terrestrial conditions to water 5 m deep[A9].

Submerged aquatics

Most of the other pondweeds have only submerged leaves. The perfoliate pondweed (*Potamogeton perfoliatus*) is a broad-leaved species, which, with other submerged aquatics, often dominates the macrophyte vegetation forming bulky masses in still or very gently flowing, mesotrophic and usually rather base-poor waters over the whole of Britain and Ireland, from Kent and Kerry to the Shetlands. The dominant *P. perfoliatus* is most often accompanied by alternate water-milfoil (*Myriophyllum alterniflorum*), a range of pondweeds including *P. gramineus*, *P. berchtoldii*, *P. obtusifolius*, *P. natans* and *P. pusillus*, Canadian waterweed, amphibious bistort, the moss *Fontinalis antipyretica* and charophyte algae of the genera *Chara* and *Nitella*[A13]. In the most base-poor and oligotrophic pools and streams alternate water-milfoil may dominate alone, with occasional *Juncus bulbosus*, intermediate water-starwort (*Callitriche hamulata*), *Potamogeton gramineus* and the moss *Fontinalis antipyretica*[A14]. In base-rich, mesotrophic to eutrophic conditions in clear, unpolluted waters, this is replaced by a community dominated by fennel pondweed (Fig. 156c) and spiked water-milfoil (*Myriophyllum spicatum*), with a long list of associates of which the most frequent are the pondweeds *Potamogeton pusillus, P. lucens* (Fig. 156b), *P. crispus, P. natans, P. filiformis, P. berchtoldii* and *P. perfoliatus,* Canadian waterweed, amphibious bistort and unbranched bur-reed (*Sparganium emersum*)[A11]. Where the water is turbid or polluted the fennel pondweed may dominate alone or with only a few associates[A12].

The white-flowered water-crowfoots, like the pondweeds, have both broad floating leaves and finely dissected submergerged leaves, and several are important in our aquatic vegetation. The common and pond water-crowfoots (*Ranunculus aquatilis* and *R. peltatus*) have been much confused; both are probably important (and interchangeable?) in the fringing vegetation of stagnant waters in ponds, ditches and slow-flowing streams along with such species as floating sweet-grass, water-cress, fool's-water-cress, broad-leaved pondweed, spiked water-milfoil, various water-starworts (*Callitriche* spp.) and the duckweeds *Lemna minor* and *L. trisulca*[A19, A20]. In coastal brackish waters these two common water-crowfoots are replaced by *R. baudotii*, typically in water half-a-metre or so deep accompanied by such species as horned pondweed (*Zannichellia palustris*), soft hornwort (*Ceratophyllum submersum*), fennel pondweed, the duckweeds *Lemna minor* and *L. gibba*, the water-starworts *Callitriche stagnalis* and *C. obtusangula*, mare's-

FIG 157. Stream water-crowfoot (*Ranunculus penicillatus* ssp. *pseudofluitans*), River Otter, East Budleigh, Devon.

tail (*Hippuris vulgaris*) and beaked tasselweed (Fig. 156d)[A21]. Two water-crowfoots
are characteristic of faster-flowing waters. The stream water-crowfoot (Fig. 157)
is a very characteristic dominant of calcareous, moderate to quite fast-flowing
rivers with sandy or gravelly bottoms, and generally draining chalk or limestone
catchments. In southern England, these include many of the classic chalk streams
and their equivalents on the Jurassic limestones of the Cotswolds – and farther
north and west rivers and streams on Devonian and Silurian rocks in Wales, on
Carboniferous limestone of Derbyshire and Yorkshire, and on Devonian rocks
in the Border counties of Scotland. Apart from the dominant water-crowfoot
(spectacular when in flower), this is a rather species-poor community; among the
commoner associates are water-cress, brooklime (*Veronica beccabunga*) and lesser
water-parsnip (*Berula erecta*)[A17]. Deep fast-flowing oligotrophic to mesotrophic
rivers with stable stony beds are the favoured habitat of the river water-crowfoot
(*Ranunculus fluitans*). Many of the British occurrences are clustered in the Welsh
Marches or around the southern Pennines; *R. fluitans* is local in Scotland, and
known only from Co. Antrim in Ireland. This is another species-poor community,
with spiked water-milfoil the most frequent associate[A18].

Free-floating aquatics

There remain only some free-floating water plants to consider (Fig. 158). At first
sight, it is paradoxical that these plants, free from the constraints of an attachment
to firm ground, are so bound to the shelter (and constraints) of small water
bodies, or of rooted aquatic vegetation on the shores of lakes. The largest of our
free-floating aquatics is the water soldier (Fig. 159b), which was formerly rather
widespread in eastern England but has declined over the years for reasons that are
not clear, and Dutch ecologists are not unanimous about its nutrient requirements.
From the available data, its surviving Broadland localities have rich aquatic floras,
with such species as greater bladderwort (*Utricularia vulgaris*), water violet (*Hottonia
palustris*) and greater spearwort[A4]. Frogbit (Fig. 158b) is much commoner and
occupies similar habitats in the clear, unshaded, mesotrophic to eutrophic (but
unpolluted) waters of ditches and ponds over the whole of lowland England. The
most constant species beside frogbit are the floating duckweeds (Fig. 159a), and
the submerged ivy-leaved duckweed (*Lemna trisulca*), Canadian waterweed and
rigid hornwort (*Ceratophyllum demersum*); the tiny rootless duckweed *Wolffia arrhiza*
occurs occasionally in this community, especially in the Somerset Levels[A3]. Several
of these free-floating species can behave as community dominants. *Ceratophyllum
demersum* often dominates still or slow-moving eutrophic waters in ditches, ponds,
sluggish streams and canals, with smaller quantities of Canadian waterweed,
common water-starwort (*Callitriche stagnalis*) and the water-crowfoot *Ranunculus*

FIG 158. Floating-leaved aquatics, a rooted and a free-floating species: (a) yellow water-lily (*Nuphar lutea*), Fiddleford, Dorset; (b) frogbit (*Hydrocharis morsus-ranae*), Shropshire Union Canal.

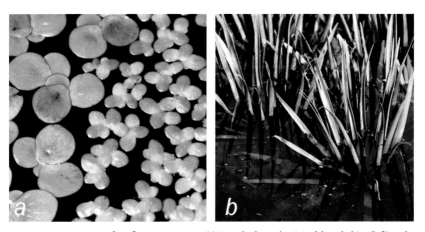

FIG 159. A contrast in free-floating aquatics. (a) Two duckweeds, *Spirodela polyrhiza* (left) and *Lemna minor* (right), Exminster, Devon; (b) water soldier (*Stratiotes aloides*), Capellen a. d. Ijssel, the Netherlands.

circinatus, and is perhaps favoured by eutrophication with agricultural run-off[A5]. By contrast, the rarer soft hornwort, with more finely divided brighter-green leaves, usually grows in company with *Potamogeton pectinatus* and *Ranunculus baudotii*; it is mainly coastal in eastern England, but comes inland in the Fens and Somerset Levels, and in the southwest Midlands[A6]. The floating *Lemna* species can both form extensive quasi-pure stands on their own, accompanied by a rather predictable list of common associates[A1, A2]. The little introduced American water-fern *Azolla filiculoides* floats like a duckweed and can cover the surface of an unshaded ditch or pond in a remarkably short space of time; its dense, buoyant and vigorous growth can crowd out virtually all competitors.

The quillwort life-form

A very distinctive 'isoetid' life-form predominates in clear, shallow oligotrophic waters (Fig. 160), a suite of unrelated plants all forming 'shuttlecock' rosettes of quill-like leaves. The most widespread is shoreweed (*Littorella uniflora*), in the plantain family. It has a typical upland distribution, with outliers in the New Forest and other heathland areas in the southeast. The two quillworts (*Isoetes lacustris* and *I. echinospora*) and water lobelia (Fig. 161a) are more closely bound to lakes and tarns in the mountains and moorlands of the north and west, and awlwort (*Subularia aquatica*, a crucifer) has a similar but slightly more restricted distribution. Lastly, pipewort (Fig. 161b) sits within the distribution of *Littorella*, *Isoetes lacustris* and the water lobelia, but is confined to Kerry, Connemara, Donegal, the Inner Hebrides and on the Scottish mainland only in Ardnamurchan. A fascinating feature of these isoetids is that they all rely largely on CO_2 taken in through their roots and diffusing through the copious air spaces of the plant to the leaves for photosynthesis. This is a habit they share with no other plants; it enables them to grow alongside algae that can deplete CO_2 in the water to very low levels.

Arguably, in their classic oligotrophic lake habitat in the north and west the isoetids could be seen as a single variable community[A22, A23]. *Littorella* has a wider distribution than the others, and can form an understorey to some of the aquatic communities already considered. In the calcareous lakes of the Burren, *Littorella* forms characteristic communities on flat surfaces of calcareous marl, flooded most of the year but dry in summer, with lesser water-plantain (*Baldellia ranunculoides*), lesser spearwort (*Ranunculus flammula*), *Juncus bulbosus*, marsh bedstraw (*Galium palustre*), *Potamogeton gramineus*, the big rich-fen moss *Scorpidium scorpioides* and species of *Chara*. On the better-drained areas of marl around some of the larger lakes, lesser water-plantain, marsh bedstraw, *Juncus bulbosus* and *Potamogeton gramineus* are missing, and their place is taken by the (normally calcifuge) spike-rush *Eleocharis multicaulis*.

LEFT: **FIG 160.** Loch nan Eilean, Skye looking towards the Cuillins, June 1981. An oligotrophic lake on hard acid rocks, with emergent bottle sedge (*Carex rostrata*) and bogbean (*Menyanthes trifoliata*), and floating-leaved white water-lily (*Nymphaea alba*) and pondweeds (*Potamogeton* spp.). The clear water allows quillwort (*Isoetes lacustris*) and plants of similar growth-form to carpet the bottom.

BELOW: **FIG 161.** Isoetids, Costelloe, Co. Galway, July 1971: (a) water lobelia (*Lobelia dortmanna*) has rosettes of blunt strap-shaped leaves on the bottom; only the flower-spikes appear above water; (b) pipewort (*Eriocaulon aquaticum*) has submerged flat rosettes of pointed leaves, and aerial flower spikes like knitting-needles.

ROCKY AND STONY STREAMS:
MOSS-DOMINATED HABITATS

Most streams in the uplands, and some in the lowlands, have rocky beds, and even slow-flowing lowland rivers have bridges, weirs and other masonry or concrete lapped by the water where aquatic mosses can get established. The trailing streamers of *Fontinalis antipyretica* (Fig. 162a) adorn probably most of the river bridges in the country, and are common in slow-flowing streams. Another big moss, *Cinclidotus fontinaloides* (Fig. 166), is characteristic of the flood zone, submerged during periods of high water level, exposed when the river is low in summer, when it forms conspicuous untidy dry blackish masses – especially on limestone, but often on concrete and sometimes brickwork. Like many mosses of the flood zone, *Cinclidotus* can tolerate severe and prolonged desiccation. Two species are common on brick or stonework in mill-leats and in streams with stony beds: *Brachythecium rivulare* (Fig. 162b) at or a little above normal water level and intermittently submerged, and *Platyhypnidium riparioides* (Fig. 162c), which forms a zone below it, exposed only when water level is particularly low. A third common (but rather nondescript) lowland species, *Leptodictyum riparium*, grows in wet places including pools, ditches and slow-flowing streams and rivers. There are a number of mosses very characteristic of riverside tree-bases in the flood zone, including *Syntrichia latifolia*, *Orthotrichum rivulare*, *O. sprucei*, *Leskea polycarpa* and *Scleropodium cespitans*. Riverbanks, particularly if they are rocky, can be very rich in bryophytes, but most of the species are not truly aquatic.

In upland streams, mosses (and sometimes liverworts) come into their own and make up most of the plant cover. *Brachythecium rivulare* and *Platyhypnidium riparioides* are still common, particularly where the water is at least reasonably base-rich, but they are often joined by *Brachythecium plumosum* and *Hygroamblystegium fluviatile* (Fig. 162d). On limestone, *Cinclidotus fontinaloides* continues to be conspicuous in the flood zone, largely replaced in fast-flowing streams by *Schistidium platyphyllum* and *Orthotrichum cupulatum*. The submerged zone is the preserve of *Platyhypnidium riparioides*, *Hygroamblystegium fluviatile* and two calcicole mosses, *Hygrohypnum luridum* and (usually where the water is saturated with calcium bicarbonate) *Palustriella falcata*.

OPPOSITE: **FIG 162.** Stream bryophytes: (a) *Fontinalis antipyretica*, common in slow-flowing rivers and streams; (b) *Brachythecium rivulare*; (c) *Platyhypnidium riparioides*, often dominant on rocks or weirs in moderately fast-flowing waters; (d) *Hygroamblystegium fluviatile*, on rocks in fast-flowing rivers and streams; (e) *Racomitrium aciculare*, common on rocks in upland streams; (f) *Scapania undulata*, a leafy liverwort often dominant in acid headwater streams.

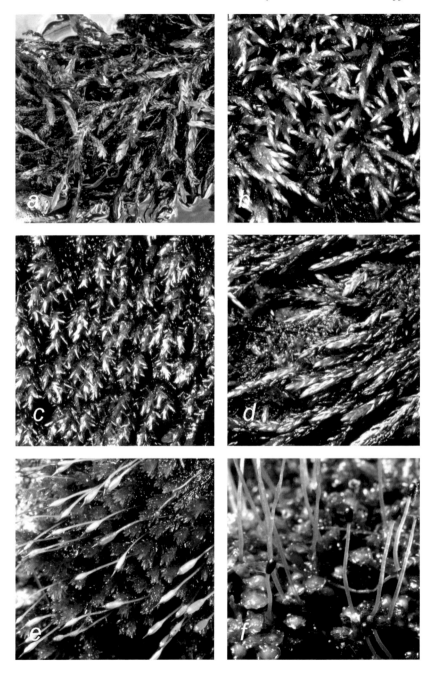

In waters draining from more acid rocks (with less calcium but pH not far below neutral), the greatest diversity is to be found. Flood-zone specialists include such species as *Racomitrium aciculare* (Fig. 162e) and *Schistidium rivulare*. The more-or-less permanently submerged species are joined by *Fontinalis squamosa* (which lacks the characteristic keeled leaves of the larger and commoner *F. antipyretica*), *Hygrohypnum ochraceum*, locally *Platyhypnidium alopecuroides*, sometimes the 'red' alga *Lemanea*, and the leafy liverworts *Marsupella emarginata* and *Nardia compressa*. Many of these species drop out in the higher reaches of upland rivers. A common pattern on Dartmoor is a threefold zonation, with sparse black cushions of *Andreaea rothii* on the granite boulders extending down into the zone of occasional flooding, a prominent belt of *Marsupella emarginata* spanning the normal variation in water level, and a dense cover of *Nardia compressa* covering the permanently submerged stream-bed.

In headwater streams draining from acid rocks the bryophyte flora is limited to a few species. *Philonotis fontana* and *Dichodontium palustre* are species characteristic of springs – waterside mosses rather than true aquatics. The same could be said of *Dichodontium pellucidum* and the common thalloid liverwort *Pellia epiphylla*, and the *Sphagnum* species and *Polytrichum commune* from the surrounding moorland (Chapter 17). True stream bryophytes are often confined to *Racomitrium aciculare* topping the boulders, and the leafy liverwort *Scapania undulata* (Fig. 162f), bright green or tinged with red, covering the wet rock surfaces.

THE PHYTOPLANKTON: SEASONAL CYCLES

The larger aquatic plants we have just considered are mostly confined to the shallow water around the margins of lakes. By contrast, the phytoplankton, the photosynthesising organisms that live suspended in the water, can be everywhere. At our latitudes, the phytoplankton is sparsest in the cold short days of winter. Diatoms, such as *Asterionella formosa* (Fig. 163a), *Stephanodiscus hantzschianus* (Fig. 163c) or *Melosira* species, with their silicified 'frustules', typically dominate the plankton in the late winter and early spring. As the days

OPPOSITE: **FIG 163.** Phase-contrast photomicrographs of some phytoplankton: (a) a colonial diatom, *Asterionella*, Malham Tarn, Yorkshire; (b) a cyanobacterium, *Anabaena* sp. (Culture coll.); (c) a centric diatom, *Stephanodiscus hantzschianus*, Slapton Ley, Devon; (d) a colonial motile chrysophyte, *Synura uvella*, Fernworthy Reservoir, Devon; (e) a motile colonial chlorophyte, *Eudorina elegans*, Slapton Ley; (f) a non-motile colonial chlorophyte, *Pediastrum boryanum*, with empty urn-shaped thecae of the colonial chrysophyte *Dinobryon*, Malham Tarn.

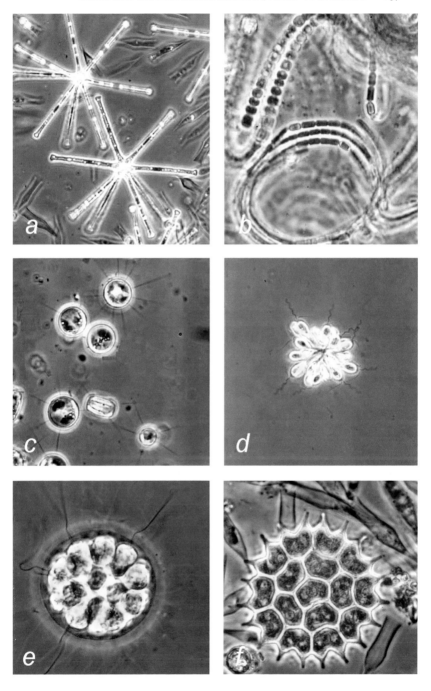

lengthen and temperatures rise, the diatom populations start to grow. At many sites the beautiful stellate colonies of *Asterionella formosa* are prominent in the phytoplankton in the early months of the year. In the Cumbrian lakes the spring rise of *Asterionella* generally begins in March or April, peaking in late May and June, but in smaller and shallower lakes this increase may begin as early as January, as at Slapton Ley in south Devon (Van Vlymen 1980) and Malham Tarn in the Pennines (Lund 1961), rising to a peak in March or April. The spring diatom maximum is brought to an end largely by depletion of essential nutrients – phosphate, and the silica essential for the diatom frustule. In Slapton Ley, a eutrophic lake, the most numerous organism in the early summer phytoplankton was the cyanobacterium *Anabaena flos-aquae*, but diatoms such as *Melosira*, *Stephanodiscus* and *Fragilaria* remained common, accompanied by green algae such as *Scenedesmus*, *Pandorina* and *Pediastrum*, and by cryptomonads whose numbers varied greatly from year to year. In late summer, with high temperatures and exhaustion of nitrogen (while phosphate remained in adequate

---

Phytoplankton organisms

Of the organisms making up the phytoplankton three groups are particularly important. The cyanobacteria ('blue-green algae', Fig. 163b), were among the first-evolved living things, at least 2500 million years ago, and for at least 1000 million years they were the principal photosynthetic organisms on our planet. The green algae (Chlorophyta), with an evolutionary history that may go back 1500 million years, include a wide range of planktonic motile (with two equal flagella to each cell) and non-motile unicells and colonial forms (Fig. 163e–f). They are the group that gave rise to the land plants some 400–500 million years ago. The silicified diatoms (Chrysophyta: Bacillariophyceae, Fig. 163a, c), which often dominate the phytoplankton, are geologically relative newcomers, which appear in the fossil record around 200 million years ago. The golden-brown algae (Chrysophyta: Chrysophyceae, Fig. 163d) are related to the diatoms (with which they share their pigmentation and ability to deposit silica in their cells), but unlike the diatoms most are motile with two unequal flagella to each cell. Some are unicellular, some colonial; they are much less important in the phytoplankton than diatoms but include some quite common planktonic organisms. Other kinds of 'algae' are frequent in the phytoplankton too, notably dinoflagellates (Pyrrhophyta), cryptomonads (Cryptophyta) and the euglenoids (Euglenophyta). Most phytoplankton organisms are somewhat heavier than water, with still-water settling velocities of some 5–15 μm a second. Cyanobacteria with 'gas vacuoles' can regulate their buoyancy, and cells with flagella can actively adjust their position in the water column.

supply), Slapton often saw dense 'blooms' of nitrogen-fixing cyanobacteria, such as *Microcystis aeruginosa, Anabaena spiroides* and *Gloeotrichia echinatula*, but the green algae such as *Pediastrum boryanum* (Fig. 163f), the diatom *Fragilaria* and cryptomonads remained a significant and sometimes major (but variable) component. As autumn leads to winter diatoms once again dominate the phytoplankton.

Shallow lakes are mixed throughout their entire depth by turbulence generated by the wind across their surfaces. Even moderate turbulent mixing ensures that *some* cells remain in suspension, and during stormy periods some re-suspension of cells from the bottom can take place, mainly in shallow water. The temperature profile of a shallow lake is gradual from top to bottom. Water expands and becomes lighter with rising temperature. In a deep lake in summer turbulence generated by the wind can no longer mix the warmer and lighter surface water into the deep body of the lake. The lake becomes *stratified*, into a well-mixed, warmer *epilimnion*, floating above a colder *hypolimnion* which extends to the depths of the lake; the two are separated by a *thermocline* (around 10 m below the surface in Windermere in summer), in which the temperature gradient (which may be around 5 °C) is concentrated. Cells that fall through the thermocline are lost to the zone where there is enough light for photosynthesis, so in periods when the lake is stratified motile cells with flagella and cyanobacteria with gas vacuoles (which can regulate their position in the water column) are at an advantage – and non-motile cells quickly disappear from the population. This results in a more complete summer crash of the diatom populations in stratified lakes and, for the same reason, stratification leads to the surface layers becoming depleted of nutrients. Declining temperature at the end of summer, along with autumn storms, brings the breakdown of stratification, and the lake again becomes turbulently mixed through the whole of its depth. For more on the complexities of phytoplankton ecology see Macan (1970) and Reynolds (1984).

COMPETITION FOR CO_2, LIGHT AND NUTRIENTS IN AQUATIC HABITATS

Photosynthesis and the acquisition of CO_2

We hear so much about the dire consequences of rising atmospheric CO_2 that it may be hard to remember that for plants CO_2 remains a scarce resource. The complex mesophyll of a flowering-plant leaf is an adaptation that can increase the area for CO_2 uptake 20-fold or more; the CO_2 compensation point (at which photosynthesis is balanced by respiration) for most temperate flowering plants (and bryophytes) is 50–100 parts per million by volume. The diffusion rate of gases in water is slower than that in air by a factor of about 10,000 to 1. Most of the phytoplankton organisms and many of the submerged macrophytes counter this by having a carbon-concentrating mechanism (CCM), enabling them to reduce the CO_2 in the water to very low levels. 'There is no such thing as a free lunch', and CCMs come at a price in energy terms, but when CO_2 is limiting an organism with a CCM can out-compete an organism without one. Emergent aquatic plants can take up CO_2 from the air in the ordinary way, but the submerged macrophytes, the phytoplankton and the algae growing on rocks on the lake bottom and around the shore, or epiphytically on the macrophytes, are all competing for the CO_2 in the water.

The vast majority of bryophytes do not have a CCM, so bryophytes are usually sparse in lakes. The exceptions are around lake shores where wave action keeps the water better aerated, in clear, unproductive oligotrophic lakes, where mosses can grow to surprising depths, and around inflow streams bringing in CO_2-rich water and organic sediment. In rivers most of the organic matter comes from the land along their banks, so CO_2 levels in rivers are generally much higher than in air (or in lakes). Bryophytes are much commoner in rivers, and become the dominant plants in mountain streams and the upper reaches of rivers with stony or rocky beds.

Light is the other *sine qua non* of photosynthesis. Even clear water absorbs light, but light attenuation in water is much increased by turbidity, by plankton in suspension, by submerged macrophytes, or by a cover of floating macrophytes at the surface. Dense growth of phytoplankton, particularly summer 'blooms' of cyanobacteria, can depress the growth and impoverish the diversity of the submerged macrophyte flora.

Limiting nutrients – and too much of a good thing

Growth in most plants and 'algae' is limited by the availability of nitrogen and phosphate, and for diatoms by silica (SiO_2). The effect of nutrient depletion on the spring diatom peaks in the phytoplankton has been considered already. In large deep lakes (particularly when they are stratified in summer) the plankton is much more nearly independent of the rest of the lake ecosystem than in smaller, shallow lakes, where the phytoplankton, the shore and the submerged macrophytes are in closer juxtaposition and more direct competition for nutrients.

Added nutrients (from treated sewage effluent or agricultural run-off) shift the trophic status in the eutrophic direction. In moderation this may do little harm, but large additions lead to excessive growth of a few vigorous species

– common reed, branched bur-reed, fennel-leaved pondweed, rigid hornwort, filamentous green algae and *Enteromorpha*, and cyanobacteria such as *Microcystis*, *Anabaena* and *Gloeotrichia* – and can be a part of the process leading to a switch from a clear-water species-rich canal or lake to one dominated by phytoplankton and a few tolerant macrophytes. Toxic pollutants, such as herbicides, pesticides, heavy metals and industrial waste products, have no redeeming features.

SEASONAL LAKES: THE IRISH TURLOUGHS

Over much of mid-western Ireland the bedrock is nearly level-bedded Carboniferous limestone, often covered by a layer of clayey glacial drift, but in which the drainage is mostly underground. The limestone is honeycombed with crevices and larger passages, through which water can move relatively freely. The scouring of the ice-sheets during the glaciations left an irregular surface in which the depressions had no relation to the surface drainage. As a consequence, the western part of the Irish midland plain is dotted with numerous seasonal lakes, *turloughs*, closed depressions that fill with water during winter or following heavy rain, but drain out more or less completely in dry summers (Figs 164, 165). Sheehy Skeffington *et al.* (2006) listed 304 turloughs in Ireland. They traditionally provide summer grazing for livestock, and they have long fascinated biologists, hydrologists and geomorphologists alike. Although some analogous sites can be found on limestones in other parts of the world (as at Pant-y-llyn in Carmarthenshire (Blackstock *et al.* 1993), in seasonal pools on the Swedish island

FIG 164. Newtown turlough, west of Gort, Co Galway, part of 194 ha liable to intermittent flooding; 180° panorama from the road, (a) 22 September 2008, after an exceptionally wet August, (b) 19 June 2010, after a long dry spring.

FIG 165. Turloughnagullaun, south of Bell Harbour, Co. Clare, a small turlough (20.4 ha): (a) 21 September 2008; (b) 20 June 2010. The lowermost hawthorns, flooded for long periods in 2008 and 2009, are dead or dying.

of Öland (Albertson 1960), and the *lacs à niveau variable* of the Île d'Anticosti in the Gulf of St Lawrence (Coté *et al.* 2006)), typical turloughs have a good claim to be seen as a distinctively Irish phenomenon. Wide expanses of flat-bedded limestone, recently glaciated, and in a rainy oceanic climate, are a rare coincidence worldwide.

Different hollows in the limestone have different patterns of seasonal variation in water level. Hollows floored by impermeable clayey glacial drift, in which water cannot fall below the level of the outflow stream, are occupied by permanent lakes, usually with more or less extensive *Schoenus* (black bog-rush) fens around them; smaller hollows have often become completely filled with *Schoenus* fen peat. These lakes and fens may flood to a depth of a metre or more in winter, but the water level during the growing season is generally stable to within half a metre or so. Hollows containing turloughs are in relatively free communication throughout their depth with the water-table in the limestone through springs, swallow-holes, or *estavelles* (openings into the underground drainage, which may function as either), so water level during the growing season may fluctuate widely (Proctor 2010). Between 2001 and 2004 water level in Skealoghan Turlough, near Balinrobe in Co. Mayo, varied over a range of 2.2 m (Moran *et al.* 2008); elsewhere the annual fluctuation may be 6–7 m or more. In July 1966, the water level in Hawkhill Lough west of Gort dropped by several metres in the course of 10 days – this is the turlough at which (in the dry summer of 1950) the sight of the aquatic moss *Fontinalis antipyretica* growing on top of a limestone field wall is said to have drawn from the eminent Swiss ecologist Josias Braun-Blanquet the astonished comment *unmöglich!* – impossible! In wet weather in summer in other turloughs I have seen autumn hawkbit (*Leontodon autumnalis*) and biting stonecrop (*Sedum acre*) still in full flower under water that had risen during the night.

The maximum winter water level in turloughs is roughly marked by the lower limit of hawthorn and hazel scrub and by the upper limit of the coarse blackish moss *Cinclidotus fontinaloides* (Fig. 166); flood debris can be left a metre higher than this. In the hollows occupied by lakes and *Schoenus* fens, the limestone grassland–fen transition is usually only a little below the lower limit of scrub. In turlough hollows there may be a zone of fen or fen-like grassland below the limit of the hazel and hawthorn bushes, especially if there is much lateral seepage of water from the surrounding country. However, in typical turloughs most of the summer grazing is provided by broad zones of pasture that recall grasslands in the flood zones of rivers (Chapter 9), or in seasonally flooded dune slacks (Chapter 19). Four species are particularly prominent and characteristic. Creeping buttercup (*Ranunculus repens*) and silverweed (*Potentilla anserina*) are almost ubiquitous in turloughs. At upper levels carnation sedge (*Carex panicea*) is abundant and often dominant in the turf along with common pasture and wet-meadow species. Common sedge (*Carex nigra*) typically forms a

FIG 166. The big desiccation-tolerant moss *Cinclidotus fontinaloides*, which grows on rocks in untidy-looking blackish masses near the flood limit in turloughs. Leaves of dewberry (*Rubus caesius*) to the right.

belt below this, still with abundant silverweed and creeping buttercup, but with a rather more 'weedy' associated flora, which commonly includes hairy sedge (*Carex hirta*) and creeping cinquefoil (*Potentilla reptans*). In shallower turloughs (as at Skealoghan) the bottom may be occupied by pools with relatively 'normal' emergent aquatic vegetation of reeds and sedges. At the lowest levels in the deeper turloughs the *Carex nigra* zone often grades down into open annual vegetation of permanently wet mud, with such species as needle spike-rush (*Eleocharis acicularis*), yellow-flowered *Rorippa* species, marsh foxtail-grass, curled dock, amphibious persicaria and other persicarias, red goosefoot (*Chenopodium rubrum*), marsh cudweed (*Gnaphalium uliginosum*), mudwort (*Limosella aquatica*), the tiny ephemeral moss *Physcomitrella patens* and the liverwort *Riccia cavernosa*.

This is a mere sketch of a complex and fascinating habitat. Turloughs have been called 'the callows of underground rivers', and they have all the complexities of river floodplains, with associated fens, marshes and aquatic habitats, plus a few of their own. The turloughs, especially those in the bare limestone country of the Burren, are home to some notable rare plants. Shrubby cinquefoil (*Potentilla fruticosa*) grows in the upper part of the fen zone around lakes and turloughs of the eastern Burren, and water germander (Fig. 167) just a little lower – both in much the same situations as in their more extensive occurrences on the bare limestone *alvar* of the Swedish island of Öland

LEFT: **FIG 167.** Water germander (*Teucrium scordium*), a rare plant of seasonally flooded calcareous places, in a turlough south of Mullagh More, Co. Clare, July 1971.

BELOW: **FIG 168.** Fen violet (*Viola persicifolia*), in a turlough near Killinaboy, Co. Clare, July 1971, with creeping bent, carnation sedge, marsh bedstraw and bird's-foot trefoil. Fen violet is characteristic of sites with large annual fluctuations in water level.

(Albertson 1960). Fen violet (Fig. 168) is another species of similar ecology and geographical distribution frequent in the *Carex panicea–Potentilla anserina* zone of the Burren turloughs, but otherwise very local in Ireland and with only a few localities in Britain.

CHAPTER 13

Peat, Peatlands and the Hydrosere

I N FORMER TIMES, FEW PARTS of northern Europe were without
waterlogged and often peat-covered tracts of country. These marshy or
boggy areas were seen by townspeople and country-dwellers alike as
desolate and even dangerous wastes to be shunned if they could not be drained
for profitable farming, although they often provided their local communities
with peat for fuel, reeds and sedge for thatching, rough hay for bedding
livestock, summer grazing, and a source of wildfowl. Nowadays we value these
areas for the richness and diversity of vegetation and wildlife they support, for
their contribution to the diversity of an increasingly urbanised and intensively
farmed landscape, and for their beneficial effects in helping to maintain the
water quality of rivers and alleviate flooding. As they accumulate, lake sediments
and peats incorporate pollen grains and larger plant remains, and sometimes
human artefacts (and even occasional humans!), as well as such other evidence
of events in their surroundings as ash from volcanic eruptions, charcoal from
burning of the surface vegetation, and pollutants such as soot, heavy metals, and
radionuclides from cold-war bomb testing and the Chernobyl disaster. Peatlands
thus provide a historical archive of extraordinary interest and value (Chapter 2),
along with a record of their own origin and development.

WETLANDS AND PEATLANDS:
A DIVERSITY OF WET HABITATS

'Wetland' is a convenient all-embracing label for wet places, familiar to everyone,
and much used by geographers, planners and conservationists. The Ramsar

conference of 1971 defined wetlands as 'areas of marsh, fen, peatland or water, whether natural or artificial, permanent or temporary, with water that is static or flowing, fresh, brackish or salt, including areas of marine water the depth of which at low tide does not exceed 6 m.' So wetland is a very broad concept; for some purposes it is *too* broad, bringing together such disparate habitats as the shallow sea (outside the scope of this book), the aquatic communities of lakes and rivers described in the previous chapter, and saltmarshes, which will be dealt with in Chapter 18.

This and the next two chapters are about the remaining wet places. These are very diverse, and the words describing them in ordinary speech have tended to be used differently, and sometimes interchangeably, in different parts of the country. The general practice among plant ecologists is to use the word *reedswamp* for reed-like vegetation in standing water, *bog* (or *moss*) for vegetation of acid peat, *fen* for vegetation on peats that are calcareous or at least influenced by water that has drained from mineral ground (bog and fen are *peatlands*), and *marsh* for vegetation of waterlogged mineral soils; the term *fen carr* (or simply *carr*) is used for wet fen scrub or woodland (Chapter 6). The term *mire* was introduced by Godwin as a general term to embrace both bogs and fens (to parallel the similar use by the Scandinavians of their word *myr*). Because the line between silty peats and highly organic wet mineral soils can only be arbitrary, and is often not reflected in the vegetation, it is convenient to extend the scope of 'mire' to include vegetation of wet sites on which there may be little or no peat developed, and that is what I shall do in this book.

PEAT AND THE ORIGIN OF PEATLANDS

Peat consists of the partly decomposed remains of the plants that once grew upon its surface. It develops in wet places where the annual input of dead organic matter from the vegetation exceeds annual breakdown, so that organic material accumulates of which the lower parts become waterlogged. The balance between the rate of input and the rate of breakdown of organic matter in the surface layers is crucial, and the rate at which peat accumulates is always much less than the rate of production of organic matter by the vegetation.

Peatlands can originate in three ways. First, and most obviously, peat-forming vegetation can develop in places where the water-table reaches the surface in low-lying hollows (*topogenous* mires), or emerges in springs or seepages on a hillside (*soligenous* mires), and the soil is permanently wet. Many of our smaller valley bogs and fens (and some larger peatlands) are of this kind. Second, peatlands have often developed through the infilling of pools, lakes and the freshwater upper reaches of estuaries by *hydrosere development*. This process took place extensively in the lake-dotted landscape left by the last glaciation. Third, in a wet enough climate, bog may grow up over previously freely drained ground by *paludification*, drainage first becoming impeded by development of an impermeable 'pan' in the soil; once significant peat growth has taken place the peat itself acts as an effectively impermeable layer. Thousands of hectares of blanket bog in the north and west of Britain and Ireland began growth in this way. All three of these processes are important. The second, hydrosere development, has some particular points of interest and we shall consider it in more detail, but that should not blind us to the importance of the other two.

EARLY HYDROSERE SUCCESSION: OPEN WATER TO FEN CARR

The hydrosere – development from open water to dry land with the progressive accumulation of sediment and peat – is a classic example of plant succession. Ideally, on a gently shelving lake shore, the successive zones of vegetation from open water to dry land can be expected to parallel the sequence preserved in the peat under the later stages in the succession. Peat borings in the larger lowland peatlands of Britain and Ireland record evidence of hydrosere development almost everywhere. It may thus seem disappointing that really good examples of hydrosere succession in action at the present day are hard to find. A major reason is that in terms of geological time lakes are short-lived, and sediment accumulation and hydrosere development are fast. Most of the hydrosere development that could readily take place in the lakes left by the retreating ice took place within the first few thousand years of the Post-glacial, and we are too late to see it. But plenty of illustrative examples of hydroseres do exist, some of long standing, some 'secondary successions' following disturbance by man.

The Esthwaite fens

A good example of hydrosere development is the fen at the north end of Esthwaite Water in the Lake District (Fig. 169), mapped by W. H. Pearsall in 1914–15 and again in 1929, as described by Tansley (1939), and re-mapped by Pigott & Wilson (1978) in 1967–69. Between the time of the Ordnance Survey of 1848 and 1968, different parts of the shore advanced between 28 and 47 m, an average of about 0.2–0.4 m a year. However, this advance has not been uniform, and Pearsall's observations seem to have spanned a period of particularly rapid

FIG 169. North Fen, Esthwaite, Cumbria, May 1981, a dozen years after Pigott & Wilson's survey. Common reed (*Phragmites*) is the main colonist of open water along most of the shoreline, supplanted within a few decades by sallow and alder carr. The large pine tree succumbed about 1990.

change. There is a striking difference between the vegetation sequence close to the mouth of the Black Beck (which brings in abundant inorganic silt) and that on the progressively less-silted shoreline farther east. At the time of Pearsall's observations, the fringing reedswamp near the mouth of the stream consisted of common reed (*Phragmites australis*) and bulrush (*Typha latifolia*), succeeded by a rich and diverse fen vegetation, in which prominent species included tufted-sedge (*Carex elata*), yellow iris (*Iris pseudacorus*) and, especially in the strip alongside the stream itself, reed canary-grass (*Phalaris arundinacea*), meadowsweet (*Filipendula ulmaria*) and purple small-reed (*Calamagrostis canescens*). The fen in turn was colonised by willows and sallows, especially purple willow (*Salix purpurea*), crack-willow (*S. fragilis*) and grey sallow (*S. cinerea*), recalling the silted alder–sallow woods of Chapter 6. Where silting was only moderate, the outer zone of the reedswamp was composed of *Schoenoplectus lacustris*, with *Phragmites* inshore. This gave way to a fen dominated by *Carex rostrata* or *C. elata*, colonised in due course by *Salix cinerea* and developing through an open carr stage with a great diversity of species to a closed *Salix* carr. On the eastern shore, farthest from the beck and with little silting, the reedswamp was again of *Schoenoplectus lacustris* and *Phragmites*, but subsequent development led quickly through a stage with *Carex rostrata* and marsh cinquefoil (*Potentilla palustris*) to a species-poor community dominated by purple moor-grass (*Molinia caerulea*), often with much bog-myrtle (*Myrica gale*).

Pearsall's maps demonstrate convincingly the advance of the successive zones of vegetation, and support the idea that the zonation in space of vegetation types from the open water up into the fen does indeed reflect succession in time at any particular point. The more recent observations confirm this picture, but also make it clear that succession has *not* consisted simply of the gradual advance of the reedswamp into the lake, followed by an orderly migration of successive zones all moving at the same rate. In fact, different species and different boundaries have often moved at different times and different rates, so that zones have contracted and expanded over the years, and different species have sometimes moved together, and sometimes not. Thus Pearsall's maps show the reedswamp vigorously colonising open water, and also extensive encroachment of fen onto former reedswamp, while *Salix cinerea* rapidly and almost alone colonised established fen. Since 1929 the reedswamp has continued to invade the open water of the northeastern corner of the lake, but there has been scarcely any movement of the reedswamp–fen boundary except close to the mouth of the Black Beck. *Salix cinerea* has continued to colonise the fen, and alder, present as only a few scattered trees in 1929, has spread throughout almost the whole of the area; the two species now form dense fen carr on ground that was fen dominated by *Carex elata* in 1929, and reedswamp dominated by *Phragmites* (but probably with some *Carex rostrata*) in 1914–15. Sedges are still abundant under the carr, now mainly lesser pond-sedge (*Carex acutiformis*) in the more silted areas, and *C. elata* and *C. rostrata* elsewhere. In fact, at the present time open fen has all but disappeared, and there is an almost direct transition from reedswamp to carr. This may reflect partly the increased supply of nitrate and phosphate to the lake and fen in recent decades, and partly the exclusion of grazing livestock.

The Norfolk Broads

The Norfolk Broads provide the most extensive examples of recent (secondary) hydrosere succession in Britain, and some of the best documented. Peat growth in the low-lying Broadland valleys (Fig. 170) began during the Atlantic period (Chapter 2), and peat borings show that the vegetation quickly developed to fen carr, which for many centuries accumulated brushwood peat, often to a depth of 3 m or more. In Romano-British times sea level rose sufficiently to bring about a return to open fen conditions in many places. Broad levées of silt were built up along the Broadland rivers. These no longer appear as levées because of subsequent peat growth, but they underlie strips of marsh, called ronds, which often separate the broads from their neighbouring rivers. The Norfolk Broads

FIG 170. Broadland: looking eastwards across the peat-filled valley of the River Ant, just downstream from Barton Broad. Part of Reedham Water in the foreground, Turf Fen drainage windmill on the riverbank, fen and a tiny broad beyond. (© Adrian Warren & Dae Sasitorn/ www.lastrefuge.co.uk)

as we know them originated as medieval peat (or 'turf') diggings cut down into the layers of peat that had filled the valleys. The earliest records of peat digging are from the mid twelfth century. For the next two centuries there is abundant documentary evidence of a flourishing peat industry in all the main areas where broads exist at the present day. In the latter half of the fourteenth century a change in the words used to describe areas that were formerly turbary suggests that the ground was becoming wetter; at the same time there is evidence of

decline in the sales of peat, and of increasing difficulty in its extraction. By the
fifteenth century the documents suggest that peat had largely given way to open
water and fen. In all, medieval peat digging had moved some 25 million cubic
metres of peat. Large as this figure is, peat winning on this scale was certainly
within the capacity of the medieval population of the area. A good day's digging
in nineteenth-century Broadland yielded 1000 turves, each of about a quarter
of a cubic foot (say 7 litres). If each of the 28 parishes that now share the Broads
within their parish boundaries had had 20 men working in the turf pits for three
weeks of the year, the volume of peat represented by the Broads could have been
removed in about three centuries (Lambert *et al.* 1960).

Because of their origin, the basins of the Norfolk Broads are shallow (seldom
more than 3 m and often less than 2 m deep), with relatively flat bottoms and
steep sides. Some silting and accumulation of nekron mud had to take place
before the water became shallow enough (c. 1 m) for reedswamp plants to become
established, but once that point was reached the development of reedswamp and
fen often took place very rapidly. This is strikingly shown by the decrease in area

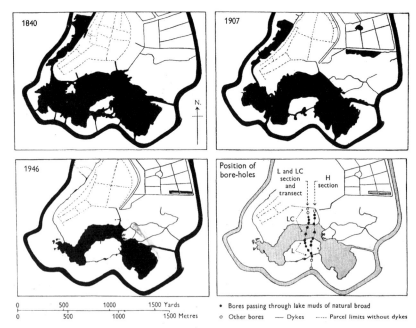

FIG 171. The outline of Hoveton Great Broad, Norfolk, from the tithe map of 1840, the
Ordnance Survey of 1907, and aerial photographs in 1946. From Lambert & Jennings (1951).

of many of the Broads since the tithe maps of 1838–41 (Fig. 171). Some former broads had already become overgrown by that time. Several, such as Strumpshaw and Hassingham Broads in the Yare valley and Woodbastwick Fen in the valley of the Bure have filled in (or practically so) since.

The first reedswamp plants to colonise the open water of the northern Norfolk Broads are the lesser bulrush (Fig. 172a), and in places the common club-rush (*Schoenoplectus lacustris*), giving way in shallower water or on wet peat to common reed (Fig. 172b). In the Yare valley these are replaced by the tall grasses *Glyceria maxima* and *Phalaris arundinacea*, corresponding with heavier silting and a richer supply of nutrients (Chapter 12). *Phragmites* maintains itself as a dominant only until the root-felt mat overlying the soft open-water mud has stabilised sufficiently to keep its surface just above water level. At that stage it is invaded by fen species, amongst which the greater tussock-sedge (Fig. 172b) is often very prominent. The tops of the *Carex* tussocks stand well above water level, and provide a dry foothold for tree and shrub seedlings. The tussock fen quickly grows up into a closed carr. With increasing shade, and increasing

FIG 172. (a) Fringing reedswamp of lesser bulrush (*Typha angustifolia*), Barton Broad, Norfolk, July 1997. (b) Common reed (*Phragmites australis*) and greater tussock-sedge (*Carex paniculata*) at margin of Upton Little Broad, Norfolk, September 1982; young sallows and downy birch rooted in the sedge tussocks. Lesser pond-sedge (*Carex acutiformis*) was prominent farther along the edge of the broad.

weight of the trees and bushes, the surface mat in which the tussocks are rooted begins to subside and break up; mature 'swamp carr' is full of rolling tussocks and a tangle of leaning trees rooted in semi-liquid peat. The surface is slowly stabilised by the gradual accumulation of brushwood peat. In sites farther from the rivers, and thus less exposed to fluctuations of water level, the *Phragmites* mat may be colonised by the lesser pond-sedge (*Carex acutiformis*). This does not form tussocks like *C. paniculata*, so shrubs cannot colonise the fen until a thicker mat of fen peat has accumulated. The unconsolidated muds under the ensuing 'semi-swamp carr' still give a quaking ground, but the surface is much more stable and most of the trees remain upright (Fig. 173). In areas still farther from the fluctuating water of the main rivers the *Phragmites* mat is often invaded by the great fen-sedge (*Cladium mariscus*, Fig. 188b), in East Anglia known simply as 'sedge'. *Cladium* has coarse strap-shaped evergreen leaves with sharp serrated edges, which may be 2 m or more long, and which live for 2–3 years. When they die they are slow to decay, remaining propped among the living leaves to form a continuous thick elastic mattress. Owing to this habit, *Cladium* excludes most other plants, though *Phragmites* maintains itself for a long time by means of its vigorous and extensive underground rhizomes. Colonisation by the fen carr

FIG 173. (a) Upton Little Broad, Norfolk, September 1982. Dense common reed and 'sedge' (*Cladium*); open water glimpsed through a gap in the sallow carr. (b) Dense fen carr of sallow (*Salix cinerea*) near Upton Great Broad, September 1982[W2]. Broad buckler-fern (*Dryopteris dilatata*) and brambles are prominent under the sallow canopy.

shrubs is long delayed, and when it ultimately takes place the carr is rooted in a firm and massive layer of *Cladium* peat.

In the Broads, as at Wicken Fen in Cambridgeshire, many of these pure dense stands of *Cladium* were regularly cut by the fenmen for thatching and kindling. If the cutting is not done oftener than once in four years the *Cladium* can maintain itself apparently indefinitely, accompanied by a scattering of plants surviving from the reedswamp, such as *Phragmites* and yellow loosestrife (*Lysimachia vulgaris*). If it is cut more frequently, say once in two years, the plant is enfeebled by the frequent loss of living leaves and *Molinia* enters the community and comes to share dominance with *Cladium*. Other plants come in at the same time, such as wild angelica (*Angelica sylvestris*), hemp agrimony (*Eupatorium cannabinum*), milk-parsley (*Peucedanum palustre*, Fig. 188a, the food-plant of the swallowtail butterfly) and marsh pennywort (*Hydrocotyle vulgaris*); a comparable mix of species can result from cutting *Phragmites* stands for thatching-reed. The accompanying plants are conspicuous after the cutting of the sedge or reed, but tend to be crowded out as the *Cladium* and *Molinia* or *Phragmites* resume dominance. If the vegetation is cut every year, *Cladium* and *Phragmites* succumb and *Molinia* remains dominant alone. Many species grow along with the *Molinia*, including meadowsweet, devil's-bit scabious (*Succisa pratensis*), marsh valerian (*Valeriana dioica*) and meadow thistle (*Cirsium dissectum*). The species composition and ecological relationships of these communities will be discussed in more detail in the next chapter.

THE LATER HYDROSERE: THE FATE OF FEN CARR

What becomes of fen carr? To answer this question we are thrown back onto the evidence of peat borings, and to inference from present vegetation. In some places the accumulation of wood peat may in due course raise the ground surface sufficiently above the water-table for sallow and alder carr to be replaced by damp oak or ash wood. This would certainly be the ultimate fate of many small pools or cut-off sections of old river channel. Godwin & Turner (1933) found evidence of development of this kind at Calthorpe Broad in Norfolk, and there are a few oak and ash trees in the older parts of the carr at Esthwaite. A classic site showing a complete zonation from *Typha* reedswamp and *Carex paniculata* fen, through sallow and alder carr, to birch and oakwood is Sweat Mere in Shropshire (Tansley 1939), but the birch and oakwood there is on drier ground that does not grade continuously with the wetter fen peats, and it is probably not part of the hydrosere (Sinker 1962). In fact, evidence of extensive hydrosere

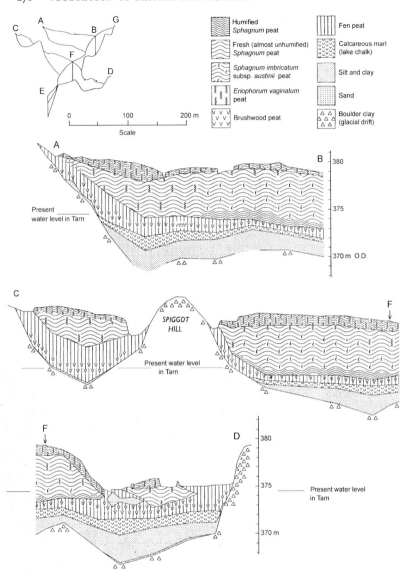

FIG 174. Sections of the raised bog and fens at Malham Tarn Moss, Yorkshire, showing succession from lake deposits, through fen and brushwood peat to ombrotrophic bog. Letters A–G correspond with the map in Fig. 200. Malham Tarn lies in a relatively infertile landscape; the main reedswamp species is *Carex rostrata*, so there is little evident differentiation of reedswamp and fen peat. Adapted from Pigott & Pigott (1959).

succession culminating in woodland is rare. In many sites the peat stratigraphy shows that brushwood peat laid down under fen carr was succeeded by the growth of thick layers of *Sphagnum* peat (Fig. 174), and observation of present-day vegetation gives some indication of how this has come about.

In the wet climate of Britain and Ireland, rainfall exceeds potential evaporation at least for most months in the year. Consequently, the continued growth of wood peat in an extensive fen basin, rather than raise the ground surface above the water-table, will tend to carry the water-table up with it. However, if peat growth makes the ground a little drier, it does have the effect of gradually raising the surface out of reach of flooding by mineral-rich drainage water from the surrounding country. The growing peat comes to depend wholly upon rainfall for its water supply. The result is that the surface layers become leached and acid. The sallows of the young carr are often replaced wholly or in part by downy birch (*Betula pubescens*), and shade- and mineral-tolerant *Sphagnum* species (such as *S. fimbriatum, S. palustre, S. fallax* and *S. squarrosum*) become established, forming a carpet over the ground in the shade of the trees and bushes (Fig. 175). Godwin & Turner described fen carr at this stage of acidification near Calthorpe Broad. The same process is taking place locally under *Betula pubescens* at Esthwaite, and similar examples of fen carr or birchwood on peat with a *Sphagnum* ground-layer can be seen in the Broads

FIG 175. *Sphagnum* carpet under downy birch and sallow, Malham Tarn, Yorkshire, July 1962[W4].

and many other places. The next stage in the succession follows opening-up of
the canopy. This may be due to the loose wet *Sphagnum* carpet waterlogging the
root systems of the woody plants, enfeebling or killing them directly, or making
them more liable to windthrow – or simply preventing regeneration as they
come to the end of their life span. Conditions are now favourable for the growth
of the common peat-forming *Sphagnum* species, such as *S. papillosum* and *S.
magellanicum*. Once a continuous bog surface has developed, dominated by these
mosses, it can continue to develop peat virtually indefinitely.

A bog of this kind, raised above the mineral-rich drainage of the surrounding
country, is called a *raised bog*. The bog is dependent for its water supply on the
rain falling directly on its surface (it is *ombrogenous* – 'rain-formed'), and the
mineral nutrient regime of the vegetation reflects the balance between additions
of solutes from rain, blown dust and other sources, and losses in drainage water
(it is *ombrotrophic* – 'rain-fed'). The same is true of the *blanket bog* that covers
wide expanses of country in the rainy west of Britain and Ireland and on the
high ground of the Pennines and elsewhere. Ombrotrophic bogs have been the
commonest end-point of large-scale hydroseres in Britain and Ireland since
the last glaciation. Ombrotrophic peats covered large tracts of ground in such
areas as the Somerset Levels, the Lancashire–Cheshire plain, the country south
of the Humber, the Solway district and (especially) the midland plain of Ireland.
However, peat cutting and drainage have changed many former raised bogs
beyond recognition, and in Britain good intact raised bogs are now rare. Some
examples are described in Chapter 15.

Directions of vegetational change, and rates of peat accumulation

The examples of early hydrosere development described above are instances
in which the course of changes in the vegetation can be demonstrated in some
detail from clear direct evidence. The probable course of succession at other sites
can often be inferred from their present vegetation. Peat borings yield a wealth
of evidence of the general course of succession from different places. With the
timescale provided by pollen analysis or (better) C-14 dating, they also give the
means to estimate rates of accumulation of peats and other sediments, and the
duration of the various phases in a hydrosere.

The sequences of hydrosere stages in the peat stratigraphy from 20 pollen
diagrams were analysed by Walker (1970). These sites would generally have
been chosen to give long undisturbed profiles near the centre of a mire basin.
The commonest sequence recorded was: open water with floating aquatics →
reedswamp → fen → fen carr → bog – the succession outlined in the preceding
pages (Fig. 176a). Walker also looked at the predominant sequences in a wider

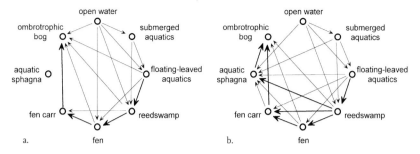

FIG 176. The frequency of transitions between vegetation stages in the hydrosere from the evidence of peat borings: (a) from the profiles used for 20 pollen diagrams (heavy lines show transitions observed six or more times); (b) from the predominant course of succession at a larger number of sites (heavy lines show transitions observed nine or more times). The arrows show only 'progressive' transitions, and the rarest transitions are omitted. Based on the data of Walker (1970).

range of sites, and in these another series of transitions emerged as important, going from open water or reedswamp to bog either direct, or through a floating *Sphagnum* stage (Fig 176b). But, as these diagrams show, a surprisingly wide range of possible transitions are recorded in the peat. Several of the common intermediate stages of the hydrosere can be passed through very quickly, or even apparently omitted altogether (but there is usually at least a brief reedswamp stage), and most of the common hydrosere transitions can be reversed. Ombrotrophic bog stands out as the one vegetation type that rarely turns into anything else; once *Sphagnum* becomes established, at whatever stage of the hydrosere, development to bog as the next stage becomes almost inevitable.

The time taken for succession from open water to fen or bog can vary widely. Peat profiles show that, once a hydrosere reaches the floating-leaved macrophyte stage, conversion to fen has frequently taken less than 1000 years – sometimes very much less (as in the Norfolk Broads), and rarely more than twice that time. Direct succession from open water through reedswamp (and sometimes swamp carr) to bog can also be rapid, rarely taking more than 2500 years. If there is a persistent fen stage the whole process takes longer; exceptionally it can be complete in less than 1000 years, but 2500–4000 years is more usual. Succession in small ponds or pools can of course be much quicker than this.

There are great variations in the rate of accumulation of almost all kinds of peats and freshwater deposits but, perhaps surprisingly, there seems to be little consistent difference in the rate of growth between one kind of deposit and another. A common rate of accumulation for fen and bog peats is 5–10 cm a century (corresponding to some 10% of the production of organic matter by the plants), but the rate of accumulation may be substantially more, or very much less than this. The actual rate of accumulation at any particular place and time represents the difference between additions to the deposit and losses due to erosion and decomposition. If conditions favour oxidation of organic matter, at least seasonally, the peat may grow slowly or not at all, and a modest amount of erosion can have the same result. The effect of shrinkage and oxidation in bringing about the wastage of peat is graphically demonstrated by the 'Holme post', close to the former Whittlesey Mere 10 km south of Peterborough. This is a cast-iron post, bolted to oak piles driven into the underlying clay so that its top was level with the peat surface at the time the surrounding fens were being drained in 1848. By 1860 the level of the peat had fallen by over 1.5 m; by the early twentieth century the post stood about 3 m above the ground, and the level of the surface changed little for several decades. With deeper drainage, the level has since fallen by a further 80–90 cm (Hutchinson 1980).

Interrelations of hydrosere succession with topography and climate

The successions that have been discussed in this chapter are typical of the hydrosere recorded in extensive peat deposits that have grown up over former lake basins. However, even a cursory inspection of a few lakes will show that hydrosere development does not take place everywhere. Lee shores of lakes are often wind-eroded, and although they may show a clear zonation of vegetation there is often no peat accumulation. Deep, steep-sided lakes often show little sign of succession. Initiation of a hydrosere typically depends in the first place on the accumulation of open-water silt or mud until the water is shallow enough for floating-leaved and emergent aquatic plants to become established. It therefore often takes place in sheltered bays or where an inflow stream brings in a supply of inorganic silt. The character, rate and extent of hydrosere development are also influenced by the nutrient supply provided by the surroundings, as we have seen at Esthwaite. A fertile lowland agricultural landscape will provide the nutrients for vigorous growth of reedswamp, fen and carr, as in the Norfolk Broads. In blanket-bog country on hard acid rocks, if hydrosere development takes place at all it is likely to be from a low open reedswamp of *Carex rostrata* to acid fen (Chapter 14) and bog. Thus there are several reasons why the vegetation of a lake may show little change, and at some sites old photographs attest remarkable stability over long spans of time.

Although different kinds of lake deposits and peats *can* accumulate at similar rates, deposits in shallow water (calcareous marl, detritus muds, reedswamp peat) often accumulate faster than those in deeper water, with the effect of steepening

FIG 177. Rinnamona Lough, near Killinaboy, Co. Clare, July 1971. The marginal vegetation is advancing over open water, leaving a deep central pool. The depth of the basin exceeds 13 m.

the underwater profile of the lake margin. There is also a common tendency for swamp vegetation to grow out as a floating raft over unconsolidated sediments or even open water, especially in small sheltered lake basins (Sinker 1962). This can be seen in places as far apart as the Norfolk Broads and the Burren. Around the margins of a small but deep lake basin close under Mullagh More, peat boring showed normal succession from calcareous marl (precipitated by the submerged green alga *Chara* in shallow water), through reedswamp, to fen dominated by black bog-rush (*Schoenus nigricans*). In the deeper centre there is little marl, and *Cladium* reedswamp is growing out as a rhizome mat over open water or very loose detritus mud. Figure 177 shows a similar site a few kilometres away, with a similar history. This pattern of development seems to have taken place in a number of lakes in the flat limestone country of the southeast of the Burren, but peat cutting and erosion have usually stripped off most of the fen peat, leaving almost level marl flanges bearing relic patches of peat, surrounding a deep centre, as at Loch Gealáin, a kilometre southwest of Mullagh More, and at Skaghard, Cooloorta (Fig. 178) and Travaun Loughs a few kilometres to the northeast. In sites poor in calcium and nutrients, floating

FIG 178. Cooloorta Lough, in the eastern Burren, July 1970. A substantial lake, with a deep centre surrounded by extensive marl flats. These were probably exposed by stripping of a former peat cover by cutting for fuel, and erosion, but are now largely vegetated with a thin community of shoreweed (*Littorella uniflora*), lesser water-plantain (*Baldellia ranunculoides*) and the moss *Scorpidium scorpioides*.

rafts of *Sphagnum*, bound together by the rhizomes of such plants as common cottongrass (*Eriophorum angustifolium*) and bogbean (*Menyanthes trifoliata*) can grow out over pools and lakelets to form a quaking mattress of the kind known by the nicely descriptive German word *Schwingmoor*; in Ireland such floating mats of vegetation are called *scragh* or *scraw*. Small-scale examples can sometimes be seen in old peat cuttings. It has been suggested that Wybunbury Moss near Crewe developed as a schwingmoor over a small deep kettlehole lake in the glacial drift which covers the Cheshire plain, but it is likely that the hollow in which it lies is at least in part due to subsidence following solution of salt deposits in the underlying Triassic rocks (Poore & Walker 1959; Green & Pearson 1977), as at Chartley Moss in Staffordshire.

In many large peat deposits there is a correlation between the stages of the hydrosere and the course of climatic change since the last glaciation. Fen and fen carr became established widely as the climate became warmer at the end of the Late-glacial period, about 11,500 years ago (Chapter 2), and were extensive and persistent through much of the Boreal period when the climate was more continental than now, and there may have been more grazing pressure from native herbivores. At many sites, the transition from fen carr to ombrogenous bog coincided roughly with the climatic change at the Boreal–Atlantic transition, about 8000 years ago, when the climate became much more oceanic in character. There is little *Sphagnum* peat anywhere before that time. In more recent periods, fen has generally given way rather quickly to ombrogenous peat except where rising sea level kept pace with the accumulating peat, as in the Broadland valleys and much of the Cambridgeshire Fenland. In the Somerset Levels, Shapwick Heath and the neighbouring moors were still part of an open estuary accumulating grey clay at the Boreal–Atlantic transition, and the whole succession from reedswamp to ombrotrophic bog took place in a relatively uniform climate (Godwin 1975).

It is obvious that the kind of succession outlined earlier in this chapter cannot go to completion over the whole of a mire system. Growth of peat in one place will affect water movement and water levels in other parts, and however extensively ombrotrophic bog develops, the mineral-rich drainage from the surroundings must go somewhere. Some of the consequences can be seen in Figure 174. As raised bog developed in the centre of the basin, the level of the marginal fens gradually rose, fen extending peripherally out over mineral ground while the bog encroached on its inner edge. In this 'lagg' fen (Chapter 14), separating the raised bog from the mineral soil, succession went sometimes from fen to carr, sometimes back again. Clearly, the more closely bog growth confines mineral-rich drainage water in this marginal belt of fen, the more will conditions inhibit further development to acid bog. The result is that the whole complex will tend to reach a steady-state pattern of ombrogenous bog and fen.

Two points may be emphasised in conclusion. First, although in a simple hydrosere there is a broad correspondence between zonation from open water to fen and the succession recorded in the fen peat, many of the changes taking place in large mire systems are not closely mirrored in zonation. In particular, the succession in the central parts of a mire complex is often different from that close to either the landward or the lakeward margin. Riverbanks and exposed lake shores often show striking zonations that are essentially stable; there is no peat accumulation and no succession – or succession, if it occurs, may go either way. Second, hydrosere succession may be seen as a process of readjustment of the vegetation to natural or artificial disturbance of the habitat. Many examples are the result of human activity – the creation of a field pond, a change in lake level, or the digging of the Norfolk Broads. The massive hydrosere development recorded in Post-glacial peats followed as a natural consequence of the massive topographic and climatic upheavals during and since the last glaciation.

CHAPTER 14

The Diversity of Peatlands: Fens and Valley Bogs

THE SORT OF CHANGE we considered in the last chapter is often too slow to be immediately perceptible. It generally takes place on a timescale of at least decades, and often centuries or even millennia, and in some places long spans of time may pass with little change at all. What is more obvious is the pattern of different kinds of plant communities, and the contrasts between the peatland vegetation in different places. Sometimes patterns in space reflect sequences in time, but often they do not.

CHEMICAL FACTORS IN PEATLAND VEGETATION: CATIONS AND pH

On a broad scale, there is a major ecological distinction between *ombrogenous* or *ombrotrophic* bogs and the rest (Fig. 179). The terms ombrogenous ('formed by rain') and ombrotrophic ('rain-nourished') are in many contexts interchangeable; these bogs depend for their water and solute supply on rain and other airborne sources alone, and are correspondingly always nutrient-poor and acid, and typically form deep peat. Ombrogenous bogs cover wide tracts of country in Britain and Ireland, and are the subject of Chapter 15.

All peatlands other than ombrogenous bogs receive water both from rain and from groundwater bearing nutrients and other solutes from neighbouring mineral ground – they are at least in some degree *minerotrophic*. What may loosely be called 'fens' in the British and Irish tradition are more or less calcareous, with a near-neutral pH (usually between 6 and 7.5). They are generally dominated

FIG 179. Malham Tarn Moss and fens, September 1959. This shows the raised bog to the left clearly above the level of the inflow stream, and the almost level surface of the fens to the right. Continued growth of fen-carr shrubs since has largely obscured this view.

by sedges (such as great fen-sedge, *Cladium mariscus*, black bog-rush, *Schoenus nigricans*, and many sedges of the genus *Carex*), reeds (such as the common reed, *Phragmites australis*, and purple small-reed, *Calamagrostis canescens*) and broad-leaved herbs. True mosses can be prominent but *Sphagnum* species are much less so and often absent altogether. By contrast, bogs (including ombrotrophic bogs) are poor in calcium and mineral nutrients, and acid, with pH often between 3.5 and 4.5. They are typically dominated by *Sphagnum* ('bog-mosses'), sedges ('cottongrasses') of the genus *Eriophorum* and dwarf shrubs of the heather family (Ericaceae) including common heather (*Calluna vulgaris*), cross-leaved heath (*Erica tetralix*) and cranberry (*Vaccinium oxycoccos*).

This bog–fen distinction reflects the water supply to the site. Measurements of pH from a large sample of peatland sites tend to show a bimodal (two-peaked) distribution, with one peak centred at about pH 4 and the other at about pH 6.5. This is because the pH of bogs is buffered by the organic acid groups on the peat colloids, whereas the pH of fens (with the ion-exchange sites on the peat saturated by calcium and other cations) is buffered by the CO_2–bicarbonate system in the water (Chapter 12). Intermediate pH values are possible (and are often associated with interesting vegetation), but they are vulnerable and easily tipped one way or the other by calcareous dust or (particularly) by 'acid rain'.

To many English people the word fen calls up a picture of the Fenland of East Anglia. But these tall reedy fens are only one corner of the range of variation

The bog–fen division

Scandinavian ecologists generally draw a primary division within mire vegetation
between ombrotrophic bogs (to which they restrict the term *bog* or *moss*) and all
more or less mineral-influenced mires, which they call *fens*. These they divide into
poor-fens (acid and species-poor) and *rich-fens* (more or less calcareous, and richer
in species). In this they follow Du Rietz (1948), who introduced the concept of the
'mineral soil-water limit', in practice recognised in the field by the limit of species
seen as indicators of mineral-enriched water. The limit of ombrotrophic bog growth
is a very important boundary ecologically, but the mineral-soil water limit can
inherently never be completely sharp, and it is reflected by different indicator species
and different water chemistry in different places. It coincides neither with a natural
floristic division nor with a natural division in terms of water chemistry – seen in the
bimodality of pH both in North America and in Britain and Ireland. There are many
'valley bogs' in southern England (and France), fed by water from decalcified rocks
and soils, which have more in common in physiognomy and species composition
with ombrotrophic bogs than with the great majority of 'fens'. For the purposes of this
book, I use the word *bog* (or wet heath) for all plant communities that Continental-
European phytosociology would place in Oxycocco-Sphagnetea, and when I am
referring specifically to ombrotrophic bogs I say so explicitly. I have used *poor-fen*
for other minerotrophic acid vegetation (largely *Caricetalia fuscae*), and *rich-fen* in the
Scandinavian sense.

in 'fen' vegetation, and a Scandinavian (or indeed an Irish) botanist would take a
very different view of what constitutes a typical fen. Much of the diversity of fens
can be thought of in terms of two major directions of variation, determined by
peat and water chemistry and the availability of plant nutrients.

Figure 180 shows pH plotted against log calcium concentration (log [Ca])
from a number of peatland sites (as pH is a logarithmic measure of acidity it is
appropriate to plot it against a logarithm). The ombrotrophic bog samples, with
low [Ca] and pH, form a cluster at the bottom end of the graph, overlapping with
the minerotrophic bog/poor-fen samples; these are concentrated in the pH range
3.5–4.5. The middle part of the graph is more thinly populated with points, but as
pH rises above 6 we are into another dense cluster of points at pH 6.5–7.5. The
pH is seldom above 8.0 in fen waters, and the highest pH values do not coincide
with the highest calcium concentrations, for reasons explained in Chapter 12.
However, while the major inorganic ions account for some of the diversity in
fens (Figs 180, 181; Table 8), they are far from explaining it all.

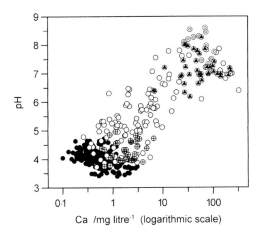

FIG 180. pH plotted against log [Ca] for 384 bog and fen water samples, 1991–7. Solid black points, ombrotrophic bogs; black triangles, typical small-sedge rich-fens; crosses are southern English valley bogs; open circles are other poor- and rich-fen sites; those enclosing a small 'o' are open-water samples.

pH

Ca /mg litre^{-1} (logarithmic scale)

TABLE 8. Some representative averages from chemical analyses of fen waters in Britain and Ireland. Figures in roman type are mg/litre; figures in *italics* are μequiv/litre (μmol ionic charge/litre). Numbers of samples in square brackets [ ].

Nature and location of site	*pH/H⁺*	*Ca²⁺*	*Mg²⁺*	*K⁺*	*Na⁺*	*Cl⁻*	*SO₄²⁻*	*HCO₃⁻*
Black bog-rush (Schoenus nigricans) fens (M13)								
Weston fen, Norfolk;	6.9	141	5.9	1.9	24	58	43	368
Schoenus fen (M13) [2]	*0.1*	*7073*	*490*	*45*	*1020*	*1631*	*970*	*6027*
Cors Erddreiniog, Anglesey;	8.1	103	3.5	0.4	11.2	28	10	370
Schoenus fen [4]	*0.0*	*5150*	*295*	*1.2*	*486*	*790*	*230*	*6063*
Burren, Co. Clare;	8.0	43	2.2	1.1	12	19	3.5	149
Schoenus fens (cf. M13) [3]	*0.0*	*2173*	*185*	*29*	*528*	*535*	*80*	*2440*
Small-sedge fens (Pinguiculo-Caricetum) (M10)								
Malham Tarn district,	7.8	72	1.7	0.3	3.8	6.3	12	146
Yorkshire [8]	*0.0*	*3588*	*138*	*8.2*	*164*	*175*	*277*	*2388*
Widdybank Fell, Teesdale,	7.5	35	0.5	0.2	3.3	5.2	7.3	106
Durham [4]	*0.0*	*1759*	*46*	*6.1*	*142*	*148*	*165*	*1745*
Yellow saxifrage (Saxifraga aizoides) mountain flushes (M11)								
Sites on Dalradian rocks,	7.5	34	2.5	1.9	8.1	11	5.3	115
Perthshire [12]	*0.0*	*1701*	*204*	*48*	*354*	*306*	*121*	*1889*
Topogenous rich-fens (Carex rostrata, C. diandra etc.) (M9 [S27])								
Southern Uplands,	6.6	38	6.7	1.0	5.5	8.4	5.0	114
northern England [5]	*0.3*	*1886*	*558*	*25*	*237*	*237*	*115*	*2354*
East Anglian-type tall-herb fens (Phragmites, Cladium etc.) (S24)								
Catfield Great Fen,	7.9	74	14	3.9	59	106	70	162
Norfolk [2]	*0.0*	*3708*	*1133*	*100*	*2503*	*2990*	*1780*	*2650*

TABLE 8. (*continued*).

Nature and location of site	pH/H⁺	Ca²⁺	Mg²⁺	K⁺	Na⁺	Cl⁻	SO₄²⁻	HCO₃⁻
Poor-fens (Caricion curto-nigrae) (M6 etc.)								
Bog below Haytor,	5.1	1.4	0.9	1.0	8.9	18	1.6	[3.4*]
Dartmoor, Devon [4]	8.4	71	76	25	386	514	37	[55*]
Southern English valley bogs (M21)								
Thursley Common,	4.0	1.4	0.9	0.4	5.3	8.7	12	0.0
Surrey [6]	104	72	75	10	229	243	276	0.0
'Fort Bog', Matley, New	4.7	0.62	1.6	0.51	12.5	24.6	1.35	0.0
Forest, Hampshire [5]	22	31	132	13	544	693	31	0.0

* including organic (humic) anions

Much of the diversity in character of fens is related to the availability of the usual trio of major plant nutrients, nitrogen (N), phosphorus (P) and potassium (K). At one extreme are fens, which may be highly calcareous, but are poor in other nutrients, especially phosphate. Their vegetation is generally short and open, and mosses are often prominent. With more N and P, the vegetation is taller but may still be rich in species. In heavily silted fen sites growth of a few bulky dominants may be so vigorous that it excludes all but a few other species. Chemical analysis of fen water samples rarely gives any indication of the availability of N and P to plants, because as soon as these limiting nutrients are released by decay of organic matter, they are immediately taken up again by plant roots or by microorganisms. However, peat contains abundant nitrogen combined in organic matter, and chemical analysis of peat can give an idea of the potential availability of N and P. What is important is the rate of 'mineralisation' of these limiting nutrients – their release in simple inorganic form, which plants can take up.

MAJOR TYPES OF FEN VEGETATION IN BRITAIN AND IRELAND: SPECIES COMPOSITION

Small-sedge rich-fens

Fens dominated by *Schoenus nigricans* are still widespread and common on the Carboniferous limestone of Ireland, and were formerly much more widespread than they are now on similar limestones in Anglesey, and on chalk and calcareous drift in East Anglia[M13] (Fig. 182). Typically, *Carex* species are

'Equivalents' and Maucha diagrams

Concentrations in Table 8 are given in mg/litre, and in μ equivalents/litre. The first is familiar (and equal to 'parts per million'), but different ions have different atomic (or molecular) weights and different valencies. Expressing all ions in μ equivalents/litre gets all ions onto a common basis relative to the hydrogen ion, which carries a single positive charge and which at pH 6.0 has a concentration of 1.0 μg/litre. In effect 'equivalents' express ionic concentration in μmol/litre ionic charge.

Maucha diagrams display the relative proportions of the eight most abundant ions in natural waters plotted in octants, on an equivalent basis; the intervening radii show the average value of all eight. The individual ions make four-sided polygons, which may vary from narrow fishtails to elongated diamonds, but whose area is proportional to concentration. If all eight ions were present at the same (equivalent) concentration, the whole plot would be a regular 16-sided polygon. The diagrams in Figures 181 and 193, from an ombrotrophic bog and a selection of British and Irish fens, are scaled so that their *areas* are proportional to their total ionic concentrations.

In Britain and Ireland, Maucha diagrams of bogs and fens are typically very different. Bog waters are generally much more dilute, and are (like rainwater) dominated by sodium and chloride, often with the hydrogen ion or sulphate in third place; the potassium concentration is typically very low, and bicarbonate is negligible. Fen waters are commonly dominated by calcium and bicarbonate, and if the groundwater drains from limestone these two ions may be present at near-saturation, and greatly outweigh all others. Sodium and chloride are everywhere present in significant quantities, and on other rocks magnesium, potassium and sulphate may appear as substantial components.

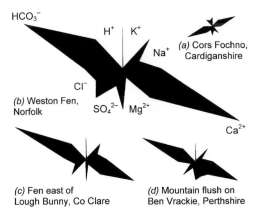

HCO_3^- H^+ K^+ Na^+ (a) Cors Fochno, Cardiganshire

Cl^- (b) Weston Fen, Norfolk SO_4^{2-} Mg^{2+}

Ca^{2+}

(c) Fen east of Lough Bunny, Co Clare

(d) Mountain flush on Ben Vrackie, Perthshire

FIG 181. Maucha diagrams comparing the water chemistry of (a) an ombrogenous bog, Cors Fochno in west Wales, and three typical rich fens: (b) Weston Fen, over the chalk in East Anglia, (c) a small fen on Carboniferous limestone east of Lough Bunny, Co. Clare, (d) a mountain flush on lime-rich schist, Ben Vrackie, Perthshire.

FIG 182. (a) Cors Erddreiniog, Anglesey, June 1990. A *Schoenus*-dominated rich-fen on Carboniferous limestone, with a rich calcareous fen flora. (b) Old peat cutting in *Schoenus* fen near Atyslany Lough, Co. Clare, September 1975. One of many such fens in the eastern Burren. Inflorescences of *Schoenus* and *Molinia* can be seen overhanging the water in the foreground.

prominent, of which *C. panicea*, *C. hostiana*, *C. viridula* and sometimes *C. nigra* are the commonest and most conspicuous, along with a very characteristic range of associated species. Of these, common butterwort (Fig. 183a), grass-of-Parnassus (Fig. 183b), dioecious sedge (Fig. 184), few-flowered spike-rush (*Eleocharis quinqueflora*), lesser clubmoss (*Selaginella selaginoides*) and the mosses *Campylium stellatum*, *Scorpidium scorpioides*, *S. revolvens* and its close relative *S. cossonii* and the thalloid liverwort *Aneura pinguis* (Fig. 186e) are particularly characteristic of this kind of vegetation. Many other species occur including purple moor-grass (*Molinia caerulea*), quaking-grass (*Briza media*), marsh valerian (*Valeriana dioica*) and bog pimpernel (*Anagallis tenella*), and in the north of England bird's-eye primrose (*Primula farinosa*). Fens of this kind provide a niche for a number of orchids, including early marsh-orchid (*Dactylorhiza incarnata*), marsh helleborine (*Epipactis palustris*, Fig. 300f) and fly orchid (*Ophrys insectifera*).

Schoenus nigricans is a characteristic dominant of calcareous fens on the Continent. That pattern is repeated with us, but here *Schoenus* is mainly a lowland plant and at altitudes much above 300 m it is generally missing from

FIG 183. Plants of small-sedge fens: (a) common butterwort (*Pinguicula vulgaris*); (b) grass-of-Parnassus (*Parnassia palustris*), Rinroe, near Corofin, Co. Clare, August 1958.

rich-fen vegetation. Dominance then falls to the *Carex* species, and small-sedge fens of this kind are widespread and locally frequent, usually in small stands, on the Carboniferous limestone of the Pennines. These upland rich-fens are usually grazed by sheep or cattle, and often have a hummocky surface with grassland species on the tops of the hummocks[M10]. Similar communities occur very widely, usually in small patches, where lime-rich water seeps out to the surface and conditions are otherwise suitable. From Teesdale and the Lake District northwards, they pass into mountain flushes and fens with which they share many species, with the conspicuous addition of yellow saxifrage (*Saxifraga aizoides*)[M11]. These are dealt with in more detail in Chapter 17.

The communities just outlined are typically *soligenous* ('soil-formed'); they are fed by springs, or more extensively where water seeps out at a spring-line on a slope. Where water accumulates in topographic hollows, or around shelving lake shores, calcareous fens occur at levels where the water-table is a few centimetres below the surface for much of the growing season. Such fens are *topogenous* ('site-formed'). They share some species (and intergrade) with the soligenous rich-fens but they are different in character, and many of the most characteristic species of the typical small-sedge fens are rare or absent. They also intergrade

FIG 184. Dioecious sedge (*Carex dioica*), a very characteristic plant of the small-sedge fens. (a) Male plant, and (b) female plant in flower; both Gordale, Malham, Yorkshire, May 1984. (c) Female plant in fruit, Widdybank Fell, Teesdale, July 1978.

FIG 185. Lesser tussock-sedge (*Carex diandra*), Malham Tarn, Yorkshire, in wet topogenous fen with marsh cinquefoil (*Potentilla palustris*), bottle sedge (*Carex rostrata*) and bogbean (*Menyanthes trifoliata*)[M9].

FIG 186. Rich-fen bryophytes: (a) *Scorpidium scorpioides*.; (b) *S. revolvens* (dark reddish brown, top) and *Campylium stellatum* (golden green, bottom); (c) *Calliergonella cuspidata*; (d) *Calliergon giganteum*; (e) the thalloid liverwort *Aneura pinguis*, with antheridial shoots; (f) *Tomenthypnum nitens*.

with emergent-aquatic communities of neighbouring waters; sometimes (apart from water level) a 'swamp' and a 'fen' differ only in the presence of a continuous bryophyte layer and the denser and more species-rich vegetation of the latter. A common kind of wet fen over much of Britain and Ireland is dominated by various combinations of bottle sedge (*Carex rostrata*), lesser tussock-sedge (Fig. 185), carnation sedge (*C. panicea*), common sedge (*C. nigra*) and slender sedge (*C. lasiocarpa*), common cottongrass (*Eriophorum angustifolium*), marsh cinquefoil, bogbean, the horsetails *Equisetum palustre* and *E. fluviatile*, water mint (*Mentha aquatica*), marsh willowherb (*Epilobium palustre*), devil's-bit scabious (*Succisa pratensis*), marsh-marigold (*Caltha palustris*), with an understorey of the mosses *Calliergon giganteum, Calliergonella cuspidata, Scorpidium scorpioides* and *S. cossonii*[M9] (Fig. 186). There are some very localised examples of essentially this community in otherwise *Phragmites*-dominated areas of the Norfolk Broads. A distinctive variant of this general kind of community, with *Sphagnum teres* and *S. warnstorfii* and often *Tomenthypnum nitens* (Fig. 186f)[M8], occurs in the central Highlands of Scotland with fragmentary outliers in the Scottish borders, northern England and northern Mayo.

Several of the species in these wet fens tolerate an extraordinarily wide range of calcium concentration and pH, notably *Carex rostrata, C. lasiocarpa* and *Menyanthes*. We are in fact at the species-rich (and calcareous) end of a range of wet fens and swamps within which there are few clear divisions, and which are largely embraced within the alliances Caricion lasiocarpae and Caricion nigrae of Continental vegetation ecology. We shall encounter them again later.

Tall-sedge and tall-herb 'fens'

These are the traditional 'fens' of East Anglia, and comparable (but usually more species-poor) vegetation elsewhere in England. They generally grow on peat with near-neutral pH, and in which nutrients are reasonably plentiful. A purist would say they are not fens at all, but reedswamps perpetuated by centuries of reed cutting (Fig. 187). Around the Norfolk Broads, and locally elsewhere in East Anglia, the fen was the essential basis for a distinctive pattern of land use, providing peat for fuel, reed (*Phragmites*) and 'sedge' (*Cladium*) for thatching, 'marsh hay' (largely *Calamagrostis canescens* and *Juncus subnodulosus* with accompanying herbs) used as livestock feed, and litter (largely *Molinia*) for bedding livestock. This traditional pattern of usage created the Broadland fens as we know them, providing a mosaic of diverse and constantly renewed habitats underlying the biodiversity we have come to value. That biodiversity, long taken for granted by visiting naturalists, was an unintended by-product of years of labour dictated by the economics of past centuries (George 1992).

FIG 187. Tall-herb fen[S24], Wicken Fen, Cambridgeshire, July 1981. Common reed dominant, with common meadow-rue (*Thalictrum flavum*), blunt-flowered rush (*Juncus subnodulosus*), yellow loosestrife (*Lysimachia vulgaris*), marsh bedstraw (*Galium palustre*) etc.

Left to itself, the fen would in due course become colonised by sallows and alder, and develop to fen carr or wet woodland. The maintenance of fen depends on cutting to prevent the growth of woody saplings. *Cladium* beds for thatching roof ridges were cut on a 3–4-year rotation. Managed reed-beds can be cut every year (in late autumn and winter), but this results in a significant drain of nutrients, and a better crop is obtained in the long run by cutting every other year. Marsh hay was cut annually from early summer to early autumn; litter was generally cut at the end of the growing season. Occasionally these areas were left uncut for a year or two.

Typical Broadland fen – the 'Peucedano-Phragmitetum' – is dominated by varying proportions of *Phragmites*, *Cladium* and *Calamagrostis canescens*. Accompanying these dominants is a very wide range of herbaceous plants, of which the commonest include marsh bedstraw, yellow loosestrife, milk-parsley (Fig. 188a, a rare species in England), hemp agrimony (*Eupatorium cannabinum*), purple-loosestrife (*Lythrum salicaria*), blunt-flowered rush, meadowsweet (*Filipendula ulmaria*), water mint, common valerian (*Valeriana officinalis*), yellow iris (*Iris pseudacorus*), gypsywort (*Lycopus europaeus*), marsh fern (*Thelypteris palustris*), wild angelica (*Angelica sylvestris*), marsh thistle (*Cirsium palustre*) and water dock (*Rumex hydrolapathum*). A long list of mainly reedswamp plants occurs less

FIG 188. (a) Milk-parsley (*Peucedanum palustre*), Catfield Fen, Norfolk, July 1997. (b) Great fen-sedge (*Cladium mariscus*), Sawston Meadows, Cambridgeshire, August 1984.

frequently[S24]. This species-rich community has its headquarters in the Broads; it also occurs in a few other places in East Anglia (notably Wicken Fen, Fig. 187) and outlying sites in Somerset and east Yorkshire. Comparable *Phragmites* communities are widely scattered, but never as rich in species. A version with rather regular occurrence of hemp agrimony, marsh bedstraw, marsh thistle, wild angelica, blunt-flowered rush, water mint and purple-loosestrife[S25] occurs in the floodplains of calcareous rivers and similar sites scattered over England and Wales, and probably Ireland too. Much more widespread are species-poor reed-beds with few associated species apart from stinging nettle, cleavers (*Galium aparine*), great willowherb (*Epilobium hirsutum*) and, in the south and west, hemlock water-dropwort (*Oenanthe crocata*)[S26].

Fen meadows
In Broadland, marsh hay and litter were cut (more or less) annually from the 'fen meadows'. After cutting, these usually served for grazing. In many parts of the country areas of fen were treated similarly, as part of the traditional agricultural landscape. Where the fen peat is base-rich these fen meadows are often very rich in species. Blunt-flowered rush is generally dominant, but such species as

Yorkshire-fog (*Holcus lanatus*), tufted-sedge (*Carex elata*), yellow iris, carnation sedge, red fescue (*Festuca rubra*), creeping bent (*Agrostis stolonifera*), water mint, marsh thistle and meadow buttercup can locally have substantial cover, and much of the bulk of the herbage is made up of a variety of broad-leaved herbs. The community usually remains open enough for the moss *Calliergonella cuspidata* and marsh pennywort (*Hydrocotyle vulgaris*) to maintain a significant presence[M22].

In western parts of Britain and in Ireland two much commoner rushes, soft-rush (*Juncus effusus*) and sharp-flowered rush (*J. acutiflorus*), are prominent in a widespread wet-meadow (or wet-pasture) community on peaty soils that are acid and base-poor. *Juncus effusus* (sometimes replaced or accompanied by *J. acutiflorus*) is almost constantly present, along with Yorkshire-fog, marsh bedstraw and greater bird's-foot trefoil. Other frequent species include marsh thistle, lesser spearwort (*Ranunculus flammula*), meadow buttercup (*R. acris*), tormentil (*Potentilla erecta*), meadowsweet (*Filipendula ulmaria*), sweet vernal-grass (*Anthoxanthum odoratum*), cuckooflower (*Cardamine pratensis*), marsh ragwort and common sorrel (*Rumex acetosa*)[M23] (Fig. 189).

A third, rather distinctive, fen meadow is dominated by *Molinia caerulea* with constant or near-constant tormentil, devil's-bit scabious, meadow thistle (*Cirsium dissectum*), greater bird's-foot trefoil and carnation sedge, with frequent marsh

FIG 189. Rushy wet pasture[M23], Hollow Moor, Devon, July 1994. Abundant soft-rush and Yorkshire-fog, with marsh ragwort (*Senecio aquaticus*), greater bird's-foot trefoil (*Lotus pedunculatus*), marsh thistle (*Cirsium palustre*), whorled caraway (*Carum verticillatum*), etc.

FIG 190. Meadow-thistle (*Cirsium dissectum*)–purple moor-grass (*Molinia caerulea*) fen-meadow[M24] on Carboniferous shales (Culm Measures), Witheridge Moor, Devon, June 1989.

FIG 191. Plants of heathy *Molinia* fen meadows: (a) wavy St John's-wort (*Hypericum undulatum*), near Folly Gate, Devon, September 1985; (b) lesser butterfly orchid (*Platanthera bifolia*) amongst *Molinia* near Tubber, Co. Clare, July 1970.

thistle, wild angelica, flea sedge (*Carex pulicaris*) and tawny sedge (*C. hostiana*)[M24]. It tolerates some variation in calcium and pH, but probably favours sites rather low in phosphorus and nitrogen, which dry out somewhat in summer. This 'Cirsio-Molinietum' is widely distributed over southern Britain and Ireland. In eastern and central-southern England it generally occurs on rather base-rich peats or peaty mineral soils, reflected by the presence of some rich-fen or base-rich grassland species, such as fen bedstraw (*Galium uliginosum*), marsh valerian and common knapweed (*Centaurea nigra*). Farther west an essentially similar community grows on peaty gley soils in mildly calcareous patches on heaths, and more widely in rough pastures on the Culm Measures of west Devon and similar situations in south Wales and Ireland (Fig. 190). This lacks the more calcicole species of its eastern counterpart, but sharp-flowered rush, compact rush (*Juncus conglomeratus*), cross-leaved heath, marsh bedstraw and heath spotted-orchid (*Dactylorhiza maculata*) are frequent[M24c]. This community is a frequent habitat of lesser butterfly orchid (Fig. 191b) and the rare southwestern wavy St John's-wort (Fig. 191a), which unaccountably seems to be absent from Ireland.

These three 'fen meadow' types are relatively widespread, but there are more local variants. Wet ungrazed hollows almost anywhere, given a reasonable supply of nutrients, can grow up to dense stands of meadowsweet, often with wild angelica (Fig. 192). Few other plants are conspicuous, but soft-rush, creeping

FIG 192. Meadowsweet–wild angelica marsh[M27], Southmoor, near Folly Gate, Devon, September 1985.

buttercup (*Ranunculus repens*), common sorrel (*Rumex acetosa*), marsh bedstraw, ragged-robin (*Lychnis flos-cuculi*), Yorkshire-fog and others often occur in smaller quantities[M27]. In principle, these meadowsweet–wild angelica stands could be replaced by wet woodland, and that would undoubtedly happen given time enough, but establishment of shrubs and trees is hindered by the density and profuse leaf-litter of the tall herbs. A distinctive northern fen meadow is much more local and has a different flavour from its southern counterparts, with abundant *Molinia* and common sedge (*Carex nigra*), frequent water avens (*Geum rivale*), and northern species such as marsh hawk's-beard (*Crepis paludosa*) and globeflower (*Trollius europaeus*)[M26].

Acid poor-fens

Poor-fens must surely be one of the Cinderellas of British plant ecology – those communities that are neither ombrogenous bogs nor 'interesting' floristically rich fens. Yet to neglect them is to leave out an important part of the mire scene, which has an interest and poses questions of its own.

Like the rich-fens, poor-fens may be roughly divided into those which occur mainly in pools or hollows or around lake shores (and in that sense are topogenous) and those which parallel the small-sedge rich-fens in that they occur on level or sloping ground wherever the soil or peat is wet enough (and in that sense are soligenous). The first group spans a continuous range from *Carex–Sphagnum* ('Caricion lasiocarpae') fens at the base-poor extreme to wet rich-fens at the other.

A number of prominent species of this first group have extremely wide tolerance of differences in water chemistry. The sedges *Carex rostrata* and *C. lasiocarpa* and bogbean are equally at home in blanket-bog pools in Sutherland and in calcareous fen in East Anglia or the Burren. Common cottongrass, bog-sedge (*Carex limosa*) and marsh cinquefoil have almost as wide an amplitude. This makes boundaries hard to draw, and to a large extent arbitrary – and yet there is a great deal of difference between the extremes. Figure 193 shows some Maucha diagrams from poor-fens (and from some transitional sites), with an ombrogenous bog for comparison. The *Carex lasiocarpa* poor-fen on Rannoch Moor (Fig. 193b) differs from neighbouring blanket bog only in its slightly higher pH, and somewhat higher calcium and magnesium levels. The valley mire near Haytor on the Dartmoor granite (Fig. 193c) has a higher pH and somewhat more magnesium and calcium than is usual in ombrogenous sites. Cockett Moss (Fig. 193d) is a *Carex rostrata–Sphagnum fallax* mire[M4] near Settle, Yorkshire. It is almost as acid as Cors Fochno, and differs mainly in the higher concentrations of calcium and sulphate. At the more acid end of Nether Whitlaw Moss, near Selkirk (Fig. 193h, upper), the only sedges were *Carex rostrata* and *Eriophorum*

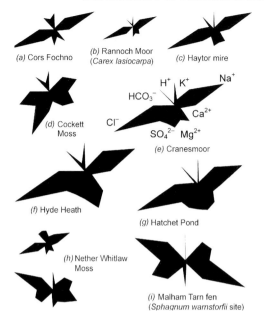

(a) Cors Fochno

(b) Rannoch Moor (*Carex lasiocarpa*)

(c) Haytor mire

(d) Cockett Moss

H^+ K^+ Na^+
HCO_3^-
Ca^{2+}
Cl^-
SO_4^{2-} Mg^{2+}

(e) Cranesmoor

(f) Hyde Heath

(g) Hatchet Pond

(h) Nether Whitlaw Moss

(i) Malham Tarn fen (*Sphagnum warnstorfii* site)

FIG 193. Maucha diagrams from a diversity of peatland sites. (a) Cors Fochno, Cardiganshire, an ombrotrophic bog; (b–d) poor-fens; (e–f) southern English 'valley bogs'; (g–i) transitional sites between rich-fen and poor-fen. For further explanation see text, and Box 7.

angustifolium, over a moss-layer mainly of *Sphagnum squarrosum*; the water analysis differed from an ombrogenous bog principally in the higher values for potassium, calcium and magnesium. With greater mineral input, calcium and bicarbonate become the dominant ions and the vegetation is richer, with lesser tussock-sedge, marsh cinquefoil, ragged-robin and *Sphagnum teres* amongst other species (Fig. 193h, lower). A step on from this is Figure 193i from a *Sphagnum*-rich transitional-fen site (with *S. teres* and *S. warnstorfii*) at Malham Tarn; pH is up in the usual rich-fen range, and the preponderance of calcium and bicarbonate is more marked. In these last two sites, the flowering plants must draw their nutrients from the groundwater permeating the peat, but the bryophytes, raised above the water-table, will be largely dependent on rainwater and other airborne sources (and in that sense are 'ombrotrophic').

The second group, the 'Caricion nigrae' small-sedge fens, are particularly associated with the old hard acid rocks of our uplands, where they are a constant part of the moorland scene, but they occur widely, if more sparsely, in wet heathy places elsewhere. A very variable community dominated by the rushes *Juncus effusus* and *J. acutiflorus*, common sedge, star sedge (*Carex echinata*), *Polytrichum commune* (our tallest moss), *Sphagnum fallax*, *S. palustre* and *S. denticulatum*[M6] occupies seepages around springs or streams in moorland, or places where water drains from blanket peat. Sometimes these are virtually three-species stands of

Juncus effusus, Polytrichum commune and *Sphagnum fallax,* especially in blanket-bog country. In more species-rich examples the sedges are typically prominent, along with tormentil, velvet bent (*Agrostis canina*), marsh violet (*Viola palustris*) and various species from surrounding communities. Numerous variants of this community have been recognised. The water chemistry of many examples of the *Carex echinata–Sphagnum recurvum/denticulatum* mire often shows little or nothing to distinguish them from ombrogenous bogs. At moderate altitudes white sedge (*Carex curta*) is often added to the mix, and becomes a regular component in the high-altitude *Carex curta–Sphagnum russowii* mire of the Highlands[M7].

Acid springs and seepages

Springs, issues of water on or at the foot of slopes, are much more prominent in the uplands (Chapter 17), but it is appropriate to say a few words about them here. A characteristic plant of neutral-to-acid springs is blinks (Fig. 194a), typically accompanied by mosses including *Philonotis fontana, Sphagnum denticulatum, Warnstorfia exannulata, Calliergonella cuspidata* and *Dichodontium palustre* and, rarely, *Hamatocaulis vernicosus.* In the mountains, starry saxifrage (*Saxifraga stellaris*) is a regular ingredient of this assemblage[M32]. At lower altitudes in western Britain it is replaced by a spring community with blinks, round-leaved crowfoot (Fig. 194b), lesser spearwort (*Ranunculus flammula*), bog pondweed (*Potamogeton polygonifolius*), bulbous rush (*Juncus bulbosus*) and the mosses *Sphagnum denticulatum* and *Philonotis fontana*[M35]. This community would be expected in southern Ireland too. These southwestern springs tend to trail off into more extensive seepages, of a kind that often fringe moorland pools, with marsh St John's-wort (Fig. 194c), bog pondweed, lesser spearwort, bulbous rush and *Sphagnum denticulatum*[M29]. This very characteristic community has a wide

FIG 194. Plants of springs, seepages and heathy pool margins: (a) blinks (*Montia fontana*), Creagan Meall Horn, Sutherland, June 1981; (b) round-leaved crowfoot (*Ranunculus omiophyllus*), Meldon, Devon, May 1991; (c) marsh St John's-wort (*Hypericum elodes*), Burley, Hampshire, July 1989.

distribution from southern England and west Wales to Kerry and Mayo, and probably more widely than that.

The southern English 'valley bogs'

These, of which the best examples are in the New Forest (Fig. 195) and the Poole Harbour area of Dorset (Fig. 196), are a provoking exception to all the rules. Scandinavian ecologists would regard them as fens. Floristically, they belong with the ombrogenous bogs in the 'Oxycocco-Sphagnetea'. Chemically (Fig. 193e, f), they share the preponderance of sodium and chloride with ombrotrophic bogs and other poor-fens, but mineral groundwater influence is betrayed by higher calcium, magnesium and potassium concentrations than could be accounted for by rainwater.

These valley bogs lie in a heathland landscape. Typically, dry heath with heather and bell heather grades down through a zone of wet heath, with cross-leaved heath, *Sphagnum compactum* and *S. tenellum* and scattered

FIG 195. New Forest valley bogs. (a) Cranesmoor, May 1996. Pools and hummock with self-sown pine on the surrounding heath; notice the 'bonsai' pine on the nearest hummock. (b) 'Fort Bog', Matley, June 1996. Pool with oblong-leaved sundew (*Drosera intermedia*) at the water's edge; the lighter red rosettes of round-leaved sundew (*D. rotundifolia*) dot the *Sphagnum* cushions farther from the water. Leaves of bog asphodel (*Narthecium ossifragum*), which is abundant here, are still only a few centimetres long, and inconspicuous.

FIG 196. Pools in valley bog^M21 on Studland Heath, Dorset, May 1973. The previous year's dead flowering stems of *Molinia* and the bleached leaves of bog asphodel are still conspicuous. *Sphagnum pulchrum* is dominant here (a rather local species), with *S. papillosum* and *S. rubellum* on the hummocks, and *S. cuspidatum* in the pools. The hummocks are often topped by grey lichens (*Cladonia* spp.).

tufts of deergrass (*Trichophorum cespitosum*), into the bog. This is most often dominated by *Sphagnum papillosum*, cross-leaved heath and common heather, with varying amounts of *Molinia* and almost always abundant bog asphodel. Almost always there are hummocks of the fine-leaved red *Sphagnum rubellum*, and usually wet hollows or pools with *S. cuspidatum*, *S. denticulatum* or in some places *S. fallax*, often with white beak-sedge (*Rhynchospora alba*). *Sphagnum papillosum* sometimes alternates with the similar-looking but deep wine-red *S. magellanicum* in the moss carpet. In the Dorset bogs but not in the New Forest wide lawns of the brownish *S. pulchrum* are the rule. Some minor components are very characteristic of the habitat. Round-leaved sundew dots the *Sphagnum* carpet almost everywhere; oblong-leaved sundew (much the more effective insect trap) is a plant of the pools, and runnels in the wet heath. A number of leafy liverworts grow amongst the *Sphagnum*, some relatively large (*Odontoschima sphagni*, *Mylia anomala*), some tiny (*Cladopodiella fluitans*, *Cephalozia connivens*, *C. macrostachya*, *Kurzia pauciflora*) but fascinating and beautiful objects with a hand-lens or microscope.

Downslope, the valley bogs grade into other communities. In some places the open hummock-and-pool *Sphagnum* carpet becomes more uniform and increasingly hidden beneath *Molinia*. Sometimes the bog grades into a strip of wet birch or alder carr in the valley bottom, with or without a zone of *Molinia* in between. These valley-bog–carr transitions can be exceedingly wet! In other places the bog becomes obviously more base-rich from the edge to the centre, as at Cranesmoor (and other places) in the New Forest and Hartland Moor in Dorset, marked by an abundance of *Schoenus* at both places. A subtler indication of base-rich influence (Fig. 197) is the occurrence of the pale butterwort and (more rarely) the bog orchid in the *Sphagnum* around bog pools, or in *S. denticulatum*-dominated seepages on the heaths.

Figure 193g is from a mildly calcareous seepage on heathland in the New Forest; a pH of 5.08 and 3.38 mg/litre of calcium are just enough to support the calcicole bog pimpernel and the mosses *Scorpidium scorpioides* and *S. revolvens*. In Devon and Cornwall *Schoenus nigricans* occurs very locally in characteristic (but variable) heathy flushes that straddle the dividing line between wet-heath and rich-fen, with *Schoenus*, *Molinia*, cross-leaved heath, bog asphodel, many-

FIG 197. Two uncommon poor-fen species widely distributed in western Britain and Ireland, which find congenial habitats in southern English valley bogs: (a) bog orchid (*Hammarbya paludosa*), Hatchet Pond, New Forest, Hampshire, August 1970; (b) pale butterwort (*Pinguicula lusitanica*), Aylesbeare Common, Devon, July 1964.

stalked spike-rush (*Eleocharis multicaulis*), bog pimpernel, the two wet-heath
sundews, *Sphagnum subnitens*, *S. denticulatum* and varying quantities of the mosses
Campylium stellatum, *Scorpidium scorpioides* and *S. revolvens*[M14].

Valley bogs were formerly widespread on the Surrey, Berkshire and north
Hampshire heaths, but apart from Thursley Common near Godalming few now
remain. Of the scattered small bogs across the Weald of Sussex and Kent, the
best that remain are at Hothfield Common in Kent and Hurston Warren in
Sussex. The most extensive lowland valley bogs outside Dorset, Hampshire and
the southeast are Dersingham Bog and Roydon Common in northwest Norfolk,
but some entirely characteristic small examples still exist in Devon.

Molinia, tussocky and otherwise

In central Europe, purple moor-grass (*Molinia caerulea*) is regarded as primarily
a 'fen meadow' plant of peaty sites long managed by traditional farming – in
short, as a plant of the 'cultural landscape'. That may still be largely true in
a broad sense in Britain and Ireland, but in the western parts of our islands
Molinia attains a prominence unknown in central Europe. It occupies boggy
valleys on heathland, fringes ombrogenous bogs, and dominates wide tracts of
damp moorland. *Molinia* favours soils poor, but not extremely poor, in nutrients;
it is tolerant of a wide range of soil moisture, calcium and pH. It does best in
full light in the open, but tolerates partial shade. In heaths, valley bogs and on
dried-out peatlands *Molinia* is favoured by disturbance and the high atmospheric
nitrogen deposition in populated or intensively farmed areas.

Molinia is often an uncompromising dominant; few other plants may be in
evidence apart from tormentil and sharp-flowered rush[M25]. In wet ground it is
often tussocky, as anyone who has taken a short-cut across a damp valley in the
New Forest will be uncomfortably aware. 'Molinieta' with us tend to fall into
three categories. Close scrutiny may reveal a considerable range of bog or poor-
fen species, such as cross-leaved heath, common cottongrass, bog-myrtle (*Myrica
gale* – this may be conspicuous among the *Molinia* tussocks), various species of
Sphagnum and the mosses *Polytrichum commune* and *Aulacomnium palustre*[M25a].
This boggy molinietum is often very tussocky. *Molinia* often dominates hill
pastures on damp slopes. These are usually less tussocky, and the associated
species reflect the surrounding acid grassland and its associated poor-fen
communities, with such species as sweet vernal-grass, Yorkshire-fog, common
bent (*Agrostis capillaris*), heath wood-rush (*Luzula multiflora*), soft-rush, marsh
violet, marsh thistle and devil's-bit scabious[M25b]. A third category has something
of the flavour of the Cirsio-Molinietum fen meadow but is dominated by dense
tall *Molinia*, often tussocky, typically with scattered wild angelica and often

hemp agrimony standing out above the *Molinia* canopy[M25c]; devil's-bit scabious, greater bird's-foot trefoil and a wide range of other species may occur in small quantities.

THE WATER SOURCES OF FENS

Fen sites can be bewilderingly complex. It is easy to say that fens receive their water and solutes from two sources, rain and water that has drained through mineral ground – but, as ever, 'the devil is in the detail.' The influx of mineral-rich water may be seasonal, as in floodplain or lake-shore fens that regularly flood in winter; flooding is usually accompanied by deposition of mineral-rich (and often nutrient-rich) silt alongside the river or stream channel. For most of the year (and often all the year), streams that run through fens drain water *from* the fen, rather than bring nutrients *into* it. Water from mineral ground may be surface run-off, or water that has percolated deep into the soil or bedrock to emerge at springs, or more diffusely, at a spring-line or seepage zone where the water-table in a porous bedrock (*aquifer*) meets the surface – a classic situation for a soligenous fen (or valley bog). On more massive rocks, such as Carboniferous limestone, springs are usually more localised. Water from calcareous springs is often saturated with calcium bicarbonate at a far higher carbon dioxide concentration than the air. The surplus is deposited as calcium carbonate, and phosphate in the water is co-precipitated as insoluble calcium phosphate with it. Consequently, spring-fed calcareous rich-fens are typically nutrient-poor. Springs may be marginal to fens, or artesian within them. A vigorous spring may support active peat growth all around it but prevent plant growth actually over the spring, leading to 'well-eyes' in the fen. Suffice it to say that bogs and fens are not simple; their possible hydrological complexities are legion.

HOW 'NATURAL' ARE OUR FENS?

Many of our fens are clearly part of the cultural landscape in that they owe their present state to traditional land use by man since he first began clearing the forest and farming some 5000 years ago. The beautiful and species-rich north fen at Malham Tarn appears on an estate map of the 1780s as a group of named meadows, 'Moss meadow', 'Long Meadow' and so on, and the area was evidently in active agricultural use. Encroachment of sallow scrub has been an increasing problem in the 60 years since the Field Studies Council took over the site. Most of the East Anglian tall-herb *Phragmites* fens, left to themselves, would in a few decades revert to wet woodland, first of sallow and in course of time to alder and ash. This is the problem of fen management almost everywhere, except in the wettest places, or in sites still open to grazing. Are any of our lowland fens naturally open and treeless? That is a question we probably cannot answer with certainty. Tree growth is very slow in some rich-fen peats, perhaps owing to phosphate deficiency. Seedlings of pine and oak are frequent in the New Forest valley bogs, but the oaks never become established and pines that do so are slow-growing and stunted (Fig. 195a). On the other hand there would have been wild herbivores – red and roe deer, elk, aurochs, wild boar, beavers – before forest clearance by man began, so there were probably openings in the forest where physical conditions were less conducive to tree growth. There are indications that some rich-fen sites have been in essentially continuous existence since the early Post-glacial (perhaps often in mire-complexes that included ombrogenous bogs), and this is consistent with evidence from peat stratigraphy and pollen analysis. A few of the larger southern valley bogs could be comparably old (and might have had local ombrotrophy in their history), but many are probably of recent origin, and some formerly open sites are now wet woodland. The balance of evidence is that most of the present extent of lowland fen is a product of human activity.

CHAPTER 15

Peatland Landscapes: Ombrogenous Bogs

EW VEGETATION TYPES make as much impact on the landscape as the ombrogenous bogs. Their expanses of wet treeless peat country, covered with cottongrasses, *Sphagnum* and heather, and summer home to curlew, dunlin and golden plover, are an essential part of the character of the windswept and rainy north and west of Britain and Ireland. The ombrogenous peatlands that formerly dominated wide low-lying tracts of flat country bordering many of our larger rivers and estuaries, in the Somerset Levels, and most extensively of all in the midland plain of Ireland, must have been just as dramatic – as indeed fragments that are left of them still are. These bogs have acquired fresh topicality with current concern over atmospheric CO_2 levels and global warming. The northern peatlands of Eurasia and North America represent one of the major stores of carbon on the planet, estimated at around 300 billion tonnes, and their fate affects us all.

THE DISTRIBUTION AND LIMITS OF OMBROGENOUS BOGS IN BRITAIN AND IRELAND

According to the Soil Survey, ombrogenous peats cover some 2.5% of the land surface area of England and Wales. Nearly 17% of Ireland is covered by peat; the figure for Scotland is around 10% (Fig. 198). These are figures for peat; the proportion carrying active ombrogenous-bog vegetation at the present day is much less. British and Irish plant ecologists have generally distinguished 'raised bogs', mainly lowland and often the product of hydrosere succession from former lakes or

FIG 198. The distribution of ombrogenous peats in Britain and Ireland, based mainly on Taylor (satellite data), the Soil Survey of England and Wales, and Doyle & O Críodáin (2003). Letters show the approximate locations of some former and extant lowland raised bogs in Britain: C, Chat Moss; F, Flanders Moss; G, Glasson Moss and Wedholme Flow; H, Humberhead peatlands (Thorne and Hatfield Moors); L, Lancashire mosses; P, former bogs south of Peterborough; S, Somerset Levels; T, Cors Tregaron; W, Fenns and Whixall Mosses.

FIG 199. Malham Tarn Moss, Yorkshire, from Chapel Fell, August 1990.

the floodplains of rivers, from 'blanket bogs', which as their name implies blanket
the landscape in high-rainfall areas. However, this is not a sharp distinction.

Malham Tarn Moss (Figs 199, 200) shows some typical features of a raised
bog and its surroundings. The gently domed area of ombrogenous peat drains
radially into a belt of minerotrophic fen, the *lagg*, separating the bog from the
surrounding mineral ground. Often the drainage in the lagg is channelled
into a stream, as it is here in the lower part of the southwest lagg. The inflow
streams from the fields to the west, and springs from the base of the limestone
to the north, skirt the bog on its northern side, meandering through a belt of
calcareous fen; some small isolated raised bogs have grown up in the meanders,
mostly now covered with birch. The margin of the acid peat, sloping down to the
lagg, is called the *rand*.

Because of their relative accessibility, lowland raised bogs have borne the
brunt of drainage and reclamation for farming. The Somerset Levels between
Street and Wedmore bore substantial raised bogs, of which little trace now
remains. The raised bogs that existed between Huntingdon and Peterborough
have all but vanished, but despite intensive exploitation in the past half-century,
more remains of the vast expanses (*c*. 3000 ha) of raised bog near the head of
the Humber estuary, and of the Fenn's and Whixall mosses near Whitchurch
on the Welsh–English border. Following drainage, ploughing and fertilising,
Chat Moss west of Manchester, which gave such trouble to George Stephenson
when he built the Liverpool & Manchester Railway in the late 1820s, has long
been a prolific market-gardening area. There are still some raised bogs around

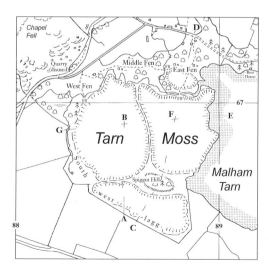

FIG 200. Malham Tarn Moss,
illustrating some features of a
raised bog. The dome of acid
peat (three coalesced domes
here) drains radially into the
lagg separating the bog from
the adjacent mineral ground, or
the Tarn to the east. The sloping
margin of the acid peat abutting
on the lagg is the *rand*. Letters
A–G correspond to the sections
in Fig. 174 (Chapter 13).

the estuaries south of the Lake District, and Glasson Moss and Wedholme
Flow south of the Solway are among the best remaining examples in Britain.
Flanders Moss in the Forth valley, 15 km west-northwest of Stirling, is another
large expanse of lowland raised bog. All these big areas of raised bog have been
encroached on at their edges by agriculture or other development, which obscure
their geographical context. Cors Caron (Tregaron Bog) in mid Wales is still
essentially intact despite some peat cutting round the edges (Fig. 218). There are
many smaller raised bogs scattered round Britain, some with their plant cover
or landscape context more or less intact. Rainfall in Britain has evidently almost
everywhere sufficed to support raised-bog growth in suitable sites.

At the close of the last glaciation, the midland plain of Ireland was left as
an undulating, lake-studded expanse of (predominantly) clayey glacial drift, a
near-perfect template for hydrosere development. By historical times, large tracts
of lowland Ireland were occupied by raised bogs, broken by hills wooded with
ash, elm and hazel (the mineral soils of the midland plain are predominantly
calcareous), and by the winding gravelly ridges of eskers (former sub-glacial
drainage channels). The bigger Irish raised bogs were on a scale that few sites in
Britain could match, but they have now mostly been cut away to fuel peat-fired
power stations, or for other uses including the horticultural industry and for

FIG 201. Clara Bog, Co. Offaly, September 1988. A view of the undisturbed western part of
the bog; a thoroughly typical piece of active raised bog surface.

FIG 202. Roundstone Bog, Co. Galway, September 1997; the Twelve Bens in the background. The grey tinge of black bog-rush (*Schoenus nigricans*) is typical of western Irish lowland blanket bogs.

peat briquettes as household fuel. Of those that remain unexploited, Clara Bog (Fig. 201) northwest of Tullamore and Mongan Bog near Clonmacnoise, both in Co. Offaly, are still impressive expanses of peatland. The rainfall over these Irish bogs is always over 800 mm and often over 1000 mm a year.

Blanket bogs occupy two at first sight rather distinct but intergrading habitats. First, they are characteristic of the high-rainfall areas of the far west of our islands, in Kerry, Galway, Mayo, Donegal, and western Scotland from Galloway, Argyll and the Hebrides to Sutherland, Caithness and Shetland (Fig. 202). In fact, lowland blanket bog is better correlated with areas experiencing more than 250 'rain days' a year than with rainfall as such (Fig. 203). Second, blanket peats cover many flattish or undulating plateaux, as on Dartmoor, much high ground in mid and north Wales, the Pennines in northern England (Fig. 204), the Southern Uplands of Scotland and similar ground in the Highlands and Ireland. These are mostly 400–600 m above sea level, with over 1500 mm annual rainfall and over 225 'rain days' a year, but the limits vary from one region to another. However, the distinction between lowland and upland blanket bogs is probably more apparent than real, and mainly reflects where vegetation ecologists have concentrated their studies, and their preconceptions.

LEFT: **FIG 203.** Mean annual number of days with at least 0.2 mm rainfall ('rain days') in Britain and Ireland, 1901–1930. The lowland blanket bogs in western Ireland occur within the area experiencing at least 250 rain days; in west and north Scotland the limiting figure is nearer 225 rain days.

BELOW: **FIG 204.** *Calluna–Eriophorum vaginatum* blanket bog in the Pennines, on the road from Swaledale to Kirkby Stephen, August 1993. Swaledale sheep are a hardy hill breed that can make a living on the meagre pasture of the bog and the mat-grass (*Nardus*) and rushy grasslands surrounding it.

Probably the same applies to the traditional distinction between 'raised bog' and 'blanket bog', of which more later. First, there is much that all ombrogenous bogs have in common.

OMBROGENOUS BOG VEGETATION: THE COMMON SPECIES

The species list on an ombrogenous bog is typically rather short, made up predominantly of a few ericaceous subshrubs, a few sedges, 'bog-mosses' of the genus *Sphagnum*, and a handful of other species. The common heather *Calluna vulgaris* is nearly ubiquitous in the bogs of north and west Europe, sometimes dominant, sometimes present only as a scatter of impoverished shoots. Towards the Atlantic seaboard of Europe, including the whole of Britain and Ireland, *Calluna* is joined by the cross-leaved heath (*Erica tetralix*). Three sedges are common in British and Irish bogs. The rhizomes of the common cottongrass (*Eriophorum angustifolium*) are virtually all-pervading in any kind of bog, ombrogenous or not, throughout most of Britain and all of Ireland. The scattered long leaves differ from most sedges in their gutter-shaped (rather than V-shaped) cross-section, and their red hue late in the season. *E. angustifolium* flowers early in spring, but the flower-spikes (several to a stem) are most

FIG 205. Hare's-tail cottongrass (*Eriophorum vaginatum*) is near-ubiquitous in ombrogenous bogs, and often dominant, as it is in the Pennines. Cottongrasses flower very early in the year, and only develop the cottony bristles in fruit. Here sunlight has picked out the inflated leaf-sheaths on the fruiting stems that give the species its name. Exmoor, June 1995.

FIG 206. Some characteristic ombrogenous bog *Sphagnum* species: (a) *S. rubellum*;
(b) *S. austinii*; (c) *S. papillosum*; (d) *S. magellanicum*; (e) *S. pulchrum*; (f) *S. cuspidatum*.

conspicuous later when the white cottony bristles elongate in fruit. Hare's-tail cottongrass (*E. vaginatum*) contrasts with its congener in every way. It grows in dense tussocks, with narrow leaves no wider than the flower stems, each of which bears only a single inflorescence at its tip (Fig. 205). Hare's-tail cottongrass dominates wide tracts of bog in the Pennines and elsewhere, and the cottony tufts dancing in the breeze are a fine sight in early summer. In many wetter bogs *E. vaginatum* is more scattered and does not form large tussocks, but it is seldom completely absent. Another tussocky sedge is deergrass (*Trichophorum cespitosum*), which again is seldom completely absent, but comes into its own when its tussocks dominate wide monotonous expanses of degraded peat in the western Highlands of Scotland and elsewhere (it seems to be favoured by trampling and compression of the peat). Its stems bear an insignificant flower-spike at their tips, and colour a yellowish tinge late in the year; the leaves are a mere few millimetres long on the sheaths at the base of the stem.

Sphagnum numbers about 30 species in Britain and Ireland, of which only some half-a-dozen are common in ombrogenous bogs (Fig. 206). The commonest is *S. papillosum*, a coarse green to brownish species forming lawns or low hummocks. In some places it is accompanied or replaced by *S. magellanicum*, of similar growth-form but deep wine-red in colour. *Sphagnum rubellum* is also pink to reddish, and typically builds taller hummocks than the last two species. *Sphagnum pulchrum* forms wide lawns in hollows; it is more local (and more oceanic) in distribution but often abundant where it occurs. *Sphagnum cuspidatum*, a common yellowish-green species with long narrow leaves, occupies wet hollows and often grows floating in bog pools. It has been likened to a 'drowned kitten' when taken out of the water. Two other hummock-forming species are of similar build to *S. rubellum*. *Sphagnum capillifolium* tends to replace *S. rubellum* on upland bogs in the north of England and in Scotland, and produces capsules much more freely. *Sphagnum fuscum* occurs mostly in Scotland and Ireland, looks like a brown *S. rubellum* or *capillifolium*, and typically forms tight cushions. Another hummock-former, *S. imbricatum* subsp. *austinii* (or *S. austinii*), was the principal peat-forming *Sphagnum* in Britain and Ireland from the 'Atlantic' period (Chapter 2) until medieval times. It is a big species, matching *S. papillosum* in size, and its easily recognised remains are abundant in the peat (Fig. 207). It remains frequent only in the north and west, and nowhere does it retain its former dominance.

A number of minor, but very characteristic, ingredients of the ombrogenous bog flora remain to be mentioned (Fig. 208). The round-leaved sundew is common in the *Sphagnum* carpet in a zone between the hummocks (too competitive) and the pools (too wet); catching and digesting small insects supplements its intake of limiting N and P. Great sundew is a bog-pool plant,

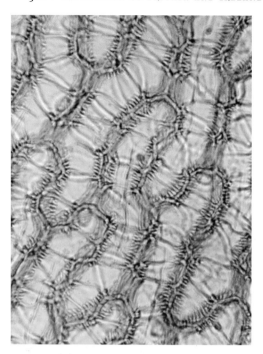

FIG 207. Leaf cells of *Sphagnum austinii* (*S. imbricatum* ssp. *austinii*) from peat, showing the characteristic 'comb fibrils' on the inner surface of the empty hyaline cells. This species was the dominant peat-former in raised bogs, but has declined greatly in recent centuries.

much more local and a much more efficient fly trap. Two pretty ericaceous plants, much smaller than the heathers and particularly characteristic of bogs, are bog rosemary and cranberry. Bog rosemary straggles among the other bog plants, its leaves recalling rosemary but broader; the flowers are like those of cross-leaved heath but much larger, borne singly and a more delicate shade of pink. The fine thread-like stems of cranberry with their half-centimetre-long leathery leaves ramify over the surface of the *Sphagnum*; they bear small pink flowers with recurved petals (like tiny *Cyclamen* flowers), which in late summer ripen to produce red-flushed cranberries, surprisingly large for the size of the plant (Fig. 209). One of the most colourful plants of our bog flora is bog asphodel (*Narthecium ossifragum*), with small iris-like leaves and spikes of bright yellow flowers at midsummer followed by orange seed capsules (Fig. 230f). In western Sweden it is generally regarded as an indicator of mineral-influenced conditions, but that is not true over most of its British and Irish range, or in blanket bog in western Norway. Lastly, the instantly recognisable white beak-sedge (*Rhynchospora alba*, Fig. 230c) is a plant of wet hollows and bog pools and can be abundant where at occurs, but wide tracts of drier bog (including most of the Pennines) are without it.

FIG 208. Ombrogenous bog plants: (a) round-leaved sundew (*Drosera rotundifolia*), the commonest sundew in bogs and wet heaths; the tiny white flowers open only in bright sunshine; (b) great sundew (*Drosera anglica*), in hollows and around pools in bogs, despite its name much commoner in Scotland and Ireland than in England; (c) bog rosemary (*Andromeda polifolia*), characteristic of lowland raised bogs; (d) cranberry (*Vaccinium oxycoccos*), often common creeping over *Sphagnum*, but rare in the south, the far west of Ireland and the north of Scotland.

FIG 209. Cranberries, Helwith Moss, Yorkshire. The berries in late summer look absurdly large for the small leathery leaves and slender wiry stems creeping over the *Sphagnum*.

RAIN-FED BOGS: THE SUPPLY OF WATER AND MINERAL NUTRIENTS

Rainwater is a meagre source of nutriment for plants, but it is not pure distilled water. The substances dissolved in rainwater come from three sources: sea-spray entrained as winds sweep over the ocean, dust of terrestrial origin, and gases absorbed from the atmosphere (CO_2, ammonia, and oxides of sulphur and nitrogen). The proportions of these three, and with them the overall chemical composition of rainwater, varies greatly from day to day, and its average over the year varies with distance from the sea and the nature of the surrounding landscape (Table 9). There are some additional airborne inputs to the bog surface, in the form of dry-deposited dust and gases from the air, and there is some interchange of nutrients in movements of animals (farm stock, deer, birds, insects) onto and off the bog.

The peat of the bog surface is an efficient cation-exchange material and, volume for volume, holds far more solutes than the water in contact with it. Therefore the water in pools on the bog surface (and permeating the peat), in effect averages the chemical composition of the rainwater (etc.) deposited on the bog surface over a period of time – somewhat concentrated by evaporation. In the long term (years, centuries), the average composition of the rain determines the composition of the peat. In the short term (days, weeks) the peat is the major determinant of the composition of the surface water. Geographical variation in the composition of bog waters reflects these factors (Fig. 210 and Table 10).

Typically, the composition of the surface water varies somewhat round the year, with pH some 0.2 units lower, and the common metallic cations (Na^+, Mg^{2+}, Ca^{2+}) somewhat higher, in summer than in winter. Summer droughts can cause 'spikes' of acidity through the oxidation of H_2S to H_2SO_4 as the water level falls

TABLE 9. Average chemical composition of rainwater at some sites in Britain and Ireland, 1986–98 (data of Hayman *et al.* 2000). The table includes the major ions except potassium (K^+), which is present in very low concentrations (*c.* 0.1 mg/litre = *c.* 3 µmol/litre). Figures in roman type are mg/litre; figures in *italics* are µequiv/litre (µmol ionic charge/litre, generally used by environmental chemists). Yarner Wood, Achanarras and Lough Navar are relatively near exposed coasts and have high values of sodium and chloride. Wardlow Hay Cop and High Muffles are the nearest to major industrial areas, and have the lowest pH and the highest values of sulphate, nitrate and ammonium nitrogen. Wardlow Hay Cop is in a limestone area and has the highest level of calcium.

Site	pH/H^+	Na^+	Mg^{2+}	Ca^{2+}	NH_4^+	Cl^-	SO_4^{2-}	NO_3^-
England and Wales								
Yarner Wood, Dartmoor, Devon	4.8	3.1	0.42	0.30	0.38	5.6	1.9	1.2
	17	*135*	*35*	*15*	*21*	*157*	*43*	*20*
Llyn Brianne, Dyfed	4.8	2.2	0.30	0.22	0.28	3.9	1.6	0.94
	17	*96*	*25*	*11*	*16*	*110*	*37*	*15*
Wardlow Hay Cop, Derbyshire	4.5	1.9	0.27	1.23	0.77	3.8	3.6	2.0
	29	*82*	*22*	*61*	*43*	*108*	*83*	*32*
High Muffles, North York Moors	4.3	2.1	0.30	0.43	0.88	4.43	3.27	2.5
	51	*93*	*25*	*22*	*49*	*125*	*75*	*40*
Cow Green Reservoir, Co. Durham	4.6	1.8	0.25	0.25	0.43	3.3	1.9	1.3
	24	*79*	*21*	*12*	*24*	*93*	*44*	*21*
Scotland								
Eskdalemuir, Borders	4.5	1.61	0.24	0.21	0.37	3.0	1.6	1.1
	34	*70*	*20*	*11*	*21*	*85*	*37*	*18*
Allt a'Mharcaidh, Cairngorms	4.7	1.6	0.22	0.18	0.10	2.9	1.1	0.65
	18	*71*	*18*	*9*	*6*	*80*	*26*	*10*
Achanarras, Caithness	4.7	4.8	0.62	0.36	0.29	8.5	2.1	1.2
	20	*208*	*52*	*18*	*16*	*239*	*49*	*20*
Ireland								
Lough Navar, Co. Fermanagh	5.1	4.0	0.55	0.42	0.20	7.3	1.6	0.58
	8	*174*	*46*	*21*	*11*	*204*	*36*	*9*

and the peat aerates. The clearest periodicity of all is shown by the brown 'peaty' colour due to dissolved organic matter, most intense in summer.

In lake or river water, the positive charges on the cations (H^+, Na^+, K^+, Mg^{2+}, Ca^{2+} etc.) are balanced by the negative charges on the inorganic anions (Cl^-, SO_4^{2-}, HCO_3^- etc.). In ombrotrophic bog waters there is typically an apparent *anion deficit*; part of the total cations is balanced by organic acids in the dissolved organic matter. Organic anions may commonly balance 40% or more of the

FIG 210. Maucha diagrams of the major ions on ombrogenous bogs in Britain and Ireland. Sodium (Na⁺) and chloride (Cl⁻) are generally the dominant ions in British and Irish ombrogenous-bog waters, but sulphate (SO₄²⁻) and acidity (hydrogen ion, H⁺) may be comparably high near industrial areas. The diagram at the centre of the key (top right) is for a bog in central Estonia, far from open ocean and from industry – the bog water is very dilute: the most abundant ions are H⁺, Ca²⁺ and SO₄²⁻. For further explanation see p. 289.

TABLE 10. Chemical composition of some bog waters in Britain and Ireland. The Malham Tarn and Plym Head data are averages of measurements from sets of six samples at monthly intervals over the five-year period 1992–98. The Mongan Bog and Roundstone data are averages from sets of samples collected in September 1997. Figures in roman type are mg/litre; figures in *italics* are μequiv/litre (μmol ionic charge/litre).

Site	pH/H⁺	Ca²⁺	Mg²⁺	K⁺	Na⁺	Cl⁻	SO₄²⁻	DEF
Malham Tarn Moss,	3.94	0.62	0.57	0.34	3.9	6.5	5.1	
Yorkshire	*115*	*31*	*48*	*9*	*168*	*182*	*118*	*71*
Plym Head, Devon	4.30	0.37	0.64	0.26	4.6	8.0	2.1	
	50	*19*	*53*	*7*	*201*	*226*	*49*	*55*
Mongan Bog, Co. Offaly	4.33	0.84	1.1	0.35	6.8	12	0.91	
(6 samples)	*48*	*42*	*90*	*9*	*296*	*329*	*21*	*135*
Roundstone Bog, Co. Galway	4.67	0.62	1.6	0.51	12.5	24.6	1.35	
(7 samples)	*22*	*31*	*132*	*13*	*544*	*693*	*31*	*18*

total cations in the Baltic countries, around 10–30% in Britain and Ireland, while in the extreme western blanket bogs of Galway and Mayo the anion deficit is very small. The organic anions are what holds the pH of ombrotrophic bogs at around 4 (Proctor 2003, 2006, 2008).

All plants need some essential elements in relatively large quantities, notably N, P and K. Of these, K is generally sufficient for the needs of the slow-growing bog plants. The indications are that, for bog plants, N is limiting in unpolluted environments, but that P soon becomes limiting as N deposition is raised by human activities (industry, vehicle exhausts, intensive farming etc.). Rainwater contains measurable amounts of both nitrate (NO_3^-) and ammonium (NH_4^+) ions, but very little of either is detectable in the surface water. *Sphagnum* uses NO_3^- from rain as a nitrogen source, but most of the NO_3^- and NH_4^+ in rain must be quickly taken up, perhaps mainly by microorganisms. Peat contains a lot of nitrogen, but mostly in organic forms inaccessible to most plants. The ericaceous plants with mycorrhizas acquire organic nitrogen through their symbiotic fungi, and the non-mycorrhizal hare's-tail cottongrass may take up most of its N requirements in the form of free amino acids. Ombrotrophic bog plants clearly exploit a diversity of N sources. Phosphorus is, as so often, the scarce resource. Phosphate is often undetectable in bog waters, because the plants (and microorganisms) take it up as soon as it is released by mineralisation. Mycorrhizal plants on bogs get much or all of their P through their mycorrhizas; the roots of non-mycorrhizal species probably rely on acid phosphatases secreted by their cell surfaces to release inorganic P from organic phosphates. Water movement can be important to the nutrition of bog plants. Where drainage from the bog surface is concentrated into 'water tracks' the nutrients from a wide area become potentially available to plants that can take them up, and slow molecular diffusion to the root surface is supplemented by much faster mass flow of the water in the peat. This often results in much greater vigour of the vegetation – especially of dominant species such as hare's-tail cottongrass, *Molinia* and black bog-rush (*Schoenus nigricans*) – in water tracks, and greater vigour of the rooted vascular plants relative to *Sphagnum*.

THE INITIATION AND GROWTH OF OMBROGENOUS PEATS

The main credit for establishing the concept of the hydrosere, with raised bog as a culminating stage, belongs to the German C. A. Weber (1902). Weber recognised, from sediments and peats in north Germany, the sequence from open water, through 'eutrophic' fen (*Niedermoor*) and 'mesotrophic' transitional communities (*Übergangsmoor*) to 'oligotrophic' bog (*Hochmoor*). His *Hochmoor* was translated in the contemporary British ecological literature as 'high moor' and caused a good deal of confusion; from the 1920s the accepted equivalent was 'raised bog'. Weber was clear that raised bog could also develop by *paludification* of forest on formerly drier ground, and that this had commonly taken place.

Ombrogenous peat growth in raised bogs most often roughly coincided with the shift to a rainier, more oceanic climate at the Boreal–Atlantic transition as the North Sea basin filled, and the sea rose to near its present level, roughly 8000 years ago. The rising sea flooded low-lying coastal expanses of land in what became the Somerset Levels, the Cambridgeshire Fenland and the lowland valley of Scotland (and elsewhere), and these places had to wait for accumulation of estuarine silt and clay (and in Scotland isostatic rise) to bring the surface above sea level before hydrosere development to bog could begin. Blanket-bog growth in the uplands began locally at much the same time as the raised bogs, but in some places (such as mid Wales, Northern Ireland, and Orkney and Shetland) it appears to have started only about 4000 years ago (Tallis 1991, 1995). Some of this variation may be more apparent than real, however, reflecting our imperfect knowledge.

FIG 211. Rannoch rush (*Scheuchzeria palustris*), common in bog pools in northern Europe and formerly widespread in Britain and Ireland, but now very rare with us. (a) Habitat of *Scheuchzeria* at Rannoch, June 1981. (b) *Scheuchzeria* inflorescences just emerging, Rannoch, June 1981. (c) Plants in fruit, Pikasaare Bog, Estonia, August 1996.

The growth of raised-bog peat, once initiated, was usually rapid over the ensuing two millennia. After that, with the climate becoming warmer and drier, the rate of peat accumulation often slowed, and the peat became darker and more humified. At the 'climatic deterioration' around 2600 years ago, the climate became cooler and wetter again, leading to a renewal of rapid growth of little-humified peat, visible as a striking 'recurrence surface' (*Grenzhorizont*) in peat cuttings. The surface of many bogs became much wetter at this time, which is often marked in the peat by the papery remains of the rhizomes of the Rannoch rush (*Scheuchzeria palustris*), a common plant of bog pools in northern Europe, which now survives with us only on Rannoch Moor (Fig. 211). The wetness and flooding caused Bronze Age people to lay down brushwood trackways across the Somerset bogs, which have been unearthed over the years by peat-digging.

Blanket bog seems almost always to have originated by paludification of pre-existing forest, as witnessed by the stump layers that often underlie blanket peats (Fig. 212). The relative importance of climate and human activity in its initiation has been much debated. It seems likely that climate was the predisposing factor, but that felling of the forest may have hastened the paludification process, or pushed the balance in the direction of bog, especially in wet periods favourable to bog growth.

FIG 212. Pine stump exposed by erosion of blanket bog, Slieve League, Co. Donegal, September 1965.

PEAT, WATER AND THE STRUTURE AND HYDROLOGY OF OMBROGENOUS BOGS

Ombrogenous peat is a remarkable substance. Fresh peat consists of plant fragments in various stages of decay in a gelatinous matrix, and about 90% of it is water. It may be thought of as rather like porridge. A recipe given for 'traditional oatmeal porridge' is 60 g medium oatmeal to 570 ml water, almost the same bulk density as ombrogenous bog peat. Peat, like porridge, is rigid up to a point, but beyond that it will give way and flow. Again like porridge, peat has a surprisingly low hydraulic conductivity: water will not flow rapidly through it. Once a layer of peat has formed over the soil surface, it tends to be self-perpetuating, a factor important in the origin of ombrogenous bogs by paludification, and the persistence of bogs once formed.

The growth and accumulation of peat: acrotelm and catotelm

Bog peat is not uniform throughout. The surface 30 cm or so of an active bog, the *acrotelm*, is where most biological activity takes place. Almost all the seasonal (and shorter-term) fluctuation of the water-table takes place in the acrotelm, which is aerobic, and the layer in which most of the bog plants are rooted. Photosynthesis and growth of vegetation on the surface adds organic material to the acrotelm. The top part of acrotelm consists of fresh and often recognisable plant remains, but this is broken down quite rapidly by aerobic decay. As new material is added year by year the residue becomes more resistant, less exposed to aerobic breakdown, and at the base of the acrotelm it has become peat, to be added to the *catotelm*, which is permanently waterlogged and anaerobic, and makes up most of the depth of the bog. The acrotelm may be thought of as a composter, fed with fresh plant material at the top, and processing it to become new catotelm peat at the bottom – or as the cooking pot to which oatmeal and water are added to make the porridge. Obviously the acrotelm is itself not uniform, and it passes insensibly into the catotelm at the bottom.

As already mentioned, catotelm peat has a very low hydraulic conductivity. The more fibrous acrotelm is much more permeable; its hydraulic conductivity is greater by a factor of 10,000 or so. Consequently, most lateral water movement – most of the radial drainage of the bog – takes place in the acrotelm. Usually, 90% or so of the organic matter produced by the vegetation is broken down in the acrotelm, leaving around 10% (or less) to be added to the store of peat; the growth-rate of an actively growing ombrogenous bog is typically 5–10 cm a century.

Anaerobic breakdown of peat and the formation of methane

Once incorporated into the catotelm, the peat is immune from further aerobic decay, but it is still subject to microbial breakdown by much slower anaerobic processes. In simple terms, this may be thought of as rearranging the atoms in carbohydrate $n(CH_2O)$ into $n/2$ molecules of methane (CH_4) and $n/2$ molecules of CO_2, with a modest release of energy. Methane is 'marsh gas', and its production is detectable in the catotelm. It rises through the peat and may escape to the atmosphere if it does not get oxidised to CO_2 and water as it passes through the aerobic surface layers. Methanogenesis leads to a slow but steady loss of mass from the deep waterlogged peat (Clymo 1984). This adds a steady anaerobic loss term, dependent on the depth of the bog, to its carbon budget. Aerobic decay in the acrotelm responds to the wetness or dryness of the surface, but anaerobic decay in the catotelm does not (Belyea & Clymo 2001).

Are ombrotrophic bogs a net sink or net source of atmospheric CO_2?

Globally, peatlands contain about the same amount of carbon as is present as CO_2 in the atmosphere. They are the single most important terrestrial carbon store, and are critical in any discussion of rising atmospheric CO_2 levels. Virtually all of the northern peatlands have grown up since the close of the last glaciation, so during that time they have been a major sink for atmospheric CO_2. It is much harder to determine whether an ombrogenous bog is a net sink or source at the present day, and sufficient data to evaluate this exist for relatively few sites worldwide. The problem is that the carbon budget of a peat bog depends on the balance between various processes, all of them hard to measure accurately. Carbon can only be added to the system by the photosynthesis of the plants growing on its surface, but a large part of that carbon is lost by aerobic respiration in the acrotelm before being transformed into peat. Some carbon is lost as methane and CO_2 in the catotelm, depending on its thickness. Those gains and losses take place all over the area of the bog. A substantial amount of carbon is lost as dissolved organic matter in the drainage water, to be oxidised to CO_2 in streams and rivers outside the bog, and this water leaves the bog already saturated with CO_2. These losses are in addition to peat lost by erosion. In Britain, the Centre for Ecology and Hydrology has research aimed at evaluating carbon budgets in four peatland catchments, at Forsinard in Sutherland, Auchencorth Moss south of Edinburgh, Moor House in the north Pennines, and in the Migneint in north Wales.

The 'hydrological mound'

Raised bogs are domed – but what determines the height of the dome? If water falls uniformly over a porous material and can drain out freely round the edges, it forms a domed water-table, gently sloping near the middle and progressively steeper towards the edges (Ingram 1983). For the case of a circular bog draining to a lagg at uniform level, the theoretical cross-section of the bog is a half-ellipse. The highest point of the dome is proportional to the radius of the bog, and to the square root of the rainfall, and inversely proportional to the square root of the hydraulic conductivity of the (catotelm) peat. The shape of the hydrological mound is a good fit to the profile of Dun Moss in Perthshire (near-parallel sided) and Ellergower Moss in Wigtownshire (nearly circular), but many bogs are of more complex shapes that defy simple mathematical analysis.

HUMMOCKS, HOLLOWS AND BOG POOLS

Often the surface of a bog has a pattern of hummocks and hollows or pools. These have long fascinated ecologists. Osvald (1923) suggested that the pattern he observed on the bog Komosse in central Sweden could be interpreted as a cyclical succession in time, the bog-pool sphagna raising the level until vigorous hummock-building *Sphagnum* species could begin a rapid phase of hummock growth, brought to an end as the hummock-top became dry enough for heath shrubs and sphagna such as *S. capillaceum* and *S. fuscum*. Other hummocks were then envisaged as growing up around it, so that in due course it was flooded and became a pool once more. Sixty years ago this elegant concept of the 'regeneration complex' was widely accepted. It became apparent, however, that there was little evidence for it in peat sections, and at many sites abundant evidence to the contrary. Since then it has been realised that many pool systems have a substantial degree of permanence, and that hummocks and hollows tend to have a regular relationship with the topography of the peat surface. If the peat surface is flat, the outline of the pools is irregular, with no preferred direction. If the surface is sloping, the pools are oriented along the contours. Many ombrogenous bog domes throughout the Boreal zone (Scandinavia, northern Russia, Canada) show similar patterns of concentrically oriented pools and hummocks, sometimes with a larger more or less circular pool at the centre. But systems of irregular deep 'mud-bottom' pools are also common. Some of the best examples of surface patterning in Britain are in the Flow Country of Caithness and Sutherland (Lindsay *et al.* 1988; Fig. 219), but there are fine examples farther south, such as Kentra Moss (Fig. 221) and Claish Moss (Moore 1977; Fig. 213) in west Argyll and the Silver Flowe in Galloway (Boatman *et al.* 1981).

There can be no doubt that patterns of hummocks and hollows stem from the interaction between water movement over the bog surface, and the growth-

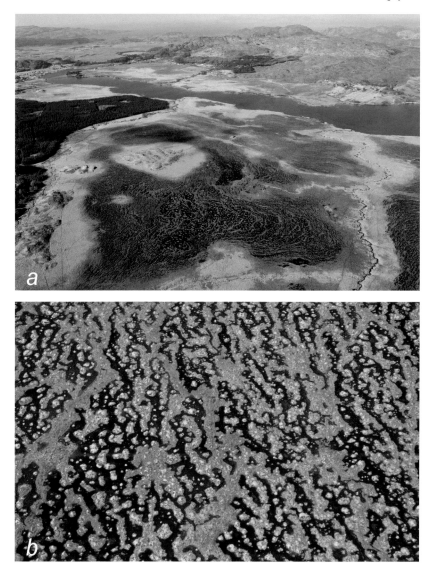

FIG 213. (a) Claish Moss, on the southern shore of Loch Shiel (of which *c.* 3 km are visible), Argyll, March 2011. The wetter ombrotrophic bogs show up as dark patches against the buff shades of the *Molinia* and deergrass of the surroundings. (b) A closer view of part of the pool system; pools oriented roughly parallel to the contours of the bog surface, separated by 'strings' of bog vegetation, with islands in many of the larger pools. The light greyish patches are *Racomitrium lanuginosum* capping the hummocks. July 2011 (Iain Thornber).

responses of the plants (especially the *Sphagnum* species) to water level. A computer model (Couwenberg & Joosten 2005), making simple (and reasonable) assumptions about rainfall, the shape of the bog, the hydraulic conductivity of the peat beneath hummocks (high) and *Sphagnum* hollows (low), and the probability of a site at any particular water level developing into either a hollow or a hummock, generates patterns reassuringly like those we see on real bogs, and we can be confident that the *initiation* of pools and hummocks depends on some such process. Once formed, large pools (as distinct from wet *Sphagnum* hollows) tend to become long-lived features; at least some of the pools on Claish Moss appear to have persisted for 5000 years. Some patterning is discernible in most bogs, and pools and hummocks are an important element in their habitat diversity, whatever the details of their origin. Needless to say, pools, hummocks and patterning on bogs have a voluminous literature, to which Standen *et al.* (1998) provide some entry points.

IS THERE A LIMIT TO BOG GROWTH?

Bog bursts

The hydrological mound obviously sets a limit beyond which a bog cannot grow, but are the surface profiles of all bogs dictated by it? Almost certainly not, if only because many bogs are still growing. What other limits are there to peat growth? Peat can only grow up to the limit set by the hydrological mound if it has the mechanical strength to resist downslope movement.

A bog-burst between Falkirk and Stirling in 1629 destroyed 16 'little farms'. On the night of 17 November 1771 peat erupted from Solway Moss, northwest of Carlisle, flowing over 400 acres of land, drowning cattle and engulfing farms, whose occupants 'passed a horrible night ... till the morning, when their neighbours with difficulty got them out through the roof.' In 1824 the Rev. Patrick Brontë wrote a graphic account of a bog-burst near Haworth on a thundery day: it was strong stuff for his sermon the following Sunday (Brontë 1824a, 1824b). A devastating burst in the early hours of 28 December 1896 high above the Ownacree valley, 15 km east of Killarney, inundated 300 acres (120 ha) of farmland with peat and swept away the house of Cornelius Donelly. He and his wife and six children all perished; part of one of their beds was picked up a few days later in Killarney Lake (Praeger 1897).

These are a few of the recorded bog bursts. There have been many others, most often in upland blanket bog. In some at least of these, the slide has taken place at the base of the peat, or in clay underlying it. Probably many minor bursts or 'peat slides' pass unrecorded, but every few years one makes at least the local headlines or for other reasons is put on record; in 2003 there were bursts in the Pennines (Teesdale), in Shetland (Channerwick) and on Slieve Aughty (Derrybrien). For more, see Tallis (2001).

Peat erosion

Almost all bogs have some erosion gullies at the edges, which may branch as miniature river systems. Sometimes, especially in upland blanket bogs, gully systems are a conspicuous feature of the mire surface, and in extreme cases they erode off the whole peat cover down to the mineral ground beneath, though usually leaving behind 'haggs' as reminders of the former bog. The southern Pennines are a classic area of peat gullying and hagging, but there are few upland

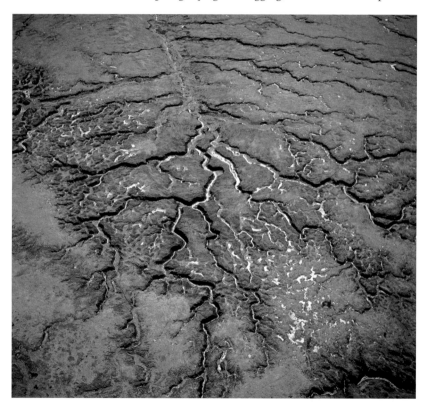

FIG 214. An erosion-gully system on blanket bog in the southern Pennines, Alport Moor, Derbyshire. Probably the finer branches develop first, often in relation to pre-existing hummocks and hollows, the parallel gullies on the steeper slopes developing later. (© Adrian Warren & Dae Sasitorn/www.lastrefuge.co.uk)

peats without at least some erosion. Erosion appears to date back to prehistoric (Bronze Age) times, and may relate to the felling of woodland in the neighbouring valleys; the climate then was warmer and drier than now. The earlier gully systems were often richly branched, creating a reticulate pattern on relatively flat ground (Fig. 214); they may have originated through coalescence and draining of pool systems. Parallel deeply-incised downslope gullies seem generally to have come later in response to more direct human pressures on the bog itself. Fire, ditching, grazing and atmospheric pollution (especially during the nineteenth and earlier twentieth centuries) have probably all played a part in accelerating erosion in recent centuries. Erosion can be rapid; exposed surfaces can lose several decimetres of peat a year. Recolonisation of eroded surfaces is often slow, from a combination of lack of propagules, poor physical and nutrient conditions for germination and establishment, disturbance by grazing sheep and the harsh climate.

Many bogs have suffered neither bog bursts nor widespread erosion of the surface, so it seems there is no hard-and-fast limit to bog growth under the present climate, or in the time span since the close of the last glaciation. Bog growth slowed (and some bogs perhaps suffered a net loss of carbon?) in the warmer, drier climate of the Bronze Age, but picked up again as the climate once more became cooler and wetter. Tantalisingly we lack fossil evidence from previous interglacials, which might give us some insight into the question. Seemingly, our northern peatlands, which bulk so large in the present landscape, are evanescent on a geological timescale!

Fire

Perhaps surprisingly, bogs are prone to fire. Even when the peat is saturated, *Calluna* and the dead previous summer's growth of *Eriophorum* and *Molinia* burn easily in dry weather. Fires under these conditions may do little lasting damage, especially if they take place in winter. Much more damaging are fires in summer droughts. If the surface layers of the acrotelm are dry enough to burn, fire can leave the bog a scene of blackened desolation. Nevertheless, the following season can bring a surprising degree of recovery. Most of the bog plants are perennials that can re-grow from underground parts, wetter spots in the hollows provide refuges for others, and *Sphagnum* can regenerate from living tissue at several centimetres depth even if the surface growth is killed. Really deep burns in severe droughts can result in the acrotelm being destroyed over wide areas, leaving a charred surface, hard to wet, and covered in ash (Fig. 215). The surface of the catotelm peat may dry out and crack irreversibly. Recovery from a severe fire like this is measured in decades. The earliest colonists are bryophytes, usually culminating after a year of two in a dense carpet of the

FIG 215. Effects of severe fire on North York Moors blanket peats. (a) Rosedale Head, October 1983, seven years after the severe fires of the dry summer of 1976. The moss *Polytrichum commune* covers wide areas of peat. Early hair-grass (*Aira praecox*) picks out areas of sandy mineral soil, and soft-rush (*Juncus effusus*) colonises moist hollows and runnels, but much bare peat remains. (b) Cockheads, 1986 burn site, February 1987. The peat surface is deeply charred, cracked and permeated with tarry material from the burnt peat above. Ash has accumulated in the hollows. Burnt heather in the foreground.

moss *Polytrichum commune*, into which rushes (mainly *Juncus effusus*) and grasses become established. Full return to normal bog vegetation takes maybe a century or two. *Polytrichum* layers are not uncommon in peat sections, and in most cases probably indicate a fire episode. Bogs can recover from occasional fires, but no bog can survive frequent burning with its vegetation intact.

Peat digging, peat exploitation

'Do you not burn turf in England?' asked Mrs Carey in reply to my compliment on her welcoming peat fire. 'Well, what do you burn then? Isn't coal terribly expensive? … And to think of it, when you can just take a donkey out on the bog, and cut as much turf as you like.' That was Co. Clare in 1953. In the 1950s many men were digging turf from the bog south of Lisdoonvarna and in the bogs west of Screeb in Connemara. There were traditional turf stacks everywhere (Fig. 216). A 1957 estimate put the annual production of hand-won peat in Ireland at about 3,500,000 tons.

Ireland has changed a lot since then. Most of the big midland raised bogs have been cut away for peat, primarily to fuel peat-fired power stations, and

FIG 216. Stacked turf ready for carting. Rinroe, near Corrofin, Co. Clare, July 1959.

FIG 217. Mechanised peat extraction, near Cloghan, Co. Offaly, September 1988. The chimneys and cooling towers of the Shannonbridge peat-fired power station can be seen in the distance; the peat-milling machinery can just be made out to the left of the power station. Many sites like this are now derelict and overgrown with birch, sallow and other vegetation.

for industrial and commercial use. Modern industrial peat extraction is highly mechanised (Fig. 217). Following drainage and stripping of the vegetation the surface peat is milled, harrowed, ridged and harvested. This sequence takes about three days, by which time moisture content has fallen to 50–55%. It can be used for power generation straightaway, but for industrial or domestic use it is dried further, to a moisture content of about 14%. Small-scale peat extraction has changed too. Many farmers now have 'sausage machines', which, drawn behind a tractor across a bog (grazed and compacted by livestock), cut a cylinder of peat below the root mat and extrude it onto the surface to dry. Small mechanical diggers are within the means of many. Some of their owners expect the same freedom to win turf as in the days when there was no alternative to a spade and muscle-power!

LOWLAND RAISED BOGS

In many respects the lowland raised bogs of Britain and Ireland are rather uniform, and the outline earlier in this chapter gives a pretty fair idea of the flora and vegetation of them all[M18]. Probably the best in Britain, judged in terms of their vegetation, are Glasson Moss (250 ha) and Wedholme Flow (270 ha), south of the Solway some 20 km west of Carlisle, but because they stand as islands in a farmed landscape and parts have been extensively exploited for peat it is not easy to appreciate their geographical context. Cors Caron (800 ha), in the Teifi valley in mid Wales is a fine group of three raised bogs set in meanders of the river. They have been cut for peat at the edges in the past, but their outlines and context are essentially intact, and they are beautiful textbook examples of the relation of raised bogs to their

FIG 218. Tregaron Bog, May 1955. The River Teifi can just be seen in the middle distance. The warm straw colour of the dead leaves of *Molinia* is prominent in the lagg, paler where it is especially luxuriant.

surroundings (Fig. 218). The river is bordered on either side by a belt of silted fen abutting on the steep rand of the bog. The rand on the landward side of the three bogs slopes gradually to a lagg fen separating the bog from the farmland on mineral soil beyond. Much of the surface of the bogs is rather dry and dominated by heather and hare's-tail cottongrass, but there are substantial areas of typical pool-and-hummock raised-bog vegetation.

A nice example of a raised bog in its context is Malham Tarn Moss (c. 30 ha) in the Yorkshire Pennines (Fig. 199). Before the Tarn Moss was acquired by the National Trust in the 1940s it had been heavily (and probably frequently) burnt; it was dominated by tussocky hare's-tail cottongrass, with wavy hair-grass (*Deschampsia flexuosa*) as the only common associate – apart from the mosses *Pohlia nutans* and *Tetraphis pellucida*, both common after burning. Since then, the bog has had 60 years to recover; there is now a good scatter of *Sphagnum rubellum* and *S. papillosum*, and most of the characteristic raised-bog species can be found somewhere, but the vegetation has a long way to go to match Tregaron or the Cumbrian mosses. As so often in Britain, *Sphagnum austinii* no longer grows on the bog surface at either site, although its highly recognisable remains make up the bulk of the peat at both. There are still many small raised bogs scattered round the north and west of Britain, particularly in Scotland.

The raised bogs of the midland plain of Ireland are more extensive than any that remain in Britain. The Irish Peatland Conservation Council listed 101 'active raised bogs of European conservation importance' of 100 ha or more, and many smaller (2000). The largest are in Roscommon, Galway and Offaly. Clara Bog, some 10 km northwest of Tullamore, is an impressively large (665 ha) expanse of raised bog south of Clara town (Fig. 201). Mongan Bog, near Clonmacnoise, is smaller (136 ha) but another good example. Both are nature reserves.

NORTHERN AND WESTERN BLANKET BOGS

Scottish bogs: the Flow Country

The Flow Country of Caithness and Sutherland is one of the wonders of the Scotland's natural heritage (Figs 219, 220). The vastness of the peat-covered terrain, broken only by a few roads, a railway, a few isolated dwellings and other buildings and occasional low rounded hills, which give a view of the patterns of the dark pools strewn over the bog, is not to everyone's taste, but for some it is addictive. There are many bogs, with magnificent pool systems. In terms of flora and vegetation the bog in the east of the Flow Country has much in common with the lowland raised bogs farther south: *Sphagnum austinii* and *S. fuscum* are

FIG 219. Towards the western end of the Flow Country, the great stretch of blanket bog that covers much of the far north of Scotland, looking towards Loch Loyal in Sutherland. The crests of the ombrogenous domes are dotted with bog pools. (Steve Moore/SNH).

FIG 220. Deep pools with black peaty bottoms and sharp vertical sides – *dubh lochan* ('little black lake'), Halsary, near Forsinard, Sutherland. Bogbean (*Menyanthes trifoliata*) abundant in the pools, many of which are ringed by flourishing deergrass (*Trichophorum cespitosum*) and bog asphodel (*Narthecium ossifragum*). September 1989.

locally frequent, and *Racomitrium lanuginosum* often caps the hummocks. *Molinia*, rare in the east, becomes commoner westward, along with the big purplish leafy liverwort *Pleurozia purpurea*. Shallow pools are dominated by *Sphagnum cuspidatum*, locally with *S. pulchrum*, white beak-sedge, great sundew (*Drosera anglica*) and bog-sedge (*Carex limosa*). Deeper pools support scattered bogbean (*Menyanthes trifoliata*), common cottongrass and loose masses of *Sphagnum*, mainly *S. cuspidatum* in the east, *S. denticulatum* becoming predominant in the west, sometimes with greater bladderwort (*Utricularia vulgaris*). The peatlands of Lewis are in effect an extension of the Flow Country westwards (Goode & Lindsay 1979).

The Moor of Rannoch

Rannoch Moor is an undulating area of desolate moorland and bog, around 360 m above sea level, ringed by mountains and covering over 250 km² almost in the centre of Scotland, with the Perthshire mountains to the east and Glencoe to the west. The vegetation is diverse, broken up by the irregular terrain; it includes areas of ombrogenous bog, with minerotrophic 'poor-fen' areas (with interesting floras) in between, and numerous lakes. The bog is generally dominated by deergrass, hare's-tail cottongrass and heather, with abundant *Molinia*[M17], great sundew, the moss *Racomitrium lanuginosum* and *Pleurozia purpurea*. Locally there are well-developed bog-pools, the habitat of the plant for which the moor is best known, the Rannoch rush. The bogs support abundant sphagna, including *S. austinii* and *S. fuscum*.

Western Scottish bogs

Ombrogenous bog is never far away in north and west Scotland, but three bogs stand out. In Argyll, Claish Moss on the south shore of Loch Shiel (Fig. 213) is a gently sloping bog with magnificent pool systems and a rich bog flora. Deergrass, hare's-tail cottongrass, heather and *Molinia* are dominant[M17], as usual in west Scotland. Other notable features include conspicuous *Racomitrium*-capped hummocks, and the common occurrence of brown beak-sedge (*Rhynchospora fusca*), bog-sedge, *Sphagnum pulchrum* and *Pleurozia purpurea*. Kentra Moss, west of Acharacle, shares many features with Claish Moss and is more accessible but has suffered more disturbance and damage. However, its southern parts still bear fine bog vegetation and splendid systems of narrow pools oriented along the contours of the gently sloping surface (Fig. 221). The Silver Flowe (Kirkcurbrightshire, 190 ha) is a chain of six more or less discrete patterned bogs on the floor of a glacial valley in the Galloway hills (Ratcliffe & Walker 1958; Boatman 1983). The southernmost major western blanket-bog site

FIG 221. Kentra Moss, Ardnamurchan, Argyll. A western Scottish bog sloping gently east towards Acharacle, with a well-developed pool system. (Top) A flattish area with large pools containing numerous islands, many topped with the moss *Racomitrium lanuginosum*. (Bottom) A gently sloping part of the bog with long pools oriented along the contours; in the left foreground is a 'spillway' from one tier of pools to the next. *Molinia* and cottongrasses are abundant, and there is a rich bog flora, including brown beak-sedge (*Rhynchospora fusca*), bog-sedge (*Carex limosa*), sundews (*Drosera* spp.) and various bladderworts (*Utricularia* spp.).

in Scotland, it is a fascinating area, offering analogies not only with the bogs of the Flow Country and other northern and western Scottish bogs, but with classic lowland raised-bog domes, and with the extensive upland blanket bogs. It is a lovely place, but could be the despair of category-minded people who seek to impose sharp boundaries on nature!

Western Irish lowland blanket bogs

In a region of near-ubiquitous blanket bog two areas stand out as classic sites: the bog between Roundstone, Ballyconneely and the Twelve Bens in Galway, and the bog around Glenamoy in northwest Mayo. The Roundstone bog has perhaps unfairly dominated our perception of western Irish bogs in general. The area is a lake-studded stretch of rocky undulating country covering some 7000 ha. Rather like Rannoch Moor, it is occupied by expanses of ombrogenous bog with mineral-influenced tracts of 'poor-fen' in between, and relatively dry heather and western gorse (*Ulex gallii*) on rocky knolls. *Molinia* and *Schoenus nigricans* are abundant almost everywhere, including areas that water analyses (and topography) show to be indisputably ombrotrophic. In species composition, the bog has much in common with the western Scottish bogs. Hare's-tail cottongrass and deergrass are usually present but seldom prominent in the plant cover. Most of the species that characteristically become commoner towards the west in Scottish bogs, such as *Pleurozia purpurea*, and bog-sedge, bladderworts (*Utricularia* spp.), and *Sphagnum denticulatum* in the pools, are in the western Irish bogs as well – with the addition of pipewort (which is in the Hebrides and just makes the Scottish mainland in Ardnamurchan). East of the Roundstone bog, there are wide tracts of blanket bog on the Connemara granite still with abundant *Schoenus* and *Molinia*, but looking much more like blanket bog elsewhere in western Britain or Ireland.

By contrast with many of the western Scottish bogs, the lowland blanket bogs of west Galway and Mayo are relatively flat. Nowhere is that more evident than in northwest Mayo, where seemingly featureless bog often stretches as far as the eye can see. The bog north of Glenamoy (nearly 5800 ha) was chosen as the site of the Irish project for the International Biological Programme in the 1960s. The Mayo bog is much less broken up by lakes than the Roundstone bog, so it is less confusing, less scenic, but it dramatically expresses how monotonously vast blanket-bog landscapes can be. There are substantial areas of lowland blanket bog in Donegal, and some in Kerry (Fig. 222), but there the mountains leave little flat land at low levels. However, the extent of turbaries on low ground around Ballinskelligs Bay and elsewhere shows that lowland bog was once widespread.

FIG 222. Blanket bog at *c.* 150 m near Kealboy Bridge, northeast of Ballaghisheen, Co. Kerry, looking towards Macgillycuddy's Reeks. Abundant *Molinia*, common cottongrass (*Eriophorum angustifolium*), black bog-rush (*Schoenus nigricans*) and sundews. As in many western bogs, *Sphagnum* is pervasive and hare's-tail cottongrass is widely scattered but neither is prominent. Large-flowered butterwort (*Pinguicula grandiflora*) present, but rare. June 2010.

MOUNTAIN BLANKET BOGS

Upland blanket bogs[M19] cover many gently domed or undulating plateaux in Wales and the Pennines, the Southern Uplands of Scotland and the eastern, central and northern Highlands. Typically they are composed largely of 'hummock' species – *Calluna, Erica tetralix* (less frequent northwards, often absent above 600 m), *Eriophorum vaginatum* and the ubiquitous *E. angustifolium*; the more northern *Sphagnum capillifolium* commonly replaces the more oceanic *S. rubellum*, and true mosses such as *Pleurozium schreberi, Dicranum scoparium* and *Hypnum jutlandicum* are prominent; *Sphagnum papillosum*, though usually present, is much less conspicuous than in lowland bogs. The heathers are often accompanied by crowberry (*Empetrum nigrum*), bilberry (*Vaccinium myrtillus*), cowberry (*V. vitis-idaea*) and in the north by bog bilberry (*V. uliginosum*). A plant particularly characteristic of these bogs is cloudberry (*Rubus chamaemorus*), and in the Highlands mountain

FIG 223. Upland blanket bog on saddle below Carn Chuinneag, some 20 km north of Dingwall, June 1981. A northeast Highland bog of rather montane character, with scattered dwarf birch (*Betula nana*) and abundant *Racomitrium lanuginosum* on the tops of the hummocks.

blanket bogs are a frequent habitat for dwarf birch (*Betula nana*) and arctic bearberry (*Arctostaphylos alpinus*). There are good areas of upland blanket bog in the Berwyns in North Wales, around upper Teesdale in the Pennines, above the Grey Mare's Tail (northeast of Moffat) in the Southern Uplands, at many places in the Highlands (Fig. 223) and on many of the gentler mountain slopes and flatter summits in Ireland. It is one of our widespread vegetation types. Pool systems can develop over flatter summits and spurs, and the vegetation then provides a similar range of habitats to lowland bogs.

Towards the west coast, mountain blanket bog takes on some of the same character as lowland bogs in the same area, notably the abundance of *Molinia*. Thus, much of the bog at 450–550 m on Dartmoor is an impoverished version of 'western blanket bog' rather than the typical 'mountain blanket bog' of high ground farther north and east, and the same is largely true of the bogs on the western mountains of Ireland.

BOUNDARIES AND TRENDS:
SOME CONCLUDING COMMENTS

Ombrogenous bogs are a remarkable ecological phenomenon. No wonder
Scandinavian ecologists see the primary division of peatlands as (ombrogenous)
bogs versus the rest! Yet they are frustratingly hard to pin down. They tend to
look recognisably the same wherever they occur, but their boundary in terms of
species composition varies from place to place; they are defined in practice by
the species that *do not* grow on them. Water chemistry is no more help; the exact
composition of their waters is everywhere different, and they overlap in pH,
Ca and any other ions with acid minerotrophic vegetation that Scandinavians
would call 'poor-fens'. The crucial test is whether they have the independence
to 'live on air' (and what it brings) and to grow and accumulate peat indefinitely
without regard to the surrounding vegetation and soils. In Estonia or Sweden
ombrogenous bogs have a short and exclusive species list. In Britain some of the
Boreal ericaceous dwarf shrubs are missing, but the regular bog flora includes
Sphagnum papillosum, common cottongrass and bog asphodel, all regarded as
'poor-fen' indicators in Sweden. In oceanic parts of Britain and Ireland *Molinia*
is added to the regular ombrogenous bog flora, and in westernmost Ireland,
Schoenus joins that exclusive assemblage. What else changes? Most conspicuously,
winter temperature, and the length of time the peat is too cold for microbial
activity; the anion deficit and no doubt other things show parallel changes.
All these are continuous trends, within which we cannot draw any natural
boundaries. The differences between 'raised bogs'[M18] and 'blanket bogs' and
between 'mountain blanket bog'[M19] and 'western blanket bog'[M17] are clear
enough (and useful) as long as you look only at their more extensive and typical
sites. But there are exceptions and intermediates, and it is in fact impossible to
define rigorously any of the categories or to specify boundaries between them.
Probably Weber would have just called them all *Hochmoor*. Precisely defined
categories, and the labels that go with them, are beloved (and needed) by lawyers
and administrators, but they do not exist in the real world. We should not be
beguiled by them. *Entia non praeter necessitatem multiplicanda sunt!*

CHAPTER 16

Heaths, Heather-Moors and Acid Grasslands

THIS CHAPTER IS ABOUT the land dominated by heathers and other heath shrubs, and the (mostly poor) grasslands that can replace heathy vegetation under grazing. Traditionally, in southeast England any land not fit for cultivation was regarded as 'heath'; in other regions 'moor' carried much the same meaning. Often these were areas on poor, infertile, acid soils dominated by dark expanses of heather giving rise to such names as Blackdown, Blackheath, Blackhill and Blackmoor. These areas were often 'commons', they were generally owned by the landowner, but his tenants (or local inhabitants) had 'common rights', which frequently included grazing.

The heaths of lowland England mark out the areas of poor, sandy or gravelly but above all acid and infertile, often podzolic, soils. They intergrade westwards with the exposed coastal heaths of southwest England, Wales and Ireland. The heather moors of hills on the harder rocks of the north and west share many species with the lowland heaths but are rather different in both flora and ecology, and in traditional land use.

LOWLAND HEATHS IN ENGLAND

The story of the English lowland heaths over the past half-century is an object-lesson for conservation. Fifty or sixty years ago much of the attention of naturalists (and biologists generally) was fixed on 'rare' or 'interesting' species or habitats. Heaths qualified for attention on neither ground; they were seen as a widespread and common vegetation type that we need not worry about.

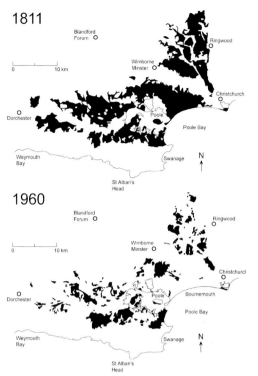

FIG 224. The extent of heathland between Dorchester and the River Avon in 1811, from the first edition of the Ordnance Survey One-Inch map, and in 1960. Re-drawn from Moore (1962).

Norman Moore in 1962 drew attention to the alarming rate of decline in the heaths of southeast Dorset (Fig. 224). In 1811, the first edition of the Ordnance Survey showed heaths covering an area of about 30,000 ha, roughly matching the area of heathland soils shown on the Soil Survey 1:250,000 map of 1983. This was well within living memory for Thomas Hardy (1840–1928) as a model for 'Egdon Heath' when he was writing his Wessex novels. The varied fortunes of farming in the nineteenth century left the overall pattern not much changed, with the major exception of the growth of Bournemouth, which did not exist in 1811. Nevertheless in 1891 the map still showed some 23,000 ha of heath. By 1934 the heathland was more fragmented, and its total area had declined to 18,000 ha. The situation when I first knew the area in 1946 was not greatly different, but heath declined very rapidly in the post-war years, and by 1960 only some 10,000 ha were left. Much of that has disappeared in the last 50 years, and the most substantial blocks of heathland that remain are now in nature reserves or army training areas.

The story has been much the same on the once-extensive heaths of the Cenozoic sands and gravels in Surrey and adjacent northeast Hampshire and

Berkshire. These heaths have been particularly hard-hit by the sprawling urban development southwest of London. The rich and diverse heathland on similar (but more varied) geology and soils in the New Forest has had better planning protection and has fared better, and it is now a National Park. The Lower Greensand at the west end of the Weald from near Dorking and Farnham in the north, southwards to Petersfield and along the foot of the South Downs to Storrington, still bears substantial areas of heath including Thursley Common and Hindhead. Heaths occur on various other substrata, including the Wealden sand of Ashdown Forest, Permo-Triassic sands and pebble beds in east Devon, but most widely on the 'clay-with-flints' overlying the chalk, and other superficial deposits of flat hill-tops, or on sandy glacial drift.

The main types of lowland heath

The heaths of lowland England intergrade in a way that defies easy classification. Most dry heaths within 100 km of London are dominated by common heather (*Calluna vulgaris*), with variable proportions of bell heather (*Erica cinerea*) and at least scattered low patches of dwarf gorse (*Ulex minor*)[H2]. Wavy hair-grass (*Deschampsia flexuosa*) is often present, but the only other vascular plants that occur at all commonly in dry heaths are bracken (mostly at the margins of the heath or where there has been disturbance in the past) and, very locally, bilberry (*Vaccinium myrtillus*). The other noticeable ingredients in the community, especially in the years following a heath fire, are a suite of calcifuge mosses, including *Ceratodon purpureus*, *Polytrichum juniperinum*, *Dicranum scoparium* and *Hypnum jutlandicum*, and lichens of the genus *Cladonia*.

Heath can grow on thin superficial deposits over chalk. Before the widespread ploughing of the 1950s and 1960s, and in the heyday of rabbits, these sites were more widespread and often closely grazed, and were very striking for their mixture of calcicole and calcifuge plants. Tansley (1939) could write: 'Thus is developed the beginnings of a vegetation which has been called chalk heath, because it is marked by a mixture of calcicolous plants rooting in calcareous soil with indifferent and calcifuge heath plants rooting in the acid surface soil. Large areas of plateau and dip slope are dominated by *Calluna* or *Erica cinerea*, accompanied by such characteristic southern heath plants as *Ulex minor*.' Most of the 'large areas of plateau and dip slope' are now under the plough, and with the relaxation of grazing, the fragments of 'chalk heath' that remain have realised Tansley's supposition that '... "chalk heaths" of the kind described develop into more typical heaths by further leaching and the accumulation of acid humus at the surface.'

Dwarf gorse is a low-growing subshrub (Fig. 225a), rather strictly confined to heathland. It has more delicate spines than common gorse (*Ulex europaeus*),

FIG 225. Heath and moorland subshrubs: (a) dwarf gorse (*Ulex minor*); (b) Dorset heath (*Erica ciliaris*); (c) Cornish heath (*Erica vagans*); (d) Mackay's heath (*Erica mackaiana*); (e) St Dabeoc's heath (*Daboecia cantabrica*); (f) Cowberry (*Vaccinium vitis-idaea*).

smaller flowers, much less hairy buds and calyces, and flowers in late summer. It is concentrated in southeast England from Surrey, Berkshire and the High Weald to the heaths of southeast Dorset, with outlying localities in Lincolnshire and Nottinghamshire and a northern outpost near Carlisle. Common gorse is an altogether more robust shrub, up to a metre or two high, with stiff sharp spines; it is common on heaths along road and track sides and in disturbed places, but also grows in scrub and open woodland on acid soils. It starts flowering in autumn, continues in mild spells through the winter, has its main flowering period from March to May, and some flowers hang on into early summer. It is common throughout Britain and Ireland.

From the southwest fringes of the Surrey–Berkshire heaths to the New Forest and the heaths of southeast Dorset, two other major players join the dry-heath community, bristle bent (*Agrostis curtisii*) and purple moor-grass (*Molinia caerulea*). Bristle bent has a compact distribution in south and southwest England and south Wales, within which it is abundant on almost every heath. It is a very distinctive grass, forming tight tufts of fine, slightly glaucous leaves amongst the heath subshrubs, quite unlike our other species of *Agrostis*. *Molinia* is a versatile grass we have already encountered in fen meadows (Chapter 14) and blanket bogs (Chapter 15). On the heaths of the New Forest–Poole Harbour area, the predominant *Ulex minor–Agrostis curtisii* heath[M3] is dominated by an intimate mixture of varying quantities of six species: heather, bell heather, cross-leaved heath (*Erica tetralix*), dwarf gorse, bristle bent and *Molinia*. In the Poole Harbour area these six are sometimes joined by Dorset heath (*E. ciliaris*, Fig. 225b), which is abundant across the Channel in the heaths of Brittany. Apart from that, the only other frequent vascular plants are tormentil (*Potentilla erecta*), heath milkwort (*Polygala serpyllifolia*) and the ubiquitous common gorse and bracken. Mosses and lichens are sometimes prominent, including *Hypnum jutlandicum*, *Campylopus brevipilus*, *Cladonia portentosa* (Fig. 235e), *C. coccifera* (Fig. 235f) and *C. crispata*. In sharply drained and exposed places, *Molinia*, cross-leaved heath and even bristle bent may drop out, leaving an impoverished dry *Calluna–Erica cinerea* heath. Conversely, with increasing wetness of the soil, bell heather and bristle bent drop out at the transition to wet heath – but that is a story that will be taken up later.

West of Dorchester, dwarf gorse gives way remarkably suddenly and completely to a second late-summer-flowering species, western gorse (*Ulex gallii*). These two species were much confused in the past, because while their most obvious differences are in size, they are both very plastic to variations in soil and exposure. They have different chromosome numbers (dwarf, 2n = 32: western, 2n = 64) and can usually be reliably told apart by the smaller flower parts of dwarf gorse (the calyces persist until well into winter).

FIG 226. 'Six-species heath' (*Ulex gallii–Agrostis curtisii* heath) on Triassic pebble beds, Aylesbeare Common, Devon, September 1987.

Heaths on the clay-with-flints and other superficial deposits that cap the chalk and the older Jurassic and Triassic rocks of west Dorset and east Devon are dominated by the same six species as the *Ulex minor–Agrostis curtisii* heath, but with western gorse substituted for dwarf gorse. This six-species *Ulex gallii–Agrostis curtisii* heath[H4] is the common dry heathland of the Southwest Peninsula, from west Dorset and the coast of south Wales to Cornwall (Fig. 226). It is more widely distributed, more variable and often richer in species than its more eastern counterpart. Tormentil is near-constant, and heath milkwort is usually present but easily overlooked. Mosses (e.g. *Dicranum scoparium, Hypnum jutlandicum*) and lichens (e.g. *Cladonia portentosa, C. floerkeana, C. coccifera, C. chlorophaea*) can be prominent, especially a few years after burning.

In southwest England the *Ulex gallii–Agrostis curtisii* heath comes into contact with other heath and moorland types, and intergrades with them. Where it grows on thin superficial deposits over limestone, it can acquire an admixture of deep-rooted calcicole species. On the upland areas of the southwest, especially Exmoor, Dartmoor and Bodmin Moor, it meets upland heather-moor and hill grasslands, and variations in geology, topography and grazing pressure can produce intricate (and sometimes baffling) gradations, mixtures and mosaics.

FIG 227. Tuddenham Heath, Suffolk. October 1980.

The heaths we have considered so far fall into a neat series of extensive heathlands on leached but stable soils across southern England. There are some lowland heaths that do not fit in with this pattern, notably the Breckland heaths of East Anglia (Fig. 227). These are on acid, sandy soils, derived from wind-borne 'coversands' deposited under dry periglacial conditions. The climate of the Breckland is dry and prone to late spring frosts. There are old records for bell heather and dwarf gorse in Breckland, but both are probably at the edge of their climatic tolerance there and no longer figure in the heathland flora. The surviving Breckland heaths are dominated by common heather, with bracken as the only other major player, but sheep's fescue (*Festuca ovina*) is usually present, along with the moss *Hypnum cupressiforme* (sensu lato) and lichens such as *Cladonia uncialis, C. portentosa, C. fimbriata, C. furcata, C. pyxidata, C. squamosa, Cetraria aculeata* and *Hypogymnia physodes*[H1]. Sites such as Cavenham Heath and Lakenheath Warren have become classic ground for British plant ecology through the researches of Farrow and Watt. Coversands were also deposited in Lincolnshire and Nottinghamshire, where until relatively recently there were quite extensive heaths, and in the Vale of York where they are now mostly farmland, and only fragments of heathland (such as Skipwith, Strensall and Allerthorpe Commons) remain. An intriguing outlier of very 'southern' heath with dwarf gorse existed until the 1960s just west of Carlisle.

The heaths of the Lizard serpentine

The Lizard district of Cornwall includes one of the largest outcrops of serpentine in Britain, and certainly the best known. Serpentine is an ultrabasic rock, rich in magnesium and heavy metals such as nickel and chromium, but poor in potassium and calcium. It is often attractively marbled and veined. Local craftsmen at the Lizard make trinkets and souvenirs out of it, and the harder forms have been valued for decorative carving and facings in building ('Connemara marble' is another serpentine rock, not a true marble). Many plants fail to thrive on serpentine soils, either because of heavy-metal toxicity, or because the ratio of calcium and potassium to magnesium is too low. But 'one man's meat is another's poison', and the Lizard is home to a remarkable concentration of plants rare in Britain, and some distinctive plant communities.

Lizard Head itself, the most southerly point in England, is on schist and has quite different vegetation from the serpentine just to the north. Any visitor driving to Kynance Cove in late summer cannot fail to notice the Cornish heath (*Erica vagans*, Fig. 225c) in flower, sometimes pink, sometimes white, in the 'mixed heath' on the roadside. Cornish heath is a southern species, occurring very locally in dry coastal heath in Brittany, and in the Basque country in France

FIG 228. The Lizard; *Erica vagans–Ulex europaeus* 'mixed heath' above Kynance Cove, (a) at the peak flowering season of Cornish heath (*Erica vagans*) in August 1970 and (b) the same vegetation in early May 1994, with common gorse in flower.

and Spain (it has one possibly native locality in Fermanagh and is naturalised on the Magilligan dunes). This *Erica vagans–Ulex europaeus* heath[H6] at the Lizard is a species-rich heathland on deep mildly acid soil, which covers large areas of the cliff-top and well-drained valley sides for a few hundred metres back from the coast (Fig. 228). Cornish heath, bell heather, common and western gorse share dominance. Glaucous sedge (*Carex flacca*), common dog-violet (*Viola riviniana*) and dropwort (*Filipendula vulgaris*) are also near-constant. Many other species are frequent in the community, including tormentil, common heather, betony (*Stachys officinalis*), common milkwort (*Polygala vulgaris*), wild thyme (*Thymus polytrichus*), spring squill (*Scilla verna*), cat's-ear (*Hypochaeris radicata*), bird's-foot trefoil (*Lotus corniculatus*), lady's bedstraw (*Galium verum*) and the grasses *Dactylis glomerata* and *Danthonia decumbens*. This heath is one of the glories of the Lizard.

On the flat poorly drained plateau over serpentine between the main Helston–Lizard road and Coverack, another distinctive community covers wide tracts of ground. This *Erica vagans–Schoenus nigricans* 'tall heath'[H5] (Fig. 229) stands at a meeting-point of heath, fen and fen meadow. Cornish heath, black bog-rush (*Schoenus nigricans*), cross-leaved heath, *Molinia* and sometimes western gorse share dominance. Other characteristic plants include devil's-bit scabious (*Succisa pratensis*), saw-wort (*Serratula tinctoria*), petty whin (*Genista anglica*), great

FIG 229. The Lizard: *Erica vagans–Schoenus nigricans* 'tall heath' on Goonhilly Downs, September 1982. Two 'dishes' of the BT Goonhilly Earth Station appear on the skyline.

burnet (*Sanguisorba officinalis*), glaucous sedge (*Carex flacca*), flea sedge (*C. pulicaris*), carnation sedge (*C. panicea*), bog pimpernel (*Anagallis tenella*) and the rich-fen mosses *Campylium stellatum* and *Scorpidium scorpioides*. Now that burning is much less frequent than 50 years ago, the smaller plants are less conspicuous than they were, and the species diversity of the heath appears to have declined.

Wet heaths

All of the dry heaths described above grade downwards with increasing wetness into a distinctive but rather uniform wet heath. Bell heather cannot grow below the level at which the soil is saturated during the winter because its roots are sensitive to the high concentrations of reduced (ferrous) iron in waterlogged soils, and the same is probably true of bristle bent. Wet heath is a widely distributed community, known as the 'Ericetum tetralicis' on the Continent[M16]. Its dominant species with us are cross-leaved heath, common heather, *Molinia* and (if the heath has been burnt in the last decade or two) *Sphagnum compactum* (Fig. 230a). Other frequent species include *Sphagnum tenellum* (Fig. 230b), deergrass (*Trichophorum cespitosum*, for which this is the main habitat in southeast England), bog asphodel (Fig. 230f), round-leaved sundew (*Drosera rotundifolia*), common cottongrass (*Eriophorum angustifolium*), *Campylopus brevipilus* (perhaps currently being ousted by the introduced *C. introflexus*) and the lichens *Cladonia portentosa* (Fig. 235e) and *C. uncialis*. Runnels through the wet-heath belt provide a habitat for oblong-leaved sundew (Fig. 230d), white beak-sedge (Fig. 230c) and, much more locally, brown beak-sedge (*Rhynchospora fusca*) and marsh clubmoss (Fig. 230e), especially a few years after a heath fire. The floristic affinities of wet heath are with bog rather than with dry heath. With still wetter conditions the wet heath grades into valley-bog[M21] and this in turn into tussocky *Molinia*[M25]. The Ericetum tetralicis wet heath can intergrade in various directions – laterally into tussocky *Molinia*, with slight base-enrichment into a version dotted with heath spotted-orchids (*Dactylorhiza maculata*, Fig. 231a) and lousewort (*Pedicularis sylvatica*), and through that into the heathy form of Cirsio-Molinietum fen meadow (M24).

Origins and conservation management of lowland heaths

Many of our heathlands are probably very old, dating back to Bronze Age or Neolithic forest clearance or maybe in some places even further back than that. How far they were ever used to grow crops is uncertain, but certainly the main traditional use of heathland was low-density grazing of stock – sheep, ponies and cattle – much as on our uplands now. This has probably been so since prehistoric times. Heath fires were a common occurrence. Some of these were accidental,

FIG 230. Wet-heath plants: (a) *Sphagnum compactum*; (b) *S. tenellum*; (c) White beak-sedge (*Rhynchospora alba*); (d) Oblong-leaved sundew (*Drosera intermedia*); (e) *Lycopodiella inundata*; (f) Bog asphodel (*Narthecium ossifragum*).

FIG 231. A common and a rare wet-heath plant: (a) heath spotted orchid (*Dactylorhiza maculata*); (b) marsh gentian (*Gentiana pneumonanthe*).

but burning was used deliberately to encourage new growth, and to get rid of old 'leggy' heather and gorse, and the mattress of dead leaves left by *Molinia* at the end of the season. The frequency of burning, and the balance between burning and grazing, has probably varied widely, but either would prevent colonisation of the inhospitable heath soils by trees. Arguably, *Calluna* and the other heathers are as fire-adapted as *Banksia* and the mallee *Eucalyptus* species in Australia. The west-European heaths could be seen as a fire-adapted ecosystem replacing oak forest, just as the fire-prone scrublands of the Mediterranean replace climax evergreen oak (*Quercus ilex*) forest.

With neither grazing nor burning, our heaths are invaded by trees, and something must replace traditional management if they are to continue as open heathland. Scots pine, introduced for landscape planting and for forestry, finds a congenial home in heathland, and many heaths in Surrey and the New Forest have been taken over by it. On slightly better soils birch (*Betula pendula* and *B. pubescens*) is the main colonist. Both pine and birch have light wind-borne seeds.

Many heathland plants are at their best 5–10 years after a heath fire. These include marsh clubmoss, the sundews, white and brown beak-sedge, the very local and beautiful marsh gentian (Fig. 231b) and many of the mosses and lichens. A heath long-unburnt tends to be a very dull place, and heathland

managers who will not use burning for management bear a heavy burden
of guilt for declining biodiversity in our heathlands. This is as true of the
heathland animals as it is of the plants. A heath fire is tough on those individual
plants or animals that happen to be above ground at the time – but occasional
fire is what has created their habitat. Burning must not be too frequent; about
every 15 years seems to be about optimal for our heaths. Mowing is no substitute,
and grazing alone does not have the same effect.

HEATHER MOORS, MAINLY UPLAND AND WESTERN

In upland areas and on rugged coasts from southwest England, Wales, Cumbria
and Galloway to Kerry and Connemara a heather moor dominated by common
heather, bell heather and western gorse is widespread on well-drained acid soils[H8]
(Fig. 232). The dominant plants leave little room for other species, but tormentil,
common dog-violet, heath bedstraw (*Galium saxatile*), fine-leaved fescues (*Festuca
ovina, F. rubra*), sweet vernal-grass (*Anthoxanthum odoratum*), common bent (*Agrostis
capillaris*) and heath-grass (*Danthonia decumbens*) are frequent. St Dabeoc's heath

FIG 232. *Calluna–Ulex gallii* heath[H8], widespread on exposed rocky ground at the coast, as
here near Land's End, September 1982, and up to moderate altitudes on our western hills.

(*Daboecia cantabrica*, Fig. 225e) is locally common in this community in Connemara, probably avoiding the peatier and most nutrient-poor sites.

On Mendip and the south Wales coast, where this community occurs on thin acid soils over limestone, transitions to limestone grassland occur, analogous to the 'chalk heaths' of the Downs, with calcicoles including salad burnet (*Sanguisorba minor*) and common rock-rose (*Helianthemum nummularium*), and species such as glaucous sedge, devil's-bit scabious, slender St John's-wort (*Hypericum pulchrum*) and betony (Etherington 1981). Western gorse does not occur in the Burren, so although there are plenty of limestone heaths on Black Head and the tops towards Ballyvaghan, none are analogous to this community.

The Midlands and North of England

From the English Midlands north to much of the Pennines and the North York Moors heaths and heather moors tend to be almost pure *Calluna*, enlivened only by a thin understorey of wavy hair-grass and mosses, mainly *Pohlia nutans* and *Hypnum jutlandicum* (Fig. 235d)[H9]. This rather dull moorland, grazed by sheep and in places managed for grouse, is a monument to the air pollution from the heavy industry that underpinned Britain's economic prosperity a century ago. It is not wholly without variety; bilberry is locally quite prominent, especially on the sharply drained scarps and crags of Millstone Grit, the 'bilberry edges', of the south Pennines. These moors are in effect an impoverished version of the next community, and show signs of recovery towards it, but it is a slow process.

The north and west of Britain and Ireland

The really widespread kind of upland heather moor from Cornwall to Shetland and on the eastern uplands of Ireland is dominated by heather with variable amounts of bilberry, with scattered wavy hair-grass and a suite of bryophytes including *Dicranum scoparium, Hypnum jutlandicum, Pleurozium schreberi* (Fig. 235a), *Hylocomium splendens* (Fig. 89d) and *Ptilidium ciliare*, and lichens such as *Cladonia portentosa* (Fig. 235e), *C. uncialis, C. pyxidata* and *C. coccifera*[H12]. Bell heather and crowberry (*Empetrum nigrum*) are frequent, as is tormentil. There is some regional variation. This community intergrades with the *Calluna–Deschampsia flexuosa* heath in the Welsh border counties and the north of England. Cowberry (*Vaccinium vitis-idaea*), rare in the south, takes an increasingly prominent role in heather moor northwards, especially in the eastern Highlands.

In the north of England and Scotland heather moorland is widely used as grouse moor. Mostly these are maintained by controlled burning in small patches, which creates a patchwork of different ages. The younger growth provides a nutritious food supply, while the taller older patches provide good

FIG 233. Muirburn pattern in heather–bilberry moorland[H12], eastern foothills of the Cairngorms near Tomintoul, September 1980.

cover for the nesting birds within a short distance. This rotational burning produces a characteristic (but hardly pretty) pattern in the grouse moors (Fig. 233). It perpetuates open heather moor, and benefits nature conservation, provided the rotation is not too short, and especially if steep rocky areas of heather are left unburnt. Nature conservation on many lowland heaths would benefit from similar rotational burning, but probably in larger patches and on a somewhat longer rotation.

On thin, acid, sharply drained soils around rock outcrops and similar dry heathy places the heather–bilberry moor gives way to a distinctive community dominated by the two heathers *Calluna* and *Erica cinerea*, with tormentil as the only other near-constant species[H10]. Heath bedstraw, sheep's fescue, sweet vernal-grass, wavy hair-grass, common bent, green-ribbed sedge (*Carex binervis*), hard fern (*Blechnum spicant*) and the mosses *Dicranum scoparium, Racomitrium lanuginosum, Pleurozium schreberi, Hypnum jutlandicum, Hylocomium splendens* and lichens such as *Cladonia portentosa* and *C. uncialis* are frequent to occasional. The influence of the wet oceanic climate is shown by occasional cross-leaved heath and deergrass. This community has a similar overall geographical distribution to the last. It is common on steep rocky ground near the west coast, but it occurs inland too, and it is well represented in the eastern Highlands.

FIG 234. Deergrass moorland[M15], showing typical autumn colour, east of Beinn Bhan, Applecross, Wester Ross, September 1974.

Deergrass moorland in the west

Wide tracts of country on peaty soils in the west of Wales, Ireland and above all Scotland are dominated by a wet heath or degraded blanket-bog community, firm underfoot, dominated by deergrass, cross-leaved heath, *Calluna* and *Molinia*[M15]. This community covers thousands of monotonous hectares in the western Highlands and islands (Fig. 234), and probably covers proportionately as much of the landscape in west Wales and in the blanket-bog districts of Ireland. Apart from the dominants the most frequent species are common cottongrass, bog asphodel, round-leaved sundew and bog-myrtle (*Myrica gale*). A long list of other bog, poor-fen and acid-grassland species occur occasionally, including several species of *Sphagnum*. The geographical limits of this community are almost the same as the heather–bilberry moor, but it covers wider expanses of ground in the west.

HEATHY GRASSLANDS ON ACID SOILS

The heath shrubs have their growing-points at the tips of their branches. Grasses keep their growing-points at ground level, protected by the bases of the leaves.

Because of this, heathers are much more severely set back by fire or grazing. In the year following burning, fire-lines can be very striking by the colour-contrast between the luxuriant growth of the grass and the unburnt heath. This may be seen by a farmer as a desirable outcome, and under favourable conditions, if grazing animals are put onto recently burnt ground, the heath can be transformed into grassland in only a few years. Most of our acid hill grasslands must have originated in essentially this way, by some combination of felling, grazing or burning on pre-existing forest or heath.

Fescue–bent grasslands

On dry, nutrient-poor soils developed over hard, acid rocks (granite, rhyolite, hard sandstones and grits) or impoverished sands, the thin turf is dominated by sheep's fescue (*Festuca ovina*) and common bent (*Agrostis capillaris*), with sheep's sorrel (*Rumex acetosella*) as the commonest associate. Some other small plants are frequent or occasional, including early hair-grass (*Aira praecox*), whitlowgrass (*Erophila verna*), stork's bill (*Erodium cicutarium*) and shepherd's cress (*Teesdalia nudicaulis*). Many other species occur occasionally. Mosses and lichens are sometimes conspicuous, the most frequent being *Dicranum scoparium* and *Brachythecium albicans*. This *Festuca–Agrostis–Rumex acetosella* grassland[U1] has a wide but scattered distribution from the south coast of England to central Scotland. Much of what was lowland 'grass-heath' in Tansley's day is now farmland.

Much more widespread is the *Festuca ovina–Agrostis capillaris–Galium saxatile* grassland[U4]. This is arguably our most important hill-pasture type. It occupies mildly acid brown-earth soils over rocks that are neither excessively base-poor nor calcareous. It is almost ubiquitous throughout our uplands from Devon and Cornwall to Shetland and the Western Isles, and similar vegetation occurs in Ireland. The turf is dominated by sheep's fescue and common bent, with abundant sweet vernal-grass; tormentil and heath bedstraw (*Galium saxatile*) are nearly always present. Other frequent species include field wood-rush (*Luzula campestris*), common dog-violet, Yorkshire-fog (*Holcus lanatus*), yarrow (*Achillea millefolium*), white clover (*Trifolium repens*), common mouse-ear (*Cerastium fontanum*), self-heal (*Prunella vulgaris*), lady's bedstraw, mountain pansy (*Viola lutea*) and bird's-foot trefoil, and mosses including *Rhytidiadelphus squarrosus* (Fig. 235b), *Pseudoscleropodium purum*, *Dicranum scoparium* and *Hypnum cupressiforme* (sensu lato). Typically this community occupies the uppermost enclosed fields on a hill farm, and the relatively steep lower slopes of the unenclosed land (Fig. 238). It provides good grazing. The enclosed fields may be ploughed and re-sown, but they revert towards the semi-natural *Festuca–Agrostis* pasture over the course of time.

FIG 235. Common mosses and lichens of heath, moorland and acidic grasslands.
(a) *Pleurozium schreberi*; (b) *Rhytidiadelphus squarrosus*; (c) *Polytrichastrum formosum*;
(d) *Hypnum jutlandicum*; (e) *Cladonia portentosa*; (f) *Cladonia coccifera* (agg.). These are two
of the commonest species of *Cladonia*; there are many others.

The *Festuca–Agrostis* grassland is variable, depending on soil fertility and pH. In upland situations where the soil parent-material provides a modicum of calcium (as on various base-rich rocks in Wales or calcareous schists in Perthshire), the grassland is richer in species. In addition to those plants already noted, wild thyme (*Thymus polytrichus*) and harebell (*Campanula rotundifolia*) are near-constant in this *Festuca–Agrostis–Thymus* grassland[CG10], white clover, bird's-foot trefoil and meadow buttercup (*Ranunculus acris*) are more frequent, and many other species occur occasionally, including quaking-grass (*Briza media*) and fairy flax (*Linum catharticum*).

Bracken

Bracken (*Pteridium aquilinum*) hardly deserves a named community of its own[U20], because it is usually either a part of a woodland understorey or an invader into *Festuca–Agrostis* grassland (Fig. 236), which, at least initially, continues as an understorey beneath the bracken canopy. As a fern, bracken reproduces by spores, but these need damp sheltered conditions to germinate and for the

FIG 236. The Llanthony valley in the Black Mountains, looking towards Darren Lwyd, August 1989. The enclosed fields on the gentle slopes are improved grassland. Bracken has spread to cover most of the potential bent–fescue grassland on the steeper slopes. There is some heather, mostly on the steeper slopes, and some mat-grass (*Nardus stricta*) on the flatter tops.

prothalli to mature, fertilise, and produce sporeling ferns. Once established, bracken can spread far and wide by means of its long-creeping underground rhizomes. In a recently invaded pasture this radial spread of bracken is often very obvious; Fig. 48), and it is easy to see the point where it became established, often a wall or a patch of disturbed ground. Bracken is intolerant of waterlogging, and needs a fairly deep soil for growth of the rhizomes. Bracken had some agricultural value in the past, when it was often cut for bedding livestock, but now it is generally seen as an unwelcome invader of pasture, added to which, it is poisonous to livestock. Bracken is hard to control because a large part of the plant's resources are underground in the rhizomes. Traditional methods of control, repeated cutting or crushing, were time-consuming and labour-intensive, and needed to be repeated year by year if they were to have an effect. Some success has been achieved in recent decades by aerial spraying with asulam, a systemic herbicide, which inhibits the development of frond initials and rhizome apices. Correctly timed and targeted application of a sufficient dose can certainly check bracken, but total eradication remains difficult.

Other upland acid grassland
Some acid grasslands are in effect grazed facies of the corresponding heath. *Deschampsia flexuosa* grassland[U2] might be derived from any one of a number of heaths in which wavy hair-grass plays a part. It has a wide but scattered distribution in England, Wales and eastern Scotland, and is probably derived from different starting points in different places.

Agrostis curtisii grassland[U3] is clearly derived from *Ulex gallii–Agrostis curtisii* heath by burning and grazing, and most of the species of the heath are still to be found in the grassland. Nevertheless, the visual difference between the grassland and the heath is very great. Much of the pale fine-leaved grass seen in late summer on Dartmoor or Bodmin Moor is bristle bent rather than the pure expanses of mat-grass (*Nardus stricta*) that would be expected on upland areas farther north.

The mat-grass-dominated *Nardus stricta–Galium saxatile* grassland (replacing damp heather moor) covers wide tracts of upland country on moist, acid peaty-gley soils. *Nardus* starts into growth and flowers early in the spring, but quickly becomes tough, wiry and unpalatable to stock. It is of much less value for grazing than the *Festuca–Agrostis* pastures of the steeper and better-drained slopes. *Nardus* generally makes up the bulk of the herbage, but heath bedstraw, tormentil, sheep's fescue, common bent and the moss *Rhytidiadelphus squarrosus* are usually present. The *Nardus* grassland often includes a seasoning of species from neighbouring vegetation – velvet bent (*Agrostis canina*), common sedge

(*Carex nigra*) and the moss *Polytrichum commune* from acid poor-fen patches, sweet vernal-grass, heath rush (*Juncus squarrosus*), deergrass and bilberry from other heaths and grasslands, and the ubiquitous upland calcifuge mosses *Dicranum scoparium, Hypnum jutlandicum, Pleurozium schreberi* and *Hylocomium splendens*. *Nardus* grassland is far from being as species-poor as its apparent uniformity might suggest.

Molinia is often prominent in the uplands, on valley floors, and on gentle slopes with damp peaty soils[M25]. The leaves die at the end of the growing season, leaving copious leaf-litter but providing no winter grazing (Fig. 237). Like *Nardus*, it starts into growth relatively early, but by the time the ewes and lambs are on the hills there is more palatable pasture available.

Heath rush is the remaining player in the hill-grassland scene. It could be said that the *Juncus squarrosus–Festuca ovina* grassland'[U6] is a grassland only in name. It grows on damp peaty soils on gentle slopes or flattish tops, often forming an irregular zone above the *Nardus* grassland. It is dominated by varying proportions of heath rush, sheep's fescue and the moss *Polytrichum commune*. The balance of the community is made up of a mixture of poor-fen, wet heath and bog species – common sedge, velvet bent, cottongrasses, *Sphagnum papillosum* and a string of other bryophytes including the mosses *Dicranum scoparium, Hypnum jutlandicum, Pleurozium schreberi, Rhytidiadelphus squarrosus, R. loreus, Hylocomium*

FIG 237. Burning damp *Molinia* pasture (here with clumps of rushes, *Juncus effusus*), near Simonsbath, Exmoor, 4 April 1980. 'Swaling' was traditional practice among upland farmers to encourage fresh green growth.

splendens and the leafy liverworts *Lophocolea bidentata*, *Lophozia ventricosa*, *Barbilophozia floerkei* and *Ptilidium ciliare*. The *Juncus squarrosus* rush-heath is a product of burning and grazing other vegetation, perhaps most often a late stage in the degradation of upland blanket bog.

Upland grazings

The heather moors and upland acid grasslands are the main ingredients of the hill-farming landscape (Fig. 238). In autumn, the various dominants with their distinctive colours make up a multi-hued patchwork covering the landscape. The improved fields in the valley[MG6] will still be bright green: if a hay crop has been taken off the grass will have had time for a fresh flush of growth. The *Festuca–Agrostis* pastures[U4] on the valley sides will be a more subdued shade of green and the bracken will already have turned red-brown. Patches of *Calluna–Erica cinerea*[H10] and *Calluna–Vaccinium* heath[H12] will be dark brownish green (with a

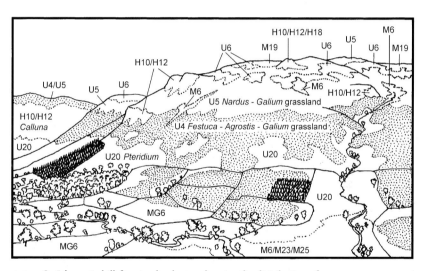

FIG 238. Schematic hill-farming landscape showing the distribution of some common upland vegetation types. MG6 is the rye-grass–crested dog's-tail improved grassland of the fields in the valley. Stippling marks the *Festuca ovina–Agrostis capillaris–Galium saxatile* pasture (U4) of the lower slopes, sometimes colonised by bracken (U20). Above this is a zone of mat-grass (*Nardus stricta*) grassland (U5) on peatier soil, with patches of heath rush (*Juncus squarrosus*) (U6). The summit is capped by upland blanket bog (M19). Sharply drained rocky terrain bears *Calluna–Erica cinerea* heath (H10) or *Calluna–Vaccinium myrtillus* heath (H12). Wet uncultivated ground in the valley is a mix of acid mire, rushy pasture and *Molinia*. Adapted from Averis *et al.* (2004).

light pinkish-brown tinge if there is a lot of *Vaccinium*). *Nardus* grassland[U5] on the gentle higher slopes will be a pale straw colour; *Molinia*[M25] will be slightly less pale and a warmer hue. *Juncus squarrosus* will be dark brownish[U6], and blanket bog[M19] on the summits will be dark brownish-green from the heather, perhaps with a reddish tinge from the autumn colour of the cottongrass leaves.

But this is farming country, and the pattern of the landscape has grown from the hard economic realities of making a living from it. The various moorland and hill-grassland types are of different value for grazing livestock. In sheepwalks in mid Wales studied in 1938, the grazing density was greatest on *Agrostis*-dominated swards at any time of year. In June, the grazing on any other type of sward was less than a fifth of the grazing on *Agrostis*, but over the year as a whole the figure was 30–40% for all but the wettest swards. In Snowdonia in June 1945, *Agrostis–Festuca* grasslands supported the highest densities of sheep, fescue–heather or fescue–bracken mixtures rather less, and other sward types with *Nardus* or *Molinia* less again. Over the year the *Agrostis–Festuca* swards supported on average 2.8 ewes per acre, *Nardus* and *Molinia* swards less than 1.4 (Hughes 1958). In the Cheviot region of southern Scotland, the year-round order of grazing intensity was bent–fescue 151, open bracken (with bent–fescue) 137, *Nardus* 72, soft-rush infested pasture 69, *Molinia* 55 (Hunter 1962). Bent–fescue (with or without an open cover of bracken) was almost equally favoured in any month of the year. *Nardus*, heather and heather–hare's-tail cottongrass (*Eriophorum vaginatum*) on deep peat were most favoured in winter, and the rush-infested pasture in February and from May to July (perhaps for shelter – 'a clump of rushes is worth a shilling'). These figures are from mid-twentieth-century sheep farming, which has changed, just as farming has changed in the past. In former centuries cattle outnumbered sheep in upland Snowdonia, and were nearly as numerous as sheep and goats combined. The cattle needed herding to take advantage of the short growing season on the high hills and were brought down to the valley in winter. Transhumance from the lowland *hendre* to the upland *hafod* (sheiling, buaile or booley) was the norm. Sheep only became predominant with the enclosures from the eighteenth century onwards. Until a century ago, mutton was in demand and the wethers overwintered on the open hill pastures, where they grazed what they could from *Nardus* and *Molinia*, and were on the ground when these grasses started into growth (Roberts 1959). Our problems with the conservation on lowland heaths in England come about because the agricultural practices that created and maintained them are no longer economically relevant and have fallen into disuse. Grouse moors and hill grasslands are part of a living cultural landscape. For that, we should cherish them more, not less.

CHAPTER 17

Mountains

WHAT CONSTITUTES A 'MOUNTAIN' depends on where you are, and who you ask. In central Europe a strict definition might be that a mountain must extend above the tree-line, so that its upper slopes and crags are naturally treeless. There the tree-line is at about 2000 m. In the Himalayas or the equatorial Andes the tree-line is at nearly double that altitude, but few would deny the status of 'mountain' to summits that fail to reach it. The highest summit of the Harz mountains in Germany, the Brocken, reaches only 1178 m so is entirely within the forest zone, and in North Wales any rocky summit can qualify, even Bangor Mountain (117 m) and Holyhead Mountain (220 m).

This chapter roughly follows the central-European definition suggested above. Our mountains are very rarely forest-covered up to the potential tree-line, and 'upland' and 'mountain' vegetation pass imperceptibly into one another. Nevertheless, there is evidence that the natural tree-line in Britain could reach around 600–650 m above sea level. Many 'hill grasslands' are on ground that would once have borne forest. So we shall take 600 m as an arbitrary (and fuzzy) line between the 'uplands' of the last chapter and the 'mountains' of this one.

The highest mountain in Britain, Ben Nevis, reaches 1343 m, but several summits in the Cairngorms are only a little lower, and the greatest extent of land above 1200 m in Britain is on the Cairngorm plateau (Fig. 239). The Highlands have 283 'Munros' – summits exceeding 3000 feet (914 m) – from Ben Lomond (973 m) 40 km northwest of Glasgow to Ben Hope (927 m) 250 km farther north. The Southern Uplands of Scotland have five summits over 800 m and another half-dozen over 700 m, but little mountain vegetation. Snowdon (1085 m) is the highest of a dozen summits, all over 920 m, in a compact area of north Wales; Aran Fawddwy (905 m) and Cadair Idris (893 m), 40 km to the south, are

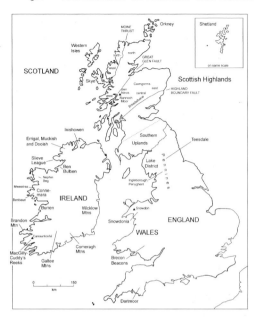

FIG 239. Location of the main mountain areas and some other localities and features mentioned in the text.

a little lower, as is Pen-y-Fan (886 m), the highest point of the Brecon Beacons in south Wales. In the north of England, mountains (and mountain plants) are concentrated in the north Pennines and the Cumbrian Lake District. The highest point on the Pennine escarpment is Cross Fell (893 m), but there are several other nearby summits that top 700 m; these have a modest arctic-alpine flora, but most of the mountain plants of the Pennines grow on limestone at rather lower altitudes. The Lake District has a good deal in common with Snowdonia. The highest summit, Scafell Pike, reaches 977 m.

Ireland has many mountains, mostly grouped around the periphery, of which only Carrauntoohil in Kerry exceeds 1000 m. There are a few more over 900 m, including Brandon Mountain and the highest points of the Galtees and the Wicklow mountains, and many over 700 m, including some of the most impressive mountains in Ireland (Errigal, Slieve League, Nephin, Croagh Patrick, Mweelrea, the 'Twelve Bens'), and many unconsidered summits equally high.

THE TREE-LINE AND ABOVE

Perhaps the only natural altitudinal tree-line in Britain or Ireland is at the northwest corner of the Cairngorms, where Scots pine forest with a fringe of

FIG 240. The Cairngorm treeline at c. 600 m on Creag Fhiaclach, June 1981. (a) Just above the forest: Scots pine, juniper, heather and bilberry. (b) In the upper fringe of the forest in Coire Buidhe; uneven-aged pine, understorey mainly of heather, juniper and bilberry, showing the coppery hue of the newly expanded leaves.

juniper scrub creeps up to an altitude of almost 600 m (Fig. 240). Taking into account the present distribution of pine and juniper, temperature, wind and other factors, Pears (1967) concluded that the potential natural tree-line in the Cairngorms lies between 610 and 685 m, depending on local topography and shelter. In Snowdonia, well-grown rowans can be found on cliff ledges inaccessible to sheep up to an altitude of about 650 m, but saplings above that level seldom attain any size. This suggests that the upper limit of forest in Britain in our present climate would probably be around 650 m. The isotherms of July maximum daily temperature suggest that the limit may be about 100 m lower in the western Highlands and much of Ireland, and 200 m lower in the northwest Highlands and northernmost Ireland. On the Continent, the upper limit of forest is generally much higher. In the Alps, coniferous forest reaches up to around 2000 m. In the high mountains of central-southern Norway birch woodland extends up to about 1200 m, while in arctic Sweden the limit of birch woodland is at about 680 m (Fig. 24). However, the tree-line drops quickly towards the Norwegian coast, over much of which the climatic tree-line is at no more than 400 m. Thus the picture in the oceanic climate of Britain and Ireland falls neatly into this Scandinavian pattern. The natural tree-line would certainly have been depressed still further in western coastal districts of Britain and Ireland by wind exposure and the cooling effect of the sea.

There is a notable difference between our mountains and those of central Europe. There, above the tree-line, we expect distinctively 'alpine' plants to be prominent everywhere. Here the general plant cover of the mountain slopes may often seem disappointingly dull, composed largely of the common calcifuge species of the hill grasslands of the last chapter. One reason for this is our mild, humid climate. This favours leaching and acidification of soils, so gley podzols and peaty soils are very widespread in our uplands. Another is the modest height and isolation of many of our mountains; any less than 600–700 m could have been forest-covered at some point in Post-glacial time.

As a consequence, many of our most characteristic mountain plants are concentrated on the immature soils of cliffs and crags, and on the associated steep slopes, especially in the glacial corries that formerly held the heads of glaciers (Fig. 244). These are mostly on the north and east sides of our mountains, and are part of the evidence that our islands were once glaciated. A corrie (*cirque* in French, *cwm* in Welsh) typically contains other evidence of glaciation, ice-smoothed rocks and moraines, and often a corrie-lake in an ice-scoured hollow or dammed by moraine.

MOUNTAIN PLANTS AND THEIR LIMITS

Arctic and alpine plants experience cooler summers than their lowland congeners, and cold and often snowy winters. Many of them grow well in cultivation in the lowlands, freed from the competition of vigorous lowland species. It is the common experience of alpine gardeners that winter is a problem time for 'difficult' alpines; our mild but moist winters favour damping-off fungi to the disadvantage of plants adapted to cold winters. Plants under snow are not necessarily particularly cold. Snow is a good insulator, so many plants beneath a blanket of snow will be at o °C, with water at their roots – the conditions of a refrigerator, not a deep freeze. Of course it is a different story for the plants of exposed places, such as trailing azalea (*Loiseleuria procumbens*), which must withstand bitterly cold air temperatures in winter. High summer temperatures mean high rates of water loss if the foliage is not to overheat in bright sunshine. They also affect the balance of photosynthesis and respiration, for which soil temperature as well as leaf temperature may be important.

The Norwegian ecologist Eilif Dahl found that, of the available climatic measures, isotherms of the average annual maximum summer temperature

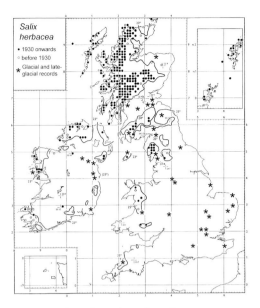

FIG 241. Dwarf willow (*Salix herbacea*, Fig. 245a) with 23 °C mean annual maximum summer isotherm for the Highlands and Ireland, the 24 °C isotherm for the Southern Uplands, and the 25 °C isotherm for England and Wales. The occurrences of the species almost all lie within these isotherms. Isolated summits are shown as 'crosswires' with their estimated summer maxima. Asterisks show sub-fossil records from Glacial and early Post-glacial deposits.

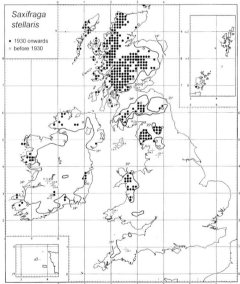

FIG 242. Starry saxifrage (*Saxifraga stellaris*) with the 24 °C mean annual maximum summer isotherm for Scotland and Ireland, and the 25 °C maximum summer isotherm for England and Wales. Conventions as in Fig. 241. The modern distribution of both species is from the first edition on the BSBI Atlas, but subsequent recording has not changed it materially. Maps simplified from Conolly & Dahl 1970.

gave the closest fit to the distributions of Scandinavian mountain plants, and he and Ann Conolly worked out the corresponding correlations for the British and Irish mountain flora (Conolly & Dahl 1970). The limiting isotherms can vary somewhat from area to area. Of the 53 species confined in the Highlands to sites with maxima in summer of 22 °C or less, only two (*Saxifraga nivalis* and *Silene acaulis*) occur in Ireland. The distributions of three-leaved rush (*Juncus trifidus*) and trailing azalea neatly fit the 22 °C isotherm in Scotland, but the area where they could potentially grow in Ireland is limited, and neither has ever been seen there. Eighteen further species occur within the 23 °C isotherm in the Highlands (Fig. 241), of which half occur in Ireland. A further 15 occur within the 24 °C or 25 °C isotherms in the Highlands, and all of those except *Vaccinium microcarpum* have at least one Irish site (Fig. 242). The 24 °C maximum summer temperature isotherm is at (or close to) sea level along most of the west coast of Scotland, and most of the north and west Irish coast from Belfast to Cork, so it is no surprise to find roseroot (*Sedum rosea*) as a sea-cliff plant on western coasts or mountain avens (*Dryas octopetala*) near sea level in western Ireland and northwest Scotland.

WALES

The Brecon Beacons

The southernmost British summits that can reasonably be called mountains are the Brecon Beacons. This Old Red Sandstone ridge rises to 886 m at Pen-y-Fan, and has a line of glacial cwms along its north side. The slopes are covered with much the same upland grasslands, heather moorland and bog as the lower neighbouring hills, but the cliffs support a modest flora of mountain plants, such as the purple saxifrage (*Saxifraga oppositifolia*), roseroot and globeflower (*Trollius europaeus*), all at the southern end of their British range. Paradoxically, the mountain flora is best developed on Craig Cerrig Gleisiad (Fig. 243a), to the west of the highest summits, at only 550–600 m. This is well below the potential tree-line, but the cliff has a nice 'mountain' feel about it, which recalls Snowdonia.

Snowdonia

The mountains of Snowdonia (Figs 243b, 244) are made up of Ordovician volcanic rocks, intruded into Cambrian and 246 grits and shales – the latter metamorphosed by pressure and time into slate. Slate-quarrying and farming were for long the twin mainstays of the economy of this part of Wales, and slate buildings, walls and roofs are everywhere. The volcanic rocks that make up the major peaks are very diverse, including hard acid rhyolite, hard but rather more base-rich dolerite, and more easily weathered and base-rich pumice tuffs.

Cadair Idris, which dominates the southern skyline from Dolgellau, is an outlier from the main mass of Snowdonia, and shares much of the character of

FIG 243. Welsh mountains. (a) Craig Cerrig Gleisiad, west of the main ridge of the Brecon Beacons (seen on the skyline), August 1989. These north-facing crags are the southernmost British locality for the arctic-alpine purple saxifrage. (b) Cwm Idwal National Nature Reserve in Snowdonia, September 1971. Glacial moraines in the foreground, the 'Nameless Cwm' in the background. The cliffs round the head of the Cwm support arctic-alpine plants including roseroot, purple saxifrage, dwarf willow, moss campion and Snowdon lily.

those mountains. It has dramatic glacially sculpted cwms, some with moraine-dammed tarns, and the cliffs have a mountain flora including purple saxifrage, moss campion (*Silene acaulis*), mountain sorrel (*Oxyria digyna*) and roseroot, with dwarf willow (*Salix herbacea*) in the sparse turf fringing the lips of the cwms (Fig. 245a). Wet mountain flushes are the habitat of greyish-green masses of the tiny liverwort *Anthelia julacea*, which grows intermingled with purplish brown *Marsupella emarginata*, and *Scapania undulata* and *Sphagnum denticulatum*, both sometimes bright green, sometimes vividly pigmented with ochre, brick-red or purplish shades, making these *Anthelia–Sphagnum* springs conspicuous and colourful[M31]. These occurrences in North Wales are their southernmost in Britain, but they are a feature of all our mountains, most common in the Scottish Highlands (Fig. 256b). Another spring community, less confined to high mountains but very much a part of the mountain scene, is characterised by the bright green moss *Philonotis fontana* (often with capsules like *petits pois* on stalks) and starry saxifrage (*Saxifraga stellaris*); other frequent species are blinks (*Montia*

FIG 244. Snowdon from the northeast: a typical piece of glacially sculpted alpine topography, with a corrie lake, Glaslyn, left of centre. The sharp ridge of Crib Goch, a classic arête, in the foreground, with a line of north-facing corries to the right, Cwm Uchaf nearest, Cwm Glas with remnants of a snow cornice on its farther lip; Clogwyn Du'r Arddu can just be glimpsed beyond. Base-rich rocks outcropping on the cwm headwalls harbour a rich mountain flora. An early spring photograph, with vegetation still dormant (© Adrian Warren & Dae Sasitorn/www.lastrefuge.co.uk).

fontana), bog stitchwort (*Stellaria uliginosa*), opposite-leaved golden-saxifrage (*Chrysosplenium oppositifolium*) and the distinctive moss *Dicranella palustris*. This *Philonotis fontana–Saxifraga stellaris* spring[M32] is common around issues of water and along streamsides in the uplands from north Wales northwards, and in most of the major mountain groups in Ireland.

Most of the slopes of Cadair Idris, Snowdon, the Glyders and the Carneddau are covered with similar upland grasslands to wide stretches of mid and north Wales – *Festuca–Agrostis* on the better soils, *Nardus* on somewhat poorer and moister soils, *Calluna* moor on the harder and more acid rocks or where grazing is less, deergrass (*Trichophorum cespitosum*) or *Molinia* where drainage is impeded and the soil is peaty, and heath rush (*Juncus squarrosus*) where a former peat cover has been persistently burnt or grazed. At higher altitudes, bilberry–wavy hair-grass heath[H18] and local patches of *Nardus* with stiff sedge (*Carex bigelowii*) [U8] probably reflect places where snow lies late, rather than burning and grazing. A distinctive plant of these acid upland communities is the fir clubmoss (Fig. 246a); it has a wide altitudinal range, but is much most conspicuous in

FIG 245. (a) Dwarf willow (*Salix herbacea*): a strict high-mountain plant and one of our most widespread, occurring from mid Wales to Shetland, and from Kerry to Donegal. (b) Snowdon lily (*Lloydia serotina*) (Chris Proctor).

FIG 246. Clubmosses: (a) fir clubmoss (*Huperzia selago*), damp heather–deergrass moor, Coire an Lochain, Cairngorms, July 1989; (b) alpine clubmoss (*Diphasiastrum alpinum*), rocky *Nardus* grassland, Cwm Idwal, Snowdonia, August 1975.

the mountains. Alpine clubmoss (Fig. 246b), more particularly an upland and mountain plant, is often common in various well-drained acid grasslands, especially over rock or scree. Bare screes in Snowdonia are often dotted with the parsley fern (*Cryptogramma crispa*)[U21]. Where grazing and disturbance allow, quite steep slopes can be occupied by a damp *Calluna–Vaccinium–Sphagnum capillifolium* moorland[H21]. Floristically like a well-drained mountain blanket bog, this community can support a rich diversity of elegant leafy liverworts, in Snowdonia including *Mylia taylori, Anastrepta orcandensis, Herbertus aduncus, Scapania gracilis* and (rarely) *S. ornithopodioides*. It comes to its best development in the west Highlands of Scotland, and locally on the western Irish mountains. The dry exposed high ridges and windswept summits, from North Wales to northernmost Scotland and the high tops of the Irish mountains, are home to a very characteristic mountain community, dominated by *Racomitrium lanuginosum* and stiff sedge[U10].

Apart from these last few communities, most of the distinctively mountain vegetation is in the steep glacial cwms north or east of the summits. Here erosion can keep pace with leaching, so the soils and the water seeping down the cliffs are more base-rich. The hardest and most acid rocks still support little more than heather and *Nardus* on open ground, but on ledges inaccessible to sheep bilberry (*Vaccinium myrtillus*) and great wood-rush (*Luzula sylvatica*) can grow luxuriantly, along with wood-sorrel (*Oxalis acetosella*), heath bedstraw (*Galium saxatile*), hard fern (*Blechnum spicant*), broad buckler-fern (*Dryopteris dilatata*), tufted hair-grass (*Deschampsia cespitosa*), sweet vernal-grass (*Anthoxanthum odoratum*) and calcifuge mosses such as *Mnium hornum, Plagiothecium undulatum, Dicranum majus* and *Rhytidiadelphus loreus*[U16]. This recalls the ground flora of acid oakwoods. On more base-rich rocks *Luzula sylvatica* remains constant and often dominant, and bilberry rather frequent, but they are joined by a string of other species, including water avens (*Geum rivale*), wild angelica (*Angelica sylvestris*), roseroot and the lady's-mantle *Alchemilla glabra*[U17]. Other frequent species include goldenrod (*Solidago virgaurea*), wood crane's-bill (*Geranium sylvaticum*), globeflower, common sorrel (*Rumex acetosa*), meadow buttercup (*Ranunculus acris*), meadowsweet (*Filipendula ulmaria*), devil's-bit scabious (*Succisa pratensis*) and marsh hawk's-beard (*Crepis paludosa*). This community is very rich, and many other species may occur. The name of the old rock-climb 'Hanging Garden Gully' on the Cwm Idwal cliffs is nicely descriptive of well-developed stands.

A rather different assemblage of small, mostly calcicolous mountain plants grows rooted on tiny ledges, broken crags or crevices in the cliff-face. Snowdonia has only rather fragmentary occurrences of a *Dryas–Silene acaulis* community[CG14], which is much better developed on calcareous schists and limestones in the Scottish Highlands. Probably the best cliffs are in Cwm Idwal below Glyder Fach, and Cwm Glas on Snowdon. Of the species that are constant in Scotland, mountain avens grows in only two localities in Snowdonia, moss campion is widespread, as are purple saxifrage, lesser clubmoss (*Selaginella selaginoides*), wild thyme (*Thymus polytrichus*), harebell (*Campanula rotundifolia*), green spleenwort (*Asplenium viride*) and the moss *Ctenidium molluscum*. In North Wales, alpine bistort (*Persicaria vivipara*), alpine meadow-rue (*Thalictrum alpinum*), alpine saw-wort (*Saussurea alpina*), Snowdon lily (*Lloydia serotina*, Fig. 245b), tufted saxifrage (*Saxifraga cespitosa*), alpine saxifrage (*S. nivalis*), hair sedge (*Carex capillaris*) and black alpine-sedge (*C. atrata*) all occur in at least one locality. *Lloydia* is circumpolar, but its localities in Snowdonia are the only occurrences in northwest Europe.

NORTHERN ENGLAND AND THE BORDERS

The Pennines

The highest summits of the Pennines are unequivocally mountains, but the most distinctive mountain plants and vegetation of the Pennines are at more modest altitudes. An exception is alpine forget-me-not (*Myosotis alpestris*), which grows in *Sesleria* grassland in several places on the high fells at 650–750 m, south of the Tees headwaters. Another is alpine foxtail (*Alopecurus borealis*), in mildly acid flushes high on Little Dun Fell, but this species has localities in the dale too, which the *Myosotis* does not.

The most famous locality in the Pennines for mountain plants and plant communities is Upper Teesdale (Pigott 1956, Clapham 1978), between High Force and Cow Green Reservoir (Fig. 247). The most noticeable of the Teesdale specialities are the shrubby cinquefoil (*Potentilla fruticosa*) and the spring gentian (Fig. 248a). *Potentilla fruticosa* grows along several kilometres of the rocky and gravelly banks of the Tees, in situations where the surface is usually dry, but the roots are within reach of water (Fig. 249). Elsewhere in our islands it is native only on damp calcareous rock ledges on Pillar and on the Wastwater screes in the Lake District, and around turloughs in the Burren. The spring gentian is locally frequent in damp calcareous grassland and tussocks in calcareous small-sedge fens in upper Teesdale, but nowhere else in Britain; it too reappears in the Burren.

Many of the notable Teesdale mountain plants grow in *Sesleria* grassland or in short-sedge fens on (or draining from) the 'sugar limestone' of Cronkley

FIG 247. Teesdale. Sugar-limestone grassland alternating with heather on Widdybank Fell, June 1970.

FIG 248. Grassland plants on Widdybank Fell, Teesdale, June 1975. (a) Spring gentian (*Gentiana verna*); in Britain restricted to Teesdale, but widespread on the Burren limestone in Ireland. (b) False-sedge (*Kobresia simpliciuscula*); Teesdale is its only locality south of the Scottish Highlands.

and Widdybank Fells, at some 520–540 m altitude. A basalt dyke – the Great Whin Sill – intruded into the Carboniferous rocks over a wide area of northern England has here metamorphosed beds of limestone, which recrystallised with the texture of lump sugar. This limestone readily breaks down to form highly calcareous permeable soils, in some places dry and sharply drained, but in some

FIG 249. Teesdale. Shrubby cinquefoil (*Potentilla fruticosa*) beside the River Tees at Holwick Head, August 1981.

places moister. Thin dry grassland on Cronkley Fell supports a form of hoary rock-rose (*Helianthemum oelandicum*, Fig. 143d) with small dark-green leaves and a small-leaved form of mountain avens. Some of the most sharply drained grassland on Widdybank Fell (Fig. 247) is home to Teesdale violet (*Viola rupestris*, Fig. 143b – rare everywhere, and more numerous in some of its other localities than here), with common dog-violet (*Viola riviniana*) and hybrids in more humus-rich grassland nearby. Moister *Sesleria* grassland (and hummocks in small-sedge fen) is home to false-sedge (Fig. 248b) and hair sedge. Flushes where calcareous water trickles over the Whin Sill are the habitat of the most inconspicuous of the Teesdale mountain plants, Teesdale sandwort (*Minuartia stricta*), which grows here with spring sandwort (*M. verna*, for which it is easily mistaken), sea plantain (*Plantago maritima*) and the cushion-forming mosses *Gymnostomum recurvirostrum* and *Catoscopium nigritum*.

The abundance of yellow saxifrage (*Saxifraga aizoides*, Fig. 259b) in springs and flushes, the flowery meadows (not as all-pervading or as flowery as they were 60 years ago), and the occurrence of *Carex rostrata–Sphagnum warnstofii* fens[M8] all add to the montane ambience of Upper Teesdale, despite its modest altitude.

Two prominent mountains in the Pennines are Ingleborough (723 m) and Pen-y-Ghent (693 m). Both have similar geology, with a Carboniferous limestone

FIG 250. (a) Limestone crags on Pen-y-Ghent, Yorkshire, home to mountain plants including roseroot and purple saxifrage. (b) Purple saxifrage in flower on Pen-y-Ghent, April 1982.

base upon which lies an almost flat-bedded succession of limestones, shales and grits (the Wensleydale series), capped by Millstone Grit. Near the top of the Wensleydale Series is a thick bed of limestone (the Main Limestone), forming a prominent cliff (Fig. 250a) some way below the summit. These cliffs have a mountain flora, which includes roseroot, purple saxifrage (Fig. 250b), mossy saxifrage (*Saxifraga hypnoides*), yellow saxifrage, alpine bistort, *Alchemilla glabra* and viviparous sheep's fescue (*Festuca vivipara*).

The Lake District

The Lake District has some fine mountains, and is geologically not unlike Snowdonia, with the great thickness of the diverse Borrowdale volcanic rocks set in a background of Ordovician slates to the north, with Silurian grits, slates and shales forming the southern fells. The mountains are a little lower than in Snowdonia, the topography more intricate, the lakes a more prominent feature, and some would say the landscape is gentler and less austere – though Wordsworth and Beatrix Potter may have something to do with that perception!

Vegetationally, the mountain slopes hold few surprises, except that alpine lady's-mantle (*Alchemilla alpina*, Fig. 265a) is common in *Festuca–Agrostis* grassland

and other habitats on the fells – especially below crags, where it benefits from a degree of flushing. The community here may be seen as an impoverished version of a herb-rich high-altitude grassland that is widespread on base-rich rocks in the Highlands, dominated by *Festuca vivipara*, *Anthoxanthum odoratum* and *Agrostis capillaris*[CG11]. On the fells, bilberry and *Nardus* communities where snow lies late are perhaps more widespread than farther south. Another conspicuous difference from Snowdonia is the abundance here of yellow saxifrage around springs, over dripping rocks and in mountain flushes[M11, U15]. As in Snowdonia, parsley fern is ubiquitous on screes, and the clubmosses *Huperzia selago* and *Diphasiastrum alpinum* are dotted through the acid grasslands on the mountain slopes.

The Lake District was heavily glaciated, and there are impressive glacial cirques on the north and east sides of Helvellyn and other high summits, with both of the *Luzula*–tall-herb communities on ledges inaccessible to sheep. All the mountain species of Snowdonia, except *Lloydia*, grow also in the Lake District, plus downy willow (*Salix lapponum*) at its southern limit in Britain, but some of them are in extremely small quantity and there is nothing to match the display of mountain plants on the cliffs of Cwm Glas on Snowdon. Most of the common alpines can be seen on the Helvellyn cliffs.

Alpine catchfly (*Lychnis alpina*) grows on a crumbling spur of Hobcarton Crag on copper-rich soil, in one of its only two localities in Britain. The other is on serpentine on Meikle Kilrannoch in Angus.

The limestone of the northern Pennines and base-rich cliffs of the Lake District mountains are the main concentration in Britain of the attractive *Asplenium viride*–*Cystopteris fragilis* community of small calcicolous ferns[OV40]. It grows on moist shady cliffs and clefts in limestone from Snowdonia to the Durness limestone in the north of Scotland. In Ireland these two species are particularly concentrated on the Sligo–Leitrim limestone, but both are widely scattered in the western Irish mountains.

The Southern Uplands of Scotland

Considering their altitude and relatively northern position, the Southern Uplands may seem an anticlimax after north Wales and northern England, but those two areas are topographically diverse, and a fairer comparison would be with the geologically similar hill country of mid Wales. Seen in this light, the Southern Uplands have a good scatter of the common mountain plants – alpine bistort, mossy saxifrage, starry saxifrage, purple saxifrage, roseroot, mountain sorrel, viviparous sheep's fescue and alpine meadow-rue – plus three outliers from the Highlands, black alpine-sedge (also in Snowdonia and the Lake

District), sheathed sedge (*Carex vaginata*) and downy willow. The most notable concentration of mountain plants is on the cliffs around the Grey Mare's Tail, 15 km northeast of Moffat. Some of the summits northwest of New Galloway are high enough to have expanses of *Carex bigelowii–Racomitrium* moss heath and even recognisable dwarf willow snow-bed communities.

THE SCOTTISH HIGHLANDS

The Scottish Highlands – beyond the Highland Boundary Fault, which runs from the Clyde at Helensburgh via Callander and Blairgowrie to the east coast at Stonehaven – are very different. Geologically they (and northwest Ireland) are a fragment of an ancient mountain range, which pre-dates the opening of the Atlantic Ocean and of which other fragments now form the Appalachians, Newfoundland, east Greenland, the mountains of Scandinavia, and Svalbard. The Highlands are a substantial mountain range even by Continental standards. Absolute altitude is no guide to effective altitude for vegetation or human habitation. The tree-line is 1300 m lower in Scotland than in the Alps, but nobody who knows them, whether hill farmer or climber, regards the Highlands as trivial.

The Cairngorms

As a starting point, the granite mass of the Cairngorms has the merit of being big, well-known and relatively simple. They have impressive corries and glaciated valleys (Fig. 251) but they are nothing like as intricately sculptured as mountains farther south and west, and the granite yields uniformly acid base-poor soils. This geological uniformity and uncomplicated topography make it easier to appreciate the broad distribution of vegetation without distracting detail (Fig. 252).

The east of Scotland is heather (*Calluna*) country. The Cairngorms are home to the most extensive remnants of the Caledonian pine forest that once covered much of the Highlands, and the heather on the lower slopes can be seen as a remnant of the heathery understorey of the forest. In freely drained and exposed places on the open moorland the heather is often accompanied by bearberry (*Arctostaphylos uva-ursi*)[H16]; in moist sheltered places deergrass and *Molinia* may get a foothold. Above about 650 m we are on ground that is naturally treeless, and the broad pattern is set by altitude, and a complex of factors that we can lump together as 'exposure'. Wind steepens the moisture and temperature gradients near the ground – so wind is both drying and cooling – and crucially, wind sweeps exposed ground free of snow, but snow accumulates in sheltered

FIG 251. The Cairngorms from the northwest. Native Scots pine in Rothiemurchus Forest, with heather in full flower in the moorland in the foreground. The Lairig Ghru is on the left, and Braeriach with its northern corries forms the middle part of the skyline, with Gleann Einich to the right of it. Creag Fhiaclach (Fig. 240) is just out of the picture to the right. (Patricia & Angus Macdonald/Aerographica).

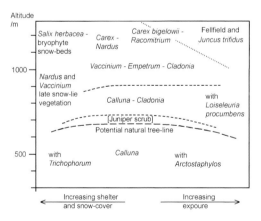

FIG 252. Schematic diagram of distribution of major vegetation types on the Cairngorms, in terms of altitude and exposure. The heather moorland on the lower slopes occupies ground that would naturally have borne pine forest, with a fringe of juniper at the tree-line. The vegetation above that level is naturally treeless. Based mainly on Watt & Jones (1948), Poore & McVean (1957) and Pears (1967).

hollows or behind obstructions. Snow has two effects. It is a good insulator, so plants under a snow cover are protected from severe frost and drying winds. But if the snow is deep and lies late into the spring, it shortens the growing season, in extreme cases to a few weeks in late summer.

Heather is dominant up to around 900 m, but above the potential tree-line it is usually no more than 10–15 cm high and often less. Crowberry (*Empetrum nigrum* ssp. *hermaphroditum*) is a near-constant associate, along with an abundance of lichens[H13]. On exposed spurs the heather is wind-cut to within a few centimetres of the ground, forming only an open cover over the surface, and bilberry and cowberry (*Vaccinium vitis-idaea*) are more frequent. Species

FIG 253. Cairngorm plants: (a) trailing azalea (*Loiseleuria procumbens*) with the lichen *Cetraria islandica* ('Iceland moss'); (b) lichen, *Thamnolia vermicularis*; (c) curved wood-rush (*Luzula arcuata*); (d) highland saxifrage (*Saxifraga rivularis*), Coire an Lochain, July 1989. These are all circumpolar plants, (a)–(c) photographed in other parts of their range, but looking very much as they do with us: (a) Valais, Switzerland, June 1974, (b) and (c) Eíriksjökull area, Iceland, August 1971.

FIG 254. The Cairngorm plateau looking towards Loch Etchachan, September 1974.

particularly characteristic of this kind of situation are trailing azalea (Fig. 253a), the white worm-like lichen *Thamnolia vermicularis* (Fig. 253b) and the crustose *Ochrolechia frigida*[H13c]. With increasing altitude, the *Calluna* heath is replaced by a community of bilberry, crowberry and *Carex bigelowii*, with frequent cowberry, sheep's fescue and wavy hair-grass (*Deschampsia flexuosa*), set in a carpet of bryophytes and (particularly) lichens[H19]. On the most sheltered slopes, heather is replaced by bilberry and wavy hair-grass, or *Nardus* grassland. The *Vaccinium–Empetrum*–lichen heath will tolerate quite prolonged snow cover, but it is replaced by *Nardus–Carex bigelowii* grassland in the most sheltered places.

The stony ground of the highest and most exposed places on the Cairngorm plateau (Fig. 254) is occupied by a sparse community of three-leaved rush (*Juncus trifidus*), *Carex bigelowii* and the moss *Racomitrium lanuginosum*[U9]. Lichens, including *Cetraria islandica*, *Ochrolechia frigida* and various *Cladonia* species, are frequent but not very conspicuous, as are starved shoots of bilberry and crowberry. The most characteristic plants of this inhospitable habitat (apart from *Juncus trifidus*) are two wood-rushes, *Luzula spicata* and *L. arcuata* (Fig. 253c). In some places high on the Cairngorms, especially on spurs towards Glen Feshie, *Racomitrium lanuginosum* grows more luxuriantly and the *Juncus trifidus* community is replaced by *Carex bigelowii–Racomitrium lanuginosum* moss heath[U10], which we first encountered on exposed summits in Snowdonia and the

Lake District. Sometimes the vegetation is so sparse that the stony summits are better described simply as 'fell-field'.

These summit communities, like the *Vaccinium–Empetrum* heaths at somewhat lower altitudes, are tolerant of quite long snow-lie, but they pass into *Nardus–Carex bigelowii* grassland where snow-lie in spring is more prolonged. The most persistent patches of snow may not melt until June or even later

FIG 255. Coire Domhain, Cairngorm: *Juncus trifidus* on the exposed southwest-facing slope in the foreground, *Nardus* grassland on the floor of the corrie, a late snow patch in the hollow on the more sheltered east-facing slope beyond, 14 June 1981.

FIG 256. The Cairngorm northern corries. (a) Looking down Coire an-t-Sneachda, September 1974. (b) *Anthelia–Scapania undulata–Sphagnum denticulatum* spring[M31], looking up Coire an Lochain; lingering snow patches on the headwall of the corrie, 17 July 1989.

(Figs 255, 256). They have very characteristic vegetation, often zoned in response to the duration of snow cover. They are often set in a background of *Nardus–Carex bigelowii* grassland, itself fairly tolerant of long snow cover. Typically an outer area of *Carex bigelowii–Polytrichum alpinum* sedge-heath[U8] encircles an inner patch of the still more tolerant *Polytrichum sexangulare–Kiaeria starkei* snow-bed[U11], dominated by its eponymous mosses, and with only a thin (if rather constant) scatter of vascular plants – tufted hair-grass, *Carex bigelowii*, dwarf cudweed (*Gnaphalium supinum*), *Nardus*, starry saxifrage and fir clubmoss, to name only the most frequent. Bryophytes make up the bulk of the community, including the mosses *Oligotrichum hercynicum*, *Conostomum tetragonum* (Fig. 262b), *Racomitrium sudeticum*, *Pohlia nutans* and liverworts including *Barbilophozia floerkei*, *Moerckia blyttii*, *Pleuroclada albescens* (Fig. 262c) and *Anthelia juratzkana*. The meltwater from the snow drains from the bottom of the patch, often marked by a patch of *Pohlia albicans* ssp. *glacialis*[M33], passing down into a *Philonotis fontana–Saxifraga stellaris* spring. The whole snow-patch may have an 'eyebrow' of *Cryptogramma crispa–Athyrium distentifolium* snow-bed[U18]. This community can occur in various situations, having in common late snow-lie and a degree of shelter. A long list of species are constant or nearly so, including *Deschampsia cespitosa*, *Alchemilla alpina* and *Saxifraga stellaris*; the rare highland saxifrage (*Saxifraga rivularis*, Fig. 253d) grows in essentially this community in shady wet gullies in north-facing corries.

Snow-bed communities in the Highlands are variable and complex, and we shall return to them later.

The eastern and central Highlands

The mountains to the south of the Cairngorms have summits almost as high, but they are more geologically varied, and generally more dissected by rivers and glaciation. For these reasons, although a similar broad pattern of zonation in relation to altitude can be discerned, they are more complicated in detail. The metamorphic Dalradian schists that make up much of the south and east Highlands generally give rise to more base-rich soils, and locally contain bands of hard limestone. The mountains east of the main Perth–Inverness road (A9), particularly the block from Lochnagar to Glen Clova, are different in character from those farther west. They are geologically very varied, granite, schist and limestone. They include some of the most vegetationally diverse and most species-rich ground in the Highlands. *Luzula*–tall-herb ledges are particularly well developed. The central Highlands to the west are mainly on mica-schists, locally base-rich. Their slopes provide good sheep pasture. The Breadalbane range from Ben Lawers to Ben Lui (Fig. 257) is outstandingly rich

FIG 257. The Breadalbane mountains; looking west from Ben Lawers, July 1989.

in mountain plants, and the less accessible Ben Alder range to the north is also good. On acid brown-earth soils *Festuca ovina–Agrostis capillaris–Galium saxatile* grassland^{U4} is widespread up to over 1000 m, alternating with the more species-rich *Festuca–Agrostis–Thymus polytrichus* grassland^{CG10} where the bedrock is more base-rich. At high altitudes this community may locally pass into patches of *Festuca–Agrostis–Alchemilla alpina* grassland^{CG11}, especially below outcrops or on steeper slopes. Very locally, on soils that are sufficiently calcareous and moist but freely drained, a *Festuca ovina–Alchemilla alpina–Silene acaulis* dwarf-herb community^{CG12} occurs, usually associated with cliffs that are themselves rich in species. This is one of the most species-rich plant communities of the Highlands, including alpine forget-me not (Fig. 258d), snow gentian (*Gentiana nivalis*), alpine cinquefoil (*Potentilla crantzii*), alpine fleabane (*Erigeron borealis*) and alpine speedwell (*Veronica alpina*). On deeper, more leached soils or on gentler slopes these *Festuca–Agrostis* grasslands give way to *Nardus–Galium saxatile* grassland^{U5}, or *Juncus squarrosus–Festuca ovina* grassland, which extend (with increasing *Racomitrium lanuginosum*) to over 800 m. At moderate altitudes, especially over the more base-poor rocks, rocky ground is often covered with grazed, rather grassy *Calluna–Vaccinium* heath. It is a fair general rule that the eastern Highlands are predominantly heathery and lichen-rich, and the western Highlands predominantly grassy and moss-rich, but these are broad trends and *Calluna* is rarely absent altogether. Above 700–800 m we are generally into the zone of *Vaccinium–Empetrum–Racomitrium* heaths[H19, H20], with *Nardus–Carex bigelowii* grassland where snow lies late.

More-calcareous bedrock has a great impact on the species diversity of the corries. The *Luzula sylvatica–Geum rivale* tall-herb community[U17] is luxuriantly developed on inaccessible ledges, and occasional patches of block-scree (and experimental exclosures) give a taste of what our mountains might look like with less ubiquitous sheep grazing. Globeflower, roseroot, arctic willows (*Salix lanata, S. lapponum, S. reticulata, S. arbuscula, S. myrsinites*), alpine saw-wort, mountain bladder-fern (*Cystopteris montana*), yellow oxytropis (*Oxytropis campestris*), *Bartsia alpina* and an array of mountain grasses and sedges (*Poa alpina, P. glauca, Carex atrata, C. vaginata*) grow in this community, with holly-fern (*Polystichum lonchitis*) in rock crevices or amongst boulders. The *Dryas–Silene acaulis* community of the small ledges and crevices[CG14] is much richer than anywhere south of the Highlands, with mountain avens, moss campion, yellow and purple saxifrages, alpine bistort, alpine mouse-ear (*Cerastium alpinum*), cyphel (*Minuartia sedoides*), hair sedge, rock sedge (*Carex rupestris*) and alpine woodsia (*Woodsia alpina*). These rich cliffs (Fig. 258a) are also a paradise for the bryologist!

The flushes and mires over base-rich rocks are also very rich in species. Typical examples of *Carex dioica–Pinguicula* mire[M10] in the Pennines, the general run of *Carex viridula* ssp. *oedocarpa–Saxifraga aizoides* mountain flushes[M11] in the Highlands (Fig. 259a), and *Carex saxatilis* mires[M12] on Ben Lawers are distinct enough, but they intergrade and it is impossible to draw sharp lines between them. They have in common an abundance of sedges, common butterwort (*Pinguicula vulgaris*, Fig. 183a), lesser clubmoss (Fig. 259e), and an abundance of rich-fen bryophytes, such as *Campylium stellatum, Scorpidium revolvens, S. scorpioides* and *Aneura pinguis*. Between them, they provide habitats for an impressive list of mountain and northern species – among vascular plants the sedges *Carex saxatilis, C. vaginata, C. atrofusca, C. microglochin*, false-sedge (*Kobresia simpliciuscula*), brown bog-rush (*Schoenus ferrugineus*), alpine rush (*Juncus alpinoarticulatus*), three-flowered rush (Fig. 259c), two-flowered rush (*J. biglumis*), chestnut rush (*J. castaneus*), Scottish asphodel (*Tofieldia pusilla*) and amongst bryophytes the mosses *Pseudocalliergon trifarium* (Fig. 259d), *Meesia uliginosa, Amblyodon dealbatus, Tayloria lingulata, Oncophorus* spp. and the liverworts *Leiocolea gillmanii, Scapania degenii, Tritomaria polita* and *Barbilophozia quadriloba*. These base-rich flushes are generally fed by *Palustriella commutata–Festuca rubra*[M37] or *P. commutata–Carex nigra*[M38] springs, depending on altitude. A distinctive *Carex rostrata–Sphagnum warnstorfii* mire[M8], with *S. teres*, frequent *Tomenthypnum nitens*, and a variety of rich-fen species, is recorded from a scatter of localities

ABOVE: **FIG 258.** Mountain-cliff plants: (a) Dalradian mica-schist cliff, Beinn Ghlas, Perthshire, with a rich flora including alpine cinquefoil (*Potentilla crantzii*) and alpine forget-me-not (*Myosotis alpestris*), July 1989; (b) moss campion (*Silene acaulis*) on Moine schist cliff, Creagan Meall Horn, Sutherland, June 1981; (c) drooping saxifrage (*Saxifraga cernua*), Ben Lawers, July 1989; (d) alpine forget-me-not, Ben Lawers, July 1981.

OPPOSITE: **FIG 259.** Base-rich mountain flushes: (a) *Saxifraga aizoides* flush, Meall Garbh (Ben Lawers), August 1981; (b) yellow saxifage (*Saxifraga aizoides*), Blà Bheinn, Skye, June 1981; (c) three-flowered rush (*Juncus triglumis*), Beinn Ghlas, July 1989; (d) moss, *Pseudocalliergon trifarium*; near Corrofin, Co. Clare, September 1975; (e) lesser clubmoss (*Selaginella selaginoides*), near Malham, Yorkshire, August 1991; (f) alpine meadow-rue (*Thalictrum alpinum*), Ben Lawers, June 1981.

at altitudes from 427 m to 833 m. In view of the much denser distribution of *S. warnstorfii* and *S. teres* in the Highlands this community must surely be more widespread. The more acid, base-poor mires also take on a more montane character with altitude. A variable *Carex curta–Sphagnum russowii* mire differs from its lowland counterpart (*Carex echinata–Sphagnum* mire[M6]) mainly in the frequency of *Carex curta* and presence of *Sphagnum russowii* and of either *S. lindbergii* or *Carex aquatilis*. They provide habitats for the sedges *Carex rariflora, C. lachenalii* and *C. aquatilis*, and again, there are surely more sites.

The western Highlands
The rainfall in the western Highlands is roughly double that in the east, and the difference is reflected in the vegetation, and indeed in the whole ambience of the landscape. From the intensity of their glacial sculpturing, and the proximity of the sea, the mountains near the west coast create a more dramatic impression than those of the eastern Highlands. There is much less *Calluna* heath. *Festuca–Agrostis* grasslands still occur, but the balance has swung in favour of *Molinia* and *Nardus*. From about 600 m upwards, *Vaccinium–[Empetrum]–Racomitrium* heath replaces the corresponding *Vaccinium–Cladonia* heath of the eastern Highlands,

FIG 260. 'Mountain-top detritus', Ben Dorain, September 1973. Many flattish mountain summits in Britain and Ireland are covered with a jumble of rock fragments, varying in size from large blocks to gravel, and usually vegetated with patchy *Racomitrium lanuginosum* and stiff sedge (*Carex bigelowii*).

and exposed summits are boulder-strewn or covered with *Carex bigelowii–Racomitrium* moss-heath (Fig. 260) (the only British locality for *Diapensia lapponica* is in this community on the top of an otherwise undistinguished mountain near Glenfinnan). We are back on hard rocks here – Moine schists and granites amongst them – but they are diverse enough to provide base-rich patches in the corries, so *Luzula*–tall-herb ledges are well represented, and the common arctic-alpines are mostly present, if not in the diversity or profusion of Breadalbane.

In these mountains south and west of the Cairngorms, late snow patches share in the greater diversity of the topography and vegetation. Figure 261 gives a picture of the situations in which the snowfall of winter persists longest. Snow lies longest of all (till late June or beyond) at the bases of cliffs in shady north- or east-facing corries. Snow-cornices at the lips of these corries can also be very persistent, and when the snow melts or slips away the cold exposed situation at the edge is less conducive to growth than the more sheltered corrie floor. These two situations are where the most extreme snow-bed communities[U11,]

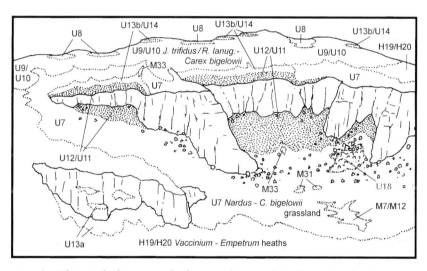

FIG 261. Schematic high-mountain landscape in the Scottish Highlands showing the distribution of snow-beds and late snow-lie communities. The snow-beds where the snow lies longest are stippled. *Vaccinium–Empetrum* heath with lichens or *Racomitrium* covers the open mountainside at the bottom of the picture; the exposed higher slopes are dominated by either *Juncus trifidus* or *Racomitrium* with *Carex bigelowii*. M7/M12 might be either an acid *Carex curta–Sphagnum russowii* mire (M7), or a calcareous *C. saxatilis* mire (M12). See text for more detail on the areas where the snow lies late (U8, U11–U14, U18) and their associated springs. Adapted from Averis *et al.* 2004.

[U12] occur. Snow is usually less persistent on open slopes or plateaux unless hollows are very deep: the picture shows patches of *Carex bigelowii–Polytrichum alpinum* sedge-heath and *Deschampsia cespitosa–Galium saxatile* grassland[U13], and *Alchemilla alpina–Sibbaldia procumbens* dwarf-herb community[U14] on the higher slopes. These communities are repeated above the corrie cliffs, a bit back from the edge. A *Cryptogramma crispa–Athyrium distentifolium* snow-bed is shown at the foot of the slope on the floor of the corrie on the right. *Pohlia wahlenbergii* ssp. *glacialis* springs[M33] are where meltwater gathers at the foot of snow patches, and where water has percolated into the ground to emerge lower down the slope there are a couple of *Anthelia julacea–Sphagnum denticulatum*

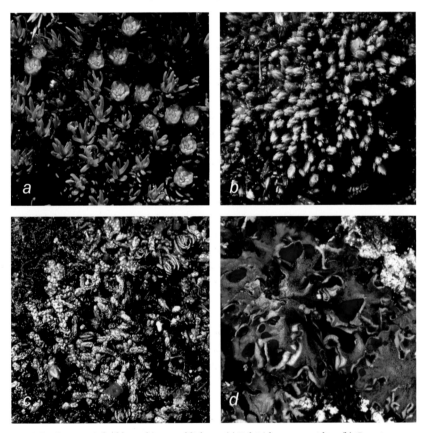

FIG 262. Late snow-bed bryophytes and lichens. (a) *Polytrichum sexangulare*; (b) *Conostomum tetragonum*; (c) the liverwort *Pleuroclada albescens* growing with a tiny black *Marsupella* species; (d) the lichen *Solorina crocea*.

springs[M31]. We have already considered *Polytrichum sexangulare–Kiaeria starkei* snow-beds[U11]. Almost as tolerant of long snow cover are *Salix herbacea–Racomitrium sudeticum* snow-beds[U12]. Like the *Polytrichum–Kiaeria* community (which forms the highest vegetation in Britain, near the summit of Ben Nevis) this snow-bed is vary variable. The dominant *Salix* and *Racomitrium* grow in a blackish crust of tiny liverworts, including *Gymnomitrion concinnatum*, *Nardia scalaris* and *Anthelia juratzkana*. As many as eight liverwort species have been found in a square centimetre of this crust! It is dotted with tiny vascular plants including *Gnaphalium supinum*, *Saxifraga stellaris*, *Alchemilla alpina* and *Diphasiastrum alpinum*. These two snow-bed communities are most extensive against the headwalls of corries, and are perhaps differentiated more by the *Salix* community's greater tolerance to instability of ground than by duration of snow cover. The striking foliose lichen *Solorina crocea* (Fig. 262d), khaki-green above but orange underneath, is characteristic of this habitat and can occur in either community. The snow-bed–meltwater-spring complex is habitat for several rare arctic-alpine plants including starwort mouse-ear (*Cerastium cerastoides*), highland saxifrage, curved wood-rush (*Luzula arcuata*) and hare's-foot sedge (*Carex lachenalii*), as well as the 'granite-mosses' *Andreaea alpestris*, *A. sinuosa*, *A. blyttii* and *A. nivalis*. *Carex bigelowii–Polytrichum alpinum* sedge-heath[U8] is lichen-rich and is mainly a community of the eastern and central Highlands, while *Deschampsia cespitosa–Galium saxatile* grassland[U13] is bryophyte-rich (usually with *Polytrichum alpinum*, *Rhytidiadelphus loreus* and *Hylocomium splendens*) and occurs on high ground throughout. Both tolerate moderately long snow cover. The mountain *Deschampsia cespitosa* grassland often includes such species as *Alchemilla alpina* and *Sibbaldia procumbens*, and sometimes it is replaced by an *Alchemilla alpina–Sibbaldia procumbens* dwarf-herb community[U14], with similar habitat preferences. The near-constant occurrence of *Gnaphalium supinum* and frequency of *Carex bigelowii* and *Luzula spicata* underline the high-mountain credentials of this community.

The western Scottish mountains are notable for their bryophyte-rich *Calluna–Vaccinium–Sphagnum capillifolium* heaths[H21]. These are generally best developed at 300–600 m on north- to east-facing slopes. The heathery canopy can be up to knee-high (but often less), over a damp layer of luxuriant bryophytes. *Calluna* is usually dominant, but bilberry is generally conspicuous, and often there is crowberry as well; bell heather (*Erica cinerea*) can be common in the drier stands. Tormentil (*Potentilla erecta*) is dotted among the heathers. The most prominent bryophytes are usually the rosy-tinged *Sphagnum capillifolium* and the glossy-green, feathery, red-stemmed *Hylocomium splendens*. Other big calcifuge mosses are common too: *Pleurozium schreberi*, *Hypnum jutlandicum*,

Rhytidiadelphus loreus, Dicranum scoparium, D. majus, Plagiothecium undulatum. The real jewels of this community are the big leafy liverworts (Fig. 263), for many of which the west Highlands are their headquarters in Europe, such as *Herbertus aduncus* ssp. *hutchinsiae, Scapania ornithopodioides, S. nimbosa, Plagiochila carringtonii, Anastrophyllum donnianum, Bazzania pearsonii* and *Mastigophora woodsii.* Some of these liverworts have remarkable disjunct world distributions. *Mastigophora woodsii* is in Scotland, Ireland, the Faroes, the Pacific Northwest of America, and central and east Asia, where it is abundant in the eastern Himalayas, largely in forest. *Anastrophyllum donnianum* and *Bazzania pearsonii* have similar

FIG 263. Liverworts of humid Atlantic-montane heaths on the mountains of west and northwest Scotland and Ireland. (a) *Herbertus aduncus* ssp. *hutchinsiae* and *Plagiochila carringtonii*; (b) *Mastigophora woodsii*; (c) *Anastrophyllum donianum*; (d) *Adelanthus lindenbergianus.*

distributions. *Plagiochila carringtonii* and the *Scapania* species are in Ireland, Scotland and the Himalayas, but not in North America.

Skye and the northwest

Skye is unique for both scenery and vegetation. It has three contrasting groups of mountains above 700 m. The Cuillins, carved from an intrusive block of gabbro, are the most famous and the highest, reaching 1014 m at Sgùrr Alasdair; they have steep cliffs and deep corries. The 'Red Hills' just to the east are of granite and are lower (to 780 m), and have few corries. The long east-facing basalt escarpment of Trotternish to the north rises to 723 m at the Storr, but landslipping over the soft Jurassic rocks below the basalt has produced a tumbled chaos of blocks and pinnacles. The 'Old Man of Storr' is a prominent landmark (Fig. 264), and the Quirang some 15 km to the north is a weird place, especially in mist!

Species-poor *Festuca–Agrostis* grassland is extensive on all three groups of mountains on suitable well-drained slopes, but the more species-rich variants are only on the Cuillins and the basalt of north Skye. *Calluna* and *Calluna–Racomitrium* heaths occur on all three but probably make up the largest

FIG 264. The Old Man of Storr and his accompanying pinnacles, Skye, June 1981. The Red Hills are visible just left of the Old Man; the cloud-capped Cuillins can be glimpsed between the two pinnacles to the right.

FIG 265. West Highland plants. (a) Alpine lady's mantle (*Alchemilla alpina*), the Storr, Skye, September 1980; (b) cyphel (*Minuartia sedoides*), Cul Mor, Sutherland, June 1981; (c) seedlings of Iceland purslane (*Koenigia islandica*), the Storr, Skye, June 1981; (d) Iceland purslane at home, Arnavatnsheiði, Iceland, August 1971.

proportion of the vegetation on the Red Hills. Bryophyte-rich *Calluna–Vaccinium* heath is significant only on the Cuillins. At high altitudes, *Racomitrium–Empetrum* heaths, the *Racomitrium–Carex bigelowii* summit heath and stony *Festuca–Luzula spicata* fellfield (in which *Juncus trifidus* occurs) are on all three mountain groups. Of the 'montane plants' that John Birks (1973) listed in his study of Skye, 40 grow on the Cuillins (including alpine rock-cress, *Arabis alpina*, at its only British locality), 16 on the Red Hills and 43 on the Storr: base-rich rock and corries and crags favour species-richness. The Storr has its speciality in the form of Iceland-purslane (*Koenigia islandica*, Fig. 265c–d), a tiny plant of the dock family (Polygonaceae), which occurs in open habitats on damp basalt gravel in Skye and Mull. On the Storr, where it was first discovered in Britain, it was misidentified as water-purslane (*Lythrum portula*) for several years. It occurs in gravelly flushes with *Carex viridula* ssp. *oedocarpa*, *Deschampsia cespitosa*, *Saxifraga stellaris* and associated bryophytes, but this is by no means its only habitat. *Koenigia* is circumpolar and is common in suitable open habitats in Iceland and Norway; there is a Norwegian specimen in the Cambridge university herbarium labelled '*In loco ubi pisces sole exsciccantur*' – 'in a place where fish are dried in the sun.'

It is noteworthy how many mountain plants descend to near sea level in Skye: alpine lady's-mantle, northern rock-cress (*Arabis petraea*), cyphel, mountain avens, mountain sorrel, holly-fern, yellow saxifrage, purple saxifrage, moss campion, roseroot and alpine meadow-rue are all recorded below 150 m.

Returning to the mainland, we are in a different scene from either Skye or the Highlands south of the Great Glen. The rocks that cover the greatest area are the Moine schists, harder and generally yielding more acid soils than the Dalradian schists of the south and east Highlands. They come to a sharp western limit at the Moine Thrust, a geological feature often very apparent in the landscape (Fig. 266). The mountains to the east are angular and 'mountain-shaped'. To the west, the mountains are of Torridonian sandstone or Cambrian quartzite, and often betray their sedimentary origin by their terraced profiles. They rise over the characteristic knobbly topography of the Lewisian gneisses, some of the oldest rocks on the surface of the earth.

FIG 266. Meall Garbh from Arkle, Sutherland, June 1981. *Calluna* on peat and bare quartzite with scattered *Racomitrium* in the foreground, grazed blanket bog covering the slope in the middle distance. Schist mountains beyond the Moine Thrust in the distance.

Blanket bog and rock are staple ingredients of the northwest Highland landscape. *Trichophorum–Erica tetralix* wet heath[M15] on shallow peat can reach 600 m on the hills, bridging the gap between lowland *Trichophorum–Eriophorum vaginatum* blanket bog[M17] of the lowlands, and the upland *Calluna–Eriophorum* blanket bog[M19] covering any flat-enough ground at higher altitudes. Upland blanket bog in the Highlands often includes cowberry, sometimes bog bilberry (*Vaccinium uliginosum*), and occasionally dwarf birch (*Betula nana*). In high-altitude blanket bog in the northern Highlands these species are more constant and are often joined by arctic bearberry (*Arctostaphylos alpinus*). This species occurs most regularly in the short wind-cut *Calluna–Arctostaphylos alpinus* heath[H17] on the thin peaty soil of stony hill sides or exposed crests and spurs across much of northern Scotland. A community of the northern Highlands generally is a *Vaccinium myrtillus–Rubus chamaemorus* heath[H22]. This intergrades with both upland blanket bog and *Calluna–Vaccinium–Sphagnum capillifolium* heath, and occurs widely where the blanket-bog peat thins off at the margins or on thin peat over rock outcrops. As ever, most exposed crests and summits are occupied by *Carex bigelowii–Racomitrium lanuginosum* moss-heath – or just bare fell-field, which is home to the few Scottish populations of Norwegian mugwort (*Artemisia norvegica*). Dwarf juniper is co-dominant with heather in a very local *Calluna–Juniperus communis* ssp. *nana* heath[H15], particularly characteristic of exposed stony ground and rock outcrops on the Torridonian and Cambrian quartzite mountains of the far west. Fragments of this community are widespread, but it is very vulnerable to burning, and well-developed stands are rare. Despite the exposed situation in which it grows, it is habitat to several rare liverworts, including *Herbertus borealis*, known only from Beinn Eighe and two localities in southwest Norway. The bryophyte-rich *Calluna–Vaccinium–Sphagnum capillifolium* heath is much more widespread, representing the major contribution of the northwest to the biodiversity of the Highlands.

The corries and cliffs of the Moine schist mountains have a respectable but not outstanding mountain flora, but in general the northwest Highlands are not rich in calcicole mountain plants. The one exception is the plants that grow on the outcrops of Cambrian limestone at Knockan, Inchnadamph and around Durness – including mountain avens, whortle-leaved willow (*Salix myrsinites*), holly-fern and arctic sandwort (*Arenaria norvegica*). The greenness and fertility of the coastal strip is in startling contrast to the rugged and austere landscape inland, and a tribute to the moderating effect on the sea upon climate. But the maritime influence that makes for the mild winters also moderates the heat of the summers, enabling mountain plants to come down to sea level. Mountain avens acts as a sand binder on the Invernaver dunes at Bettyhill, and purple oxytropis (*Oxytropis halleri*) grows in the machair and yellow saxifrage in dune slacks; I have seen purple saxifrage growing around the bases of marram grass there.

IRELAND

Ireland has many mountains, but only a relatively modest arctic-alpine flora and little overtly 'mountain' vegetation. In part this is a reflection of the fact the Ireland has fewer vascular plants than Britain, just as Britain has fewer than a comparable area on the Continent. In part it is a reflection of climate. Our cool, moist oceanic climate favours leaching and acidification of soils, so that conditions for most mountain plants deteriorate westwards. On the other hand, bryophyte diversity benefits from oceanic conditions, and the Irish mountains share with the Scottish and Welsh mountains a rich moss and liverwort flora. A distinctively Irish plant, frequent on many Irish mountains, is St Patrick's cabbage (Fig. 267).

Table 11 summarises the distribution of 24 mountain plants that reach Ireland, and shows how widely dispersed these species are. The clear hotspot is the limestone massif that includes Ben Bulben (Figs 148, 268), with 83% of the list. But six areas ranging from Donegal to Kerry have 45–70% of the list. The richest sites for mountain plants, such as Slieve League, tend to be in the

FIG 267. A distinctively Irish mountain plant, St Patrick's cabbage (*Saxifraga spathularis*), found on most Irish mountains, but known as a native nowhere in Britain. Croagh Patrick, Co. Mayo, July 1970.

Species	Inishowen	Errigal	Slieve League	Ben Bulben	Nephin Beg	Connemara	Burren
Highest summit (m)	615	752	601	644	629	819	327
Alchemilla alpina	–	–	–	–	–	–	–
Alchemilla glabra	*	*	*	*	–	*	*
Arabis petraea	–	–	–	*	–	–	–
Arctostaphylos uva–ursi	–	*	*	–	*	*	*
Arenaria ciliata	–	–	–	*	–	–	–
Carex bigelowii	*	*	*	*	*	*	–
Dryas octopetala	*	–	*	*	–	*	*
Oxyria digyna	–	*	*	*	*	*	–
Persicaria vivipara	*	–	*	*	–	–	–
Poa alpina	–	–	–	*	–	–	–
Polystichum lonchitis	*	*	*	*	–	*	–
Salix herbacea	*	*	*	*	*	*	–
Saussurea alpina	*	*	*	*	*	*	–
Saxifraga hypnoides	*	–	–	*	–	–	*
Saxifraga aizoides	–	–	*	*	–	–	–
Saxifraga nivalis	–	–	–	*	–	–	–
Saxifraga oppositifolia	*	*	*	*	*	*	
Saxifraga spathularis	–	*	–	–	*	*	
Saxifraga stellaris	–	*	*	–	*	*	
Sedum rosea	*	*	*	*	*	*	*
Silene acaulis	*	–	–	*	–	–	–
Thalictrum alpinum	–	*	*	*	*	*	–
Trollius europaeus	–	*	–	*	–	–	–
Vaccinium vitis-idaea	*	*	*	*	*	*	
Number of species /24	12	14	15	20	11	14	5

northwest, while the high mountains of the south are generally poorer. Latitude and summer temperature may be partly responsible, but the occurrence of dwarf willow on almost all high Irish mountains suggests that much of the variation is due to factors of bedrock and soil. In the Twelve Bens of Connemara, Muckanaght has been called 'an oasis of schist in a Sahara of quartz', and it does indeed have a richer arctic-alpine flora than the quartzite mountains around it – as does Lissoughter, southeast of the Twelve Bens, despite its modest altitude (400 m).

Most of Donegal is made up of either metamorphic schist, gneiss or quartzite, or granite. Ben Bulben in Co. Sligo is a large block of Carboniferous limestone, with a capping of peat. Nephin and Nephin Beg are largely Cambrian quartzite, but there are other metamorphic rocks in the vicinity as well. Mweelrea is Ordovician slate, but the Connemara mountains south of Killary

Brandon	Iveragh	Galtee	Comeragh	Wicklow
953	1041	920	795	926
*	*	–	–	*
–	–	–	–	*
–	–	*	–	–
–	–	–	–	–
–	–	–	–	–
*	*	*	*	*
–	–	–	–	–
*	*	*	–	–
*	–	–	–	–
*	–	–	–	–
*	*	–	–	–
*	*	*	*	*
*	*	*	–	*
		*	*	*
–	–	–	–	–
–	–	–	–	–
–	–	–	–	–
*	*	*	*	*
*	*	*	*	*
*	*	*	*	*
–	–	–	–	–
–	–	–	–	–
–	–	–	–	–
11	9	9	6	9

TABLE 11. The distribution of mountain plants in the main mountain areas of Ireland. The headings of the columns should be interpreted 'and district' in each case. Inishowen includes Benevenagh; Errigal includes Muckish, Dooish etc.; Ben Bulben includes the whole massif; Burren inclues Aran and the Cliffs of Moher; Comeragh includes the Knockmealdown and Monaullagh mountains. Compiled from Praeger (1934) and Preston *et al.* (2002).

Harbour are a complex mix including schists and quartzites. These (except the limestone) are predominantly hard, acid rocks, which might be seen as a southward outpost of the western Highlands. As there, they carry infertile peaty soils. The Wicklow mountains are of granite, schists and slates, and invite comparisons with the granite dome of Dartmoor, the rocks of north Wales and the Lake District, and the Southern Uplands of Scotland. Most of the other upland areas from Slieve Aughty and Slieve Bloom southwards are of Old Red Sandstone age, and this includes all the high summits from Kerry to Waterford; they could be regarded as a large-scale extension westwards of the geology of the Brecon Beacons.

The Wicklow mountains, above the level of *Festuca–Agrostis* pasture, are largely covered with heather–bilberry moorland, with western gorse (*Ulex gallii*)

FIG 268. The limestone cliffs on the north side of Ben Bulben, Co. Sligo, seen from the Horseshoe Road in Gleniff, June 2010.

up to about 400 m, and grading above into the upland blanket bog that covers wide tracts of the high ground – with heather, bell heather, deergrass, hare's-tail cottongrass (*Eriophorum vaginatum*), *Sphagnum* spp. etc. *Racomitrium lanuginosum* becomes prominent on the crests and summits. Cliffs and outcrops that would give a footing to anything other than peat-loving plants are few, and the mountain flora is correspondingly modest (Table 11). The Old Red Sandstone yields better soils, and grasslands are correspondingly more widespread. Heather and bilberry are still prominent, and western gorse reaches its altitudinal limit (670 m) in Macgillycuddy's Reeks, but crowberry is much less common than in the north of the country, or in Wales or Scotland. There is bryophyte-rich *Calluna–Vaccinium–Sphagnum capillifolium* heath on shady slopes, in which gingery cushions of *Herbertus aduncus* are conspicuous from Kerry to the Comeraghs (Fig. 269); the rarer *Mastigophora*, *Bazzania pearsonii* and *Scapania ornithopodioides* occur only in the high western mountains. *Racomitrium lanuginosum* is prominent on high exposed ground everywhere, joined by stiff sedge and dwarf willow on a few high summits. Brandon on the Dingle Peninsula (Fig. 270) has a fair mountain flora in its magnificent corries, but it is no match for Connemara or Slieve League (Table 11). The mountains of Connemara are still in the zone of abundance of western gorse, which grows everywhere with heather and bell heather on well-drained rocky ground. Any ground that is not well drained is

LEFT: **FIG 269.** Coumshingaun, Comeragh mountains, Co. Waterford, looking down the northeastern corrie to Coumshingaun Lough and the fertile farming country beyond, August 1974.

BELOW: **FIG 270.** Brandon Mountain, with stepped corrie lakes, from Conor's Pass, Co. Kerry, September 2008.

FIG 271. View from the Twelve Bens looking towards Mweelrea, July 1966; foreground quartzite with *Calluna* and *Racomitrium lanuginosum*.

FIG 272. Errigal from near Gweedore, Co. Donegal, September 1965. Blanket bog in the foreground, haycocks and stooked corn in the middle distance.

occupied by blanket bog, *Calluna–Trichophorum* wet heath, or wet *Molinia*. These give way with increasing altitude in turn to heather moorland, and windswept *Racomitrium* summit heath with stiff sedge and, locally, dwarf willow. Bryophyte-rich *Calluna* heath was better developed and richer on the quartzite hills of Connemara (Fig. 271) than on the Old Red Sandstone of Kerry, with *Adelanthus lindenbergianus* (also in Donegal and on Islay) and *Plagiochila carringtonii* to add to the species that reach Kerry, but it suffered severely from overgrazing in the 1990s. As an Irish friend of mine said, 'People used to raise wool and mutton on these hills; now they raise European Community subsidies.' That was in the worst days of sheep-headage payments. It remains to be seen whether the hills recover. The mountains farther north (Fig. 272) are outside the range of any but local western gorse (but within the area in which crowberry is common), so their slopes mostly ring the changes on blanket bog or *Calluna–Trichophorum–Molinia* wet heath, and *Calluna–Vaccinium* moorland with some mat-grass (*Nardus*) and heath rush (*Juncus squarrosus*); *Racomitrium* covers exposed outcrops, crests and summits often with stiff sedge and dwarf willow. Slieve League, on the north side of Donegal Bay, deserves particular mention. It is not particularly high (601 m), but looks higher because its southern face drops 'into the Atlantic in a magnificent precipice, one of the finest things of its kind in Ireland. The northern face, which drops with equal steepness down towards the little Lough Agh, is the home of a remarkable assemblage of Alpine plants' (Praeger 1934). Slieve League is one of the richest sites for mountain plants in Ireland.

The Irish mountains may seem rather an anticlimax after the riches of the Scottish Highlands. In view of their modest altitude, it is perhaps a matter for agreeable surprise that the Irish mountain flora is as rich and widely dispersed as it is. On his map of the potential natural vegetation of Ireland, John Cross (2006) shows the highest summits as acid birchwood, but at the scale of the map any naturally treeless areas would probably have been too small to show. There is no evidence in the present vegetation where the tree-line might have been. Analogy with Britain suggests a tree-line in northwest Ireland at about 450–500 m, and 500–600 m in the southern counties. Knud Jessen (1952) considered that during the Bronze Age 'in the Wicklow mountains the timber-line lay at about 600 metres.' At the Post-glacial climatic optimum the tree-line might have been somewhat higher than these estimates, which would have left only scattered summits and little (and fragmented) land area above the trees – islands in a sea of forest, subject to all the vicissitudes of island life. But floristic austerity cannot detract from the individuality, fascination and remarkable beauty of the Irish and northwest Scottish mountains, and the landscapes in which they lie.

CHAPTER 18

The Sea Coast: Saltmarshes

CCORDING TO THE ORDNANCE SURVEY'S 1:10 000 map, the length of
the coastline of Britain is about 17,820 km or, if the larger off-shore
islands are taken into account, 31,370 km. The length of the Irish
coast has been estimated at 5631 and 6437 km. We should not worry about these
discrepancies, because the 'length' of any irregular boundary depends on the
scale at which you measure it, a well-known mathematical problem (Mandelbrot
1967). By any measure, our two islands have a lot of coast, and our coasts account
for some very distinctive vegetation and a significant part of the biodiversity
of Britain and Ireland. But there are different kinds of coast: 'soft' coastlines
– saltmarshes, sand-dunes and shingle – and rocky shores and cliffs, each with
its own suite of characteristic species. These are the subject of the next three
chapters.

Saltmarshes, sand-dunes and shingle are all 'accretional' habitats, built up
from material water-borne by coastal tides and currents, with a continuous range
of particle size from the finest clay particles of saltmarsh mud to beach pebbles.
The clay and silt particles (less than 0.05 mm diameter) that predominate in
saltmarsh mud are kept in suspension by even slight turbulence and only settle
out in estuaries and sheltered bays. Sand particles (0.05–2 mm) settle out quickly
but are readily moved by waves and currents, and of course dry sand is picked up
and carried by wind. Shingle (2–200 mm) is too large to be carried by wind but
small enough to be moved by the combination of waves and longshore currents.
Each kind of coast has its own suite of land-forms and distinctive vegetation
and species. Rocky shores and sea-cliffs can only be eroded by the elements;
they have a degree of permanence, but for plants they can be among the most
stressful of habitats.

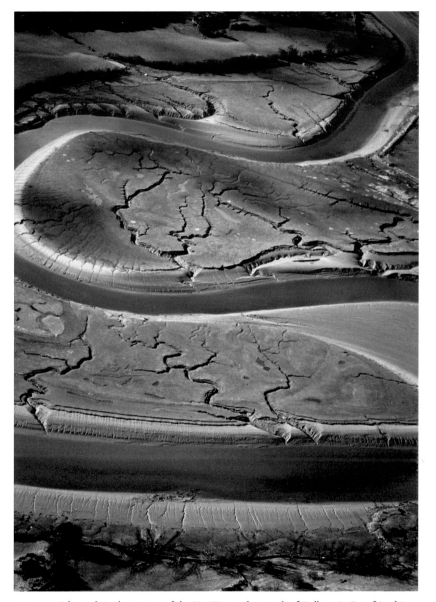

FIG 273. Saltmarsh in the estuary of the Urr Water, 5 km south of Dalbeattie, Dumfriesshire. Sediment erodes from the outer and downstream sides of the bends of the meandering river channel, and is deposited on the inner, lower sides of the loops. Various stages in the development of saltmarsh channels can be seen. (© Patricia & Angus Macdonald/SNH).

SALTMARSHES

Saltmarshes sometimes develop on open coasts, provided these are not too exposed, but more usually in estuaries (Fig. 273), at the head of sheltered bays or inlets, or in the shelter of shingle spits. They occur all round our coasts below high tide mark, with particular concentrations in the Poole Harbour–Solent–Chichester Harbour area, the north Kent, Essex and south Suffolk estuaries, northwestern Norfolk and the Wash, and the Humber estuary on the east coast, and the Bristol Channel, the northern Welsh and Lancashire estuaries and the Solway on the west. There are numerous saltmarshes, all of them small, in the Scottish sea-lochs, and many, mostly small, in Ireland.

The tides and saltmarsh zonation

The tides are caused by the gravitational pull of the sun and moon on the water of the oceans. There are generally two high tides and two low tides each day. When the sun, earth and moon are lined up (new and full moon) the tides are at a maximum; these are *spring tides*. When the moon is at right angles to the sun (half moon) tides are at a minimum, giving *neap tides*. Tides are about 51 minutes later each day because the moon is orbiting round the earth, but spring and neap tides always occur at the same time of day at any particular place. Figure 274 shows the twice-daily rise and fall of the tides through a spring–neap cycle, with the mean high water of spring tides (MHWS), mean high water of neap tides (MHWN), and mean low water of neaps (MLWN) and springs (MLWS) –

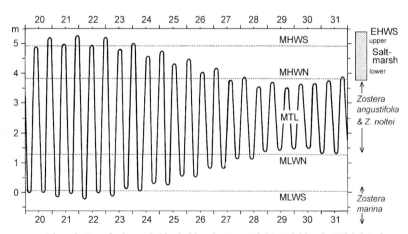

FIG 274. The relation of saltmarsh (shaded band at top right) to tidal levels. Tidal data for Holyhead, July 1959, from Lewis 1964. See text for further explanation.

calculated on tides for the whole year, and based on 'Chart datum', which is lower than mean sea level. The exact height of the tide varies with the relative position of the sun and moon and with the seasons, so additional levels can be defined, such as extreme high and low water of spring tides (EHWS and ELWS) and mean tide level (MTL). These are all based on astronomical predictions. Weather also influences tidal level. The sea acts as a 'water barometer', so if atmospheric pressure is low, sea level is raised by about 1.3 cm per millibar, and vice versa. Onshore winds also raise tidal level, especially when they funnel water into confined spaces such as the English Channel or the southern North Sea.

Saltmarshes as we usually think of them occupy only the shore between MHWN and EHWS – roughly the top quarter of the whole tidal range. This means that they spend a large proportion of time exposed to the air. At mean tide level a mud-flat (or beach) is submerged (and exposed) twice every (tidal) day – about 630 submergences a year. At about mean high water level the marsh is submerged 360 times a year, for an average of 1.2 hours a day (c. 440 hours a year), and never exposed for more than nine days at a stretch. In his work on the saltmarsh at Scolt Head Island on the Norfolk coast, V. J. Chapman took this as the boundary between lower marsh, dominated by time of *submergence*, and upper marsh, in which factors related to *emergence* are more important. Clearly this boundary cannot be sharp, but it enshrines a useful concept in thinking about saltmarsh ecology. Above mean high water level the saltmarsh can be exposed continuously to the air for many days on end, until at EHWS it is inundated only on rare occasions.

Below mean high water, the mud- or sand-flats may be colonised by eelgrasses (*Zostera* spp.), of which we have three species around our coasts. *Zostera marina*, the largest, is not generally part of the saltmarsh zonation at all, but is usually to be found on sandy substrata at or below low water of spring tides, to a depth of 3 or 4 m. The other two species, narrow-leaved eelgrass (*Z. angustifolia*, Fig. 275) and dwarf eelgrass (*Z. noltei*) are smaller, and *Z. noltei* is distinctive in being unbranched. These two species can form quite dense stands in suitable sites (Fig. 276), and are an important food resource for migrant brent geese and wigeon. Where the two grow together, as they commonly do, *Z. noltei* tends to grow on the slightly raised areas of the undulating mud, while *Z. angustifolia* prefers the slight depressions. These two have similar geographical distributions with us, strikingly different from the big *Z. marina*.

Above MHWN there is a steep rise in the average number of days the surface is undisturbed by tidal ebb and flow. This is crucial for establishment of seedlings. On the Dovey marshes in Wales, Wiehe (1935) found a dramatic drop in the density of mature glasswort (*Salicornia*) plants in midsummer at about high water of neap

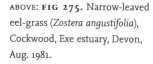

ABOVE: **FIG 275.** Narrow-leaved eel-grass (*Zostera angustifolia*), Cockwood, Exe estuary, Devon, Aug. 1981.

LEFT: **FIG 276.** Eelgrass (*Zostera angustifolia*) beds: (a) Lympstone, Exe estuary, Devon, May 1980; (b) Oranmore, Co. Galway, with *Zostera* rooted in the mud, seaweeds (*Fucus* spp.) attached to the stones and an eroded edge of saltmarsh in the middle distance on the right. September 1959.

tides, and that the mortality of seedlings between April and July was 70–90% within reach of neap tides, but abruptly fell to around 30% in the zone between high water of neaps and high water of springs (Fig. 277). The two plants that

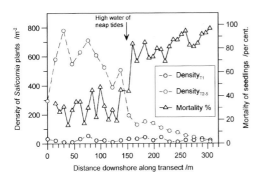

FIG 277. Establishment of glasswort (*Salicornia*) seedlings along transects in the Dyfi estuary, North Wales. Transect 1 (T1) was entirely covered at every tide; transects 2–5 (T2–5) were half above, half below high water of neap tides. Triangles and the full line show mortality of seedlings between April and July on transect 2. (Data from Wiehe 1935).

generally form the lower fringe of the saltmarsh proper, the annual *Salicornia* species (Fig. 278a) and the perennial grass *Spartina anglica* (Fig. 278d), share a need for two or three days free of tidal disturbance for seedlings to establish themselves. Two other short-lived species that often become established with the *Salicornia* are annual sea-blite (*Suaeda maritima*) and sea aster (*Aster tripolium*), often in its rayless form. Typically there follows a broad zone dominated by common saltmarsh-grass (Fig. 279), often with sea-purslane (Fig. 280) and sea-lavenders (*Limonium vulgare* and *L. humile*; only the latter in Ireland), greater sea-spurrey (Fig. 278c) and English scurvygrass (*Cochlearia anglica*). This gives way upwards to a zone in which the saltmarsh-grass (with a prominent ligule) is replaced by red fescue (*Festuca rubra*, without a ligule) with a very characteristic list of associates including saltmarsh rush (Fig. 281c), sea-milkwort (Fig. 281a), sea plantain (*Plantago maritima*), sea arrowgrass (*Triglochin maritimum*) and thrift (*Armeria maritima*).

At its upper edge the saltmarsh passes, gradually or abruptly, into whatever vegetation lies behind it, and as this neighbouring vegetation can be diverse, upper saltmarsh vegetation is very variable. Common transitions are to grassland, freshwater marsh, sand-dune and shingle.

Saltmarsh accretion and erosion

Saltmarshes grow by accretion of silt, but this is by no means inevitable. Rivers bring down silt in suspension, and if they debouch into a wide sheltered estuary they deposit a large part of this. The consequence is that the estuary silts up, creating mud-flats, which sooner or later reach a level at which the pioneer plants of the lower saltmarsh can become established. Further accretion of silt raises the surface, so that plants of successively higher levels take over, and this process continues until the marsh is inundated by only the highest spring tides. Upper saltmarsh turf affords good grazing, so by this stage the marsh may

FIG 278. Saltmarsh plants: (a) an annual glasswort, *Salicornia dolichostachya*, Budleigh
Salterton, Devon; (b) common sea-lavender (*Limonium vulgare*), Thornham, Norfolk, July 1971;
(c) greater sea-spurrey (*Spergularia media*), Budleigh Salterton, Devon, June 1988; (d) common
cord-grass (*Spartina anglica*), Cockwood, Devon, July 1980.

LEFT: **FIG 279.** Salt-marsh grass (*Puccinellia maritima*) colonising a dense stand of *Salicornia* with scattered sea aster (*Aster tripolium*), Blakeney, Norfolk, August 1970.

BELOW: **FIG 280.** Sea-lavender (*Limonium vulgare*) marsh; with sea-purslane (*Atriplex portulacoides*) lining the creeks, Thornham, Norfolk, July 1971.

well have been enclosed from the sea and added, behind an embankment, to the farmland to landward. This process ('inning') has obviously continued for millennia in the past, and the lower courses of many of our rivers flow through tracts of flat farmland that were formerly saltmarsh. The evidence for this is often apparent in maps and aerial photographs. The river becomes confined to a channel a fraction of the width of the former estuary. It may still support saltmarshes, but deposition and erosion of sediment are likely to be much more nearly in balance. The other situation in which saltmarsh can develop *de novo* is in sheltered inlets and behind growing shingle spits, of which Scolt Head Island in Norfolk provides classic examples (Figs 310, 311). Wave action on exposed shores stirs up sand and silt, which then settles in less turbulent places.

Accretion and erosion of silt, and their interplay, are complex. Maximum accretion of silt in a growing marsh can reach 3 cm a year in the *Salicornia* zone,

FIG 281. Upper saltmarsh plants. (a) Sea-milkwort (*Glaux maritima*), Cuckmere Haven, Sussex, June 1986; (b) lesser sea-spurrey (*Spergularia marina*), Turf Lock, Devon, May 1977; (c) saltmarsh rush (*Juncus gerardii*), Holme-next-the-Sea, Norfolk, July 1981; (d) parsley water-dropwort (*Oenanthe lachenalii*), Cuckmere Haven, July 1989.

and 10 cm a year or more in a vigorous closed turf of *Puccinellia* or *Spartina*, but these are maxima and 10–15% of these figures would be more usual. Accretion on emergence marshes is commonly 0.2–1 cm a year. In relatively young and uniform marsh, most accretion takes place near the landward margin, but the zone of maximum accretion tends to move seaward as the marsh builds up and matures, so the marsh becomes flatter (approaching high water of spring tides) as it ages. Accretion is not uniform round the year. Fine clayey mud tends to be deposited more around the seaward edges of the marsh during spring and summer. In autumn, with the vegetation full grown and at maximum trapping capacity, silt mobilised by autumn storms is deposited generally over the marsh; there may be a small net loss of sediment during the winter.

Changing current patterns may lead to a switch from accretion to erosion – which typically takes place at the seaward edge of the marsh, creating a small cliff. If the sediment regime switches back to accretion, saltmarsh development begins anew on the eroded surface. Erosion features are very common on long-established saltmarshes, often manifested as a series of 'steps' as one walks down the marsh. But the possibilities are almost endless. A common pattern in estuaries is for the river channel to follow a meandering course through the deposited silt, eroding the outer and downstream sides of the meanders, and depositing silt on the more sheltered inner and lower side. The result is a series of lobes of saltmarsh in successive loops of the river, each one youngest on the downstream side, and progressively truncated on the upstream side. Thus the whole pattern slowly moves downstream. This process is well illustrated in Figure 273, which will repay close scrutiny.

Pans and creeks, and aeration

Even if the original surface is level, small irregularities inevitably arise as the plant cover develops. More silt tends to be deposited where plants slow the flow of the water, and minor depressions can become accentuated as *primary pans*. As accretion goes on, the water covering the marsh at flood tide becomes concentrated in regular creeks (or channels) as the water drains off at the ebb. These often become intricately branched and meandering, especially if the marsh is extensive and flat. The creeks are erosion features subject to tidal scour on the ebb, even on an accreting marsh, but they are not necessarily permanent. The vegetation on either side of the creek continues to trap silt, so the surface of the marsh remains flat. Narrow creeks can be roofed over by encroachment of the turf on either side, or blocked by collapse of their walls. The former creek becomes stagnant and detached from the drainage, and it can break up into a string of *channel pans*, whose origin is often obvious from their shape and orientation. Most pans in a mature saltmarsh have originated in this way. Some pans may be formed where persistent masses of tidal debris on the marsh have killed the underlying vegetation, initiating a *trash pan*. Whatever their origin, pans are subject to wide variations in temperature and salinity, many are permanently waterlogged, and they are generally an unfavourable environment for plant growth (Fig. 282).

FIG 282. Saltmarsh pans, Malltraeth, Anglesey. The pans are dry in a spell of fine weather; a more normal water level is marked by the line of bleached algae.

The better-aerated strip a few metres wide along the margin of a channel is often marked by the abundance of the grey-leaved sea-purslane. Even below the *Salicornia* zone, the surface is constantly burrowed by lugworms, cockles and the small crustacean *Corophium volutator*, but the saltmarsh contains a great deal of organic matter needing oxygen for its breakdown. Chapman found that when he poked a hole through the mud of a flooded saltmarsh, bubbles of air were released. At low levels on the marsh (*Aster* zone) the oxygen concentration in this air was only around 1%, but in sea-lavender marsh it was 10.5–17.5%. Measurement of oxidation–reduction (redox) potential in a Humber saltmarsh showed that low on the marsh under *Spartina*, even at 5 cm depth the soil was permanently waterlogged and highly reducing for all but a brief period at neap tides in midsummer (Armstrong *et al.* 1985). In *Puccinellia* marsh, conditions were only permanently reducing at 30 cm depth; at 10 cm or less high spring tides caused episodes of waterlogging and lowered redox potential. Rather higher, in the 'general saltmarsh', the soil at all levels was oxidising most of the time, broken only by periods of waterlogging following unusually high spring tides. Pans remain waterlogged and reducing most of the time. Highly reducing conditions favour anaerobic bacteria, which reduce iron from the ferric to the ferrous state, and sulphate to sulphide. Anyone incautious enough to go up to their thighs in a saltmarsh pan is likely to emerge covered in stinking black mud!

Geographical (and other) variation in British and Irish saltmarshes

The details of saltmarsh zonation and of individual saltmarshes vary greatly from place to place, depending on the geographic setting of the marsh, its exposure to wave action and currents, the nature of the substratum (muddy, sandy or peaty), whether the marsh is grazed or not, the age of the marsh and whether it is accreting or eroding. These paragraphs give only the barest outline.

The classical studies on British saltmarsh vegetation were concentrated on Blakeney Point and Scolt Head Island in north Norfolk, and the Dyfi estuary in Cardigan Bay, in the years before the Second World War. A much broader and more representative picture was provided in the 1970s by Paul Adam's detailed survey of 133 saltmarshes all round the coasts of Britain (Adam 1978, 1981). He suggested a division of saltmarsh sites round the British coast into three rough categories. Type A marshes, in which communities of the lower and middle marsh, dominated by such plants as *Spartina anglica*[SM6], *Salicornia* spp.[SM8], the rayless form of sea aster[SM11], common saltmarsh-grass[SM13] and sea-purslane[SM14] are particularly prominent, are concentrated along the south and east coasts of England. They are seldom grazed, are typically backed by low coasts (sometimes dunes), and are the habitat of some southern species that are scarce or absent elsewhere on our coasts, notably matted sea-lavender (*Limonium bellidifolium*, confined with us to north Norfolk), shrubby sea-blite (*Suaeda vera*) at the upper fringe of the marsh[SM21] and sea-heath (*Frankenia laevis*) often growing with sea-purslane[SM22]. Golden-samphire (*Inula crithmoides*) is more widespread than the last two (often on sea-cliffs) but grows in saltmarshes only from Essex to Hampshire[SM26]. Sea wormwood (*Seriphidium maritimum*) occurs more widely again but is commonest in Type A saltmarshes in southeast England.

In type B marshes the emphasis shifts towards the upper marsh and to communities in which such species as red fescue, sea-milkwort, thrift, saltmarsh rush[SM16], sea rush (*Juncus maritimus*)[SM15,18], autumn hawkbit (*Leontodon autumnalis*) and parsley water-dropwort (Fig. 281d) are prominent. Most of the lower-marsh species are still present, but zonation tends to be less clearly marked than in type A marshes. Type B marshes are often grazed, usually by sheep, less often by cattle, and are particularly concentrated on the Irish Sea coastline of Wales, northern England and Galloway. They typically combine an abundance of the common saltmarsh species – thrift can make spectacular displays of pink in early summer (Fig. 283) – with rich upper-marsh transitions to non-saline wet or dry grasslands. These are marked by some characteristic species, such as autumn hawkbit, parsley water-dropwort and silverweed (*Potentilla anserina*), but nothing particularly notable or rare.

FIG 283. Thrift (*Armeria maritima*) in flower on upper saltmarsh, Holy Island, Northumberland, June 1999.

The saltmarshes in the Scottish sea-lochs are a category of their own (type C), characterised as much as anything by absences. Common saltmarsh-grass is still there, but the turf is largely dominated by the ubiquitous red fescue. Sea-milkwort, thrift, sea arrowgrass, sea plantain and buck's-horn plantain (*Plantago coronopus*) are abundant. *Salicornia* and annual sea-blite are sparse, and *Spartina* is absent altogether. Features very characteristic of these Scottish saltmarshes are the 'turf fucoids', depauperate brown seaweeds growing mixed with *Puccinellia*, the detached floating masses of seaweeds (usually *Ascophyllum nodosum*) in the most sheltered sea-lochs (Fig. 284), and the yellow iris (*Iris pseudacorus*) and meadowsweet (*Filipendula ulmaria*) in the fringe at the head of the saltmarsh[M28] (Fig. 285). Western Scottish saltmarshes are the headquarters in Britain and Ireland for saltmarsh flat-sedge (*Blysmus rufus*)[SM19] and slender spike-rush (*Eleocharis uniglumis*)[SM20], both regularly with sea-milkwort, saltmarsh rush, creeping bent (*Agrostis stolonifera*), sea arrowgrass and red fescue, and often accompanied by long-bracted sedge (*Carex extensa*).

Tom Curtis and Micheline Sheehy Skeffington (1998) made an inventory of the saltmarshes of Ireland, noting their substrata (sand, mud or peat), and whether they were ungrazed or grazed, and if grazed, by what kind of livestock. Generally, marshes in estuaries or sheltered bays are on mud, and these predominate, especially on the south and east coasts. The saltmarshes on the

FIG 284. Free-floating masses of a finely-divided form of knotted wrack (*Ascophyllum nodosum*) in the very sheltered water of Loch Sunart, Strontian, September 1973. The inset shows a close-up of the seaweed with a clump of red fescue and sea-plantain for comparison. 'Turf fucoids' are similarly depauperate fucoid seaweeds amongst the saltmarsh plants.

FIG 285. Upper saltmarsh with yellow iris (*Iris pseudacorus*), Drynoch, Skye, June 1981.

east coast of Ireland would pass with little remark in southeast England, and most are ungrazed. Sea-purslane and *Spartina* are generally present, as they are in saltmarshes of most of the south-coast estuaries; they too are mostly ungrazed. On the west coast the marshes are generally grazed, mostly by cattle. A particular feature of the west, especially in west Galway and Mayo, is the number of saltmarshes overlying freshwater peat – a testimony to Post-glacial changes of land and sea level. These west-coast Irish saltmarshes are mostly small, and have much in common with similar marshes in western Britain, at low levels dominated by common saltmarsh-grass, with sea plantain, sea aster, sea-milkwort, thrift, greater sea-spurrey and, locally, lax-flowered sea-lavender (*Limonium humile*). 'Turf fucoids' are often prominent, as in the western Scottish sea-lochs. *Salicornia* and annual sea-blite are scattered among the other saltmarsh plants but, as in British west-coast marshes, seldom form extensive pioneer stands. At higher levels red fescue and saltmarsh rush become dominant, accompanied by all the common mid- and upper-saltmarsh plants, and such species as long-bracted sedge, sea rush and brookweed (*Samolus valerandi*).

The south- and east-coast marshes stand in a similar relationship to the west-coast marshes as the southeastern (type A) marshes do to the west-coast (type B and C) marshes in Britain. The distributions of thrift and sea-milkwort are noticeably biased toward the west side of Ireland, while sea-purslane is heavily concentrated along the south and east coasts – its distribution is nicely symmetrical around the Irish Sea. *Spartina anglica* is mainly a south- and east-coast plant in Ireland, with the conspicuous exception of the Shannon estuary. Lax-flowered sea-lavender, greater sea-spurrey, annual sea-blite and *Salicornia*, all lower-marsh species, all show a bias towards the estuarine marshes of the south and east.

SPARTINA IN BRITAIN AND IRELAND

The long-standing native small cord-grass (*Spartina maritima*), once recorded all along the south and east coasts of England from Devon to Lincolnshire, is now a very local plant with its headquarters in the Essex and Suffolk estuaries, and outlying localities on the Lincolnshire shore of the Wash and in the Chichester Harbour–Solent area. The American smooth cord-grass (*S. alterniflora*) was first recorded in the River Itchen near Southampton in 1829, having presumably arrived accidentally with shipping. By the 1870s it had spread widely and become abundant in the Itchen and as far down Southampton Water as Hamble. About 1870 a sterile hybrid between *S. maritima* and *S. alterniflora* was noticed, and in 1880 it was described by H. & J. Groves as a new hybrid species under the name

Spartina × *townsendii*. The sterile hybrid has a chromosome number of 2n = 62; *S. maritima* has 2n = 60 and *S. alterniflora* 2n = 62. The sterile hybrid spread by rhizome growth and fragmentation, and by the early 1900s was established throughout the Solent area and had reached Poole Harbour. Probably about 1890 a fertile amphidiploid plant appeared, with a chromosome number of 2n = 122. This was vegetatively as vigorous as the sterile hybrid, but it could spread by seed. In the early decades of the twentieth century both continued to spread, and '*Spartina townsendii* sensu lato' (including *S. anglica*) was widely planted for reclamation work from 1907 onwards. Goodman *et al.* (1959) estimated that by then the plant covered at least 12,000 ha of tidal flats in Britain and Ireland. Patches of die-back in formerly vigorous *Spartina*, with death and decay of the rhizomes (Fig. 286), are perhaps due to the growth of the plant itself along with accumulation of fine sediment leading to waterlogging and locally toxic anaerobic conditions, because no possibly causal disease organism has been found.

Spartina anglica is notable for the wide range of conditions it can tolerate. Its lower limit is normally about (or just below) MHWN, but in Poole Harbour, in sheltered conditions and with an unusually narrow tidal range, it grows almost down to MLWN. It can grow as high as MHWS, but is a poor competitor in dry soils, so its upper limit is usually 20–30 cm lower. It can thus cover a large part of the normal saltmarsh range. It is now generally distributed round the shores of Britain and Ireland north to Galloway and the Firth of Tay.

FIG 286. *Spartina* dieback. Exe estuary, behind Dawlish Warren, Devon, July 1980.

SALINITY

Saltmarsh plants are *halophytes*, that is, they can tolerate relatively high concentrations of salt (sodium chloride, NaCl). Halophytes face twin problems. To take up water, the cells of a plant must be at a lower water potential than the soil water; that means in practice that the cell sap has to be more concentrated than the soil water. Many of the enzymes essential for normal metabolism are inhibited by raised concentrations of NaCl, and in a saline habitat the soil water is at a lower water potential (more concentrated) than in normal soils, and contains a large excess of salt. This means that the plant's cell-sap must be more concentrated too, but at the same time the plant must keep the sodium and chloride concentrations in its cells within bounds. Selective uptake of potassium (K) and exclusion of sodium (Na) is important to all halophytes, as it is in a lesser degree to non-halophytes. Grasses and other monocotyledonous halophytes tend to rely mainly on excluding Na, but many, including *Spartina*, have glands excreting salt to the exterior; their (molar) ratio of Na to K is commonly around 1:1. Many dicotyledonous halophytes take up salt, and have higher Na:K ratios, often around 10:1. They rely on various combinations of secreting salt through specialised glands into hairs or bladders on the leaf surface, or segregating it into the cell vacuoles, and balancing the osmotic potential of the vacuoles and cytoplasm by synthesising *compatible solutes* in the cytoplasm – substances like proline, glycine betaine and sugars – which can adjust osmotic potential but are benign to enzyme activity. In plants with this mechanism the vacuoles often make up a large proportion of their bulk, which is largely why succulence is so common in saline habitats. These two mechanisms are not mutually exclusive, and any simple explanation sidesteps a great deal of variation and physiological detail.

The salinity at the seaward edge of a saltmarsh or at the mouth of an estuary is roughly that of the open sea, about 35 g per litre. This is around 500 times the concentration of sodium and chloride in an average fresh water. Measurements of the salinity of the soil-water at 10 cm depth in the saltmarshes of the Exe estuary are plotted in relation to distance from the mouth in Figure 287. The effect of the river in diluting the seawater entering the estuary at every tide is immediately apparent, and minor inflows into the sheltered bay behind the sand-spit of Dawlish Warren result in a good deal of variation even close to the mouth. However, for the first 5 or 6 km, where the estuary remains a broad expanse of water at high tide, salinity does not fall greatly below the open-sea concentration. It is only as the estuary begins to narrow above Turf Lock and the confluence with the River Clyst at Riversmeet that the influence of fresh water is felt strongly. From here, there is a steep gradient of salinity to fully fresh water at Countess Wear on

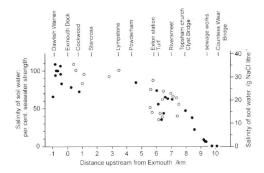

FIG 287. Salinity at 10 cm depth in muds in the Exe estuary from its mouth to near the tidal limit, projected onto a straight line through Exmouth dock and Topsham church. Solid dots, west shore; open circles, east shore. From Proctor (1980).

the Exe, and about 1 km below Clyst St Mary on the Clyst. Of the dominant species in the Exe saltmarshes, the eelgrasses (*Zostera angustifolia* and *Z. noltei*) occur only where salinity is above about 25 g per litre – about 70% of seawater. Sea-purslane occurs down to about half seawater, and *Spartina anglica* and *Puccinellia maritima* down to about a third of seawater strength. Sea club-rush (*Bolboschoenus maritimus*) and common reed (*Phragmites australis*) by contrast are restricted to the less saline parts of the estuary; sea club-rush was found at salinities from 7% to 65% of seawater and common reed up to just over 35% seawater.

The picture is essentially similar in the saltmarsh in the neighbouring estuary of the River Otter at Budleigh Salterton – a small estuary now, but larger and longer in the past. When the first edition of the Ordnance Survey was surveyed early in the nineteenth century, unreclaimed saltmarsh occupied much of the width of the valley for almost 2 km, and the river followed the parish boundary west of the embankment constraining the river to its present course. The present saltmarsh has grown up in the last 150 years. The first edition of the 25-inch map, surveyed in 1888, shows 6.26 ha of 'saltings' above high water mark of ordinary tides. By 1933 this had increased to 8.90 ha, and by 1955 to 10.45 ha. The map surveyed in 1970 gives the area of saltings as 12.13 ha, largely filling the estuary, and a recent satellite image shows little further change.

The lower-marsh species are frequently covered by the tide and are seldom exposed for more than a week at a stretch. The salinity around their roots (Fig. 288) thus mostly stays close to that of the flooding water – the central trend of the salinity curve of Figure 287. Salinity is much more variable in the upper marsh, subject both to seepage of fresh water and leaching by rain, and to concentration of the water of the high spring tides during spells of hot fine weather in summer. Hence the highest salinities are encountered between high water of neaps and high water of springs, and are seasonal and erratic. Five samples of the mud under *Zostera angustifolia* in the Exe estuary spanned a range of salinities of only 26–34

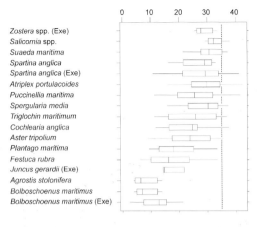

FIG 288. Box plot of salinity at 10 cm depth under saltmarsh plant species in 39 1 m² quadrats from the Otter estuary, Devon, with some comparative data from the Exe estuary. The line dividing each box is the median value. The dotted line shows the composition of average seawater. Data from Proctor (1980) and Brooks *et al.* (1988).

g per litre, with a median of 28. The common lower and middle-marsh species generally give median salinities of 25–30 g per litre, with maxima around or a little above the salinity of seawater. The few upper-saltmarsh species listed in Figure 288 all have lower median salinities, but red fescue (*Festuca rubra*) and sea plantain (*Plantago maritima*) stand out for the wide range of salinities they can tolerate. This can lead to the apparent paradox, in upper reaches of estuaries, of the frequently flooded lower levels being occupied by brackish-water species, while the more salt-tolerant species are confined to the higher parts of the zonation (Fig. 289).

The variation in salinity of upper saltmarshes is one factor in their great diversity of species composition. The widespread *Festuca rubra–Juncus gerardii*[SM16] and *Juncus maritimus*[SM18] communities are never very saline in their upper parts and intergrade without a break into non-saline pastures and meadows[MG6,11,13]. In sites on clayey or compacted soils, subject to tidal flooding and often to a degree of disturbance, conditions can sometimes become very saline; this is the preferred habitat of reflexed saltmarsh-grass (*Puccinellia distans*) and lesser sea-spurrey (Fig. 281b) – both species that (with Danish scurvygrass, *Cochlearia danica*) have spread inland along the verges of salt-treated roads. Many of the common middle- and upper-saltmarsh species still occur in this community[SM23], including common saltmarsh-grass, sea-milkwort, sea-purslane, sea arrowgrass, sea plantain, as well as the very characteristic little hard-grass (*Parapholis strigosa*).

Where a saltmarsh abuts on a shingle ridge or dune, or an embankment or sea-wall, the back of the marsh is often marked by a strand-line community. Usually these involve one or other of two species of *Elytrigia*, sea couch (*E. atherica*) or common couch (*E. repens*). Much the commonest is dominated by sea couch, with frequent red fescue, sea-milkwort, spear-leaved orache (*Atriplex*

FIG 289. Inverted zonation: saltmarsh grass, sea aster, sea plantain, and English scurvygrass above, sea club-rush nearest the river channel. River Exe, Topsham, Devon, May 1980.

prostrata) and a scatter of other common saltmarsh plants[SM24]. In the north-Norfolk saltmarshes the rare shrubby sea-blite (*Suaeda vera*) is constant in a community of this kind, with sea couch and sea-purslane[SM25]. This is related to a *Suaeda vera–Limonium binervosum* (rock sea-lavender) saltmarsh community from the same marshes[SM21]. Common couch dominates a comparable community, mainly on the west coast[SM28]. It is less tied to freely drained soils, and often subject to some disturbance and deposition of drift-line debris. Sea couch, red fescue, creeping bent and spear-leaved orache are constant or nearly so, often accompanied by silverweed, parsley water-dropwort, curled dock (*Rumex crispus*), perennial sow-thistle (*Sonchus arvensis*) and creeping thistle (*Cirsium arvense*). Similar vegetation occurs in north Germany and Scandinavia.

The saline–freshwater transition: brackish habitats

Halophytes tend to be confined to the top of the tidal range near their upstream limits (e.g. sea aster, English scurvygrass), while fresh- and brackish-water species are confined to a narrow zone at the top of the tidal range near the estuary mouth, but extend down to lower levels farther upstream (e.g. sea club-rush, parsley water-dropwort). As already noted, this can lead to inversions of zonation from one part of an estuary to another. Thus sea club-rush is confined to freshwater seepage areas at the back of the more saline marshes (as in the Cefni marsh in Anglesey) but is the lowermost member of the zonation in brackish marshes.

Many species can grow in brackish habitats, but most of them have an alternative niche, either in saltmarsh or in non-saline habitats. Thus sea aster and sea plantain can grow at salinities a quarter of seawater or less, and common reed is primarily a freshwater plant, but has been found growing at a soil salinity more than half seawater strength in Poole Harbour. Parsley water-dropwort, distant sedge (*Carex distans*), false fox-sedge (*Carex otrubae*), grey club-rush (*Schoenoplectus tabernaemontani*) and brookweed (Fig. 300b) are other species often associated with salt-influenced habitats, but with inland freshwater habitats as well. And of course many common grassland species can grow in the less saline upper saltmarshes, including bird's-foot trefoil (*Lotus corniculatus*), white clover (*Trifolium repens*), autumn hawkbit (*Leontodon autumnalis*), creeping thistle and silverweed.

Saltmarshes are fertile habitats, and estuaries tend to be hubs of human activity. Upper saltmarshes are embanked and enclosed as farmland, river channels are straightened and dredged for navigation, or for control of flooding upstream. All of these processes tend to sharpen the boundary between saline or brackish estuarine water and the freshwater drainage of the surrounding country. The species and communities of the transition are the most vulnerable to this process, and they come increasingly under pressure with improvements of the sea-walls and more sophisticated engineering of the drainage. Of the specialist brackish-water plants, the widely tolerant and versatile sea club-rush remains abundant. Other brackish-water specialists with narrower ecological niches have fared less well. Divided sedge (*Carex divisa*) and marsh-mallow (*Althaea officinalis*), both plants of brackish places, have declined over the past century and are now very local. The tasselweeds (*Ruppia*) are usually found in brackish ditches and pools, often behind sea-walls (see also Chapter 12). Their main habitat in the past may have been in shallow pools on mud-flats in the upper reaches of estuaries[SM2], and *R. maritima* grows abundantly in this situation in the Cromarty Firth. It occurs in similar situations elsewhere, but not commonly. Triangular club-rush (*Schoenoplectus triqueter*) is a plant of mud-banks in brackish reaches of strongly flowing rivers, whose decline has certainly been due to development of the riverside. It now survives only in a single locality on the Tamar in southwest England, and by the Shannon and its tributaries in Ireland. A very different brackish-mud specialist is dwarf spike-rush (*Eleocharis parvula*). It occurs close to the upper tidal limit on firm estuarine mud, and in tidal pans on brackish grazing marshes. It has very few recorded localities widely scattered over England, Wales, Scotland and Northern Ireland, but it is so inconspicuous, and seems confined to so narrow an ecological niche, that it may well have been overlooked elsewhere[SM3].

CHAPTER 19

Sand-dunes and Shingle

S ALTMARSH IS ONE ACCRETIONAL HABITAT; sand-dunes and shingle are the other two. Sand-dunes occur all round our coasts. The only long segments of coastline without dunes are stretches of unbroken steep cliff, or unbroken wet muddy foreshore; even modest breaks in the continuity of a cliffed coast provide toeholds for small dunes to build up. Shingle beaches often fringe cliffed coasts, but the shingle is often not derived from the cliffs behind them. More distinctive (and often vegetationally more interesting) shingle features are much more local, and generally their shingle has clearly been carried by wave and current action from elsewhere. Nevertheless, sea-cliffs, dune and saltmarsh coexist in close proximity in many places.

COASTAL SAND-DUNES

For a dune system to develop, three things are necessary. The first is a source of sand on the foreshore – a wide sandy beach, sometimes the sandy shore of an estuary, which the local tides and currents keep replenished with sand. The second is at least reasonably frequent onshore winds – which do not have to be the prevailing winds, as shown by the numerous dune systems on our east and north coasts. The third is space to develop. That space may be landwards, as at Braunton Burrows (Fig. 290), Newborough Warren (Fig. 291) and many other west-coast 'backshore' dune systems, seawards over an accreting beach or 'ness', as at Tentsmuir, laterally along the shore or over a growing shingle spit or offshore island. In some places, dunes have overwhelmed farmland and settlements in historic times, as at Culbin and Forvie on the east coast of Scotland.

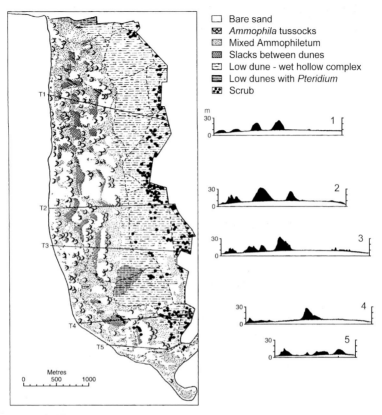

FIG 290. The dune system at Braunton Burrows, Devon, with levelled profiles across the dunes. The domed water-table is shown in white; the freely drained sand of the dune ridges above it in black. Compiled from Willis *et al.* (1959).

The course of dune development: embryo dunes and foredunes

The highest strand-line left after the winter storms is made up of seaweed, dead crabs and the odd dead bird and all sorts of flotsam and jetsam. It includes also seeds of the strand-line plants, which germinate in spring, and flower and fruit from midsummer onwards. They are a limited but very characteristic assemblage – prickly saltwort (*Salsola kali*), sea rocket (*Cakile maritima*), Babington's orache (*Atriplex glabriuscula*), frosted orache (*Atriplex laciniata*) and often the common weedy spear-leaved orache (*Atriplex prostrata*) (Fig. 291a). These are all annuals, but they are commonly joined by two perennials inseparable from the early stages of dune formation, sand couch (*Elytrigia juncea*) and sea sandwort (*Honckenya peploides*)[SD2]. This strand-line community provides the nuclei for embryo dunes to develop.

FIG 291. Newborough Warren, Anglesey. (a) Sea rocket–prickly saltwort–frosted orache strand-line community on the foreshore. In the middle distance sand couch is just beginning to colonise the strand-line, and to form a small embryo dune; the first ridge of high marram dunes to the left. (b) Embryo dune with sand couch.

Sand couch can withstand occasional immersion in seawater, but once it is firmly established it can build up substantial foredunes a metre or so high, and out of reach of the tides[SD4] (Fig. 291b). The only other species at all frequent on these foredunes are *Cakile* and *Honckenya*, along with a thin scatter of strand-line annuals, and pioneer plants of species, like sea-holly (*Eryngium maritimum*), more typical of the high marram dunes. Sand couch cannot withstand burial under more than 15–20 cm of sand, and cannot sustain indefinite upward growth, but it raises the foredunes to a level where the more vigorous dune-building grasses, marram (*Ammophila arenaria*) and lyme-grass (*Leymus arenarius*) can become established.

Dune ridges and mobile dunes

Marram is intolerant of even brief immersion in seawater, but with its vigorous spreading growth it is very effective at intercepting and trapping blown sand; it can keep pace with burial under a metre or so of sand a year. Marram quickly builds up dunes 5 m or more high. Trapping of sand is most obvious on the lee side of individual marram tussocks; on the windward face an equilibrium is soon established between trapping of sand by the marram and its tendency either to be blown over the top or to slide downhill again (Fig. 292a). Behind the dune crest there is a modicum of shelter, and a great deal of eddying, and sand is deposited patchily, in drifts (Fig. 292b). The detailed physics of wind-transport of sand is quite complex; a good introduction is given by Packham & Willis (1997).

A dune ridge of any substantial height is liable to wind erosion of its windward side, the sand being carried over the crest of the dune and deposited on the lee side. Hence marram dunes tend to be mobile. Sometimes erosion gets the upper hand locally, often through disturbance, creating a blow-out. If this is extensive, it may initiate a 'parabolic dune' moving downwind at a few metres or tens of metres a year (Fig. 293). Aerial photographs show that this has been a common process

LEFT: **FIG 292.** Mobile marram dune ridge at Newborough Warren. (a) Eroding windward face of dune; (b) Accreting sand in the lee of the dune crest.

BELOW: **FIG 293.** Newborough Warren. A parabolic dune; the darker moist sand of a newly created slack can be seen in front of the eroding face, already being invaded by the mobile sand of the next dune ridge. May 1956.

on active west-coast dune systems such as Braunton Burrows and Newborough Warren. Parabolic dunes move apex-foremost (contrary to the unvegetated *barchans* of sandy deserts), their arms trailing back upwind; the whole system is commonly 100 m or so across. Erosion continues down to the level at which the sand is moist close to the water-table. Actively growing or mobile marram dunes generally leave a good deal of sand exposed; they have been called 'yellow dunes'. A mobile ridge ceases its downwind travel as other ridges build up in front of it, erosion becomes less active, and the sand surface between the marram plants becomes vegetated and stabilised. The dune becomes 'fixed'.

The marram community is not rich in species, but three are frequent and particularly characteristic: sea-holly (Fig. 294a), sea bindweed (Fig. 294b) and sea

FIG 294. Plants of the mobile and semi-fixed dunes: (a) sea-holly (*Eryngium maritimum*); (b) sea bindweed (*Calystegia soldanella*); (c) dune pansy (*Viola tricolor* ssp. *curtisii*); (d) common restharrow (*Ononis repens*).

FIG 295. Newborough Warren. Sand on the foreshore contains a high proportion of broken-down seashells, and is highly calcareous.

FIG 296. Tentsmuir dunes, Fife, September 1968. (a) Lyme-grass (*Leymus arenarius*) forming the first line of dunes; sand-couch embryo dunes can be seen on the foreshore below. (b) The dunes are growing seawards; this line of concrete blocks was placed along the shore-line in 1940 as an anti-invasion defence. (c) A 'grey dune'. The sand has become leached and acid; the ground between the residual marram tufts is covered with grey lichens (*Cladonia* spp.), with scattered heather (*Calluna*) bushes.

spurge (*Euphorbia paralias*). These are all common on dunes from Norfolk to Galloway and all round the coast of Ireland; sea-holly extends up the east coast to Northumberland, and sea bindweed reaches Tayside, the southern Hebrides and Orkney. From the Isle of Wight westwards, these three are generally joined by Portland spurge (*Euphorbia portlandica*), which also grows on fixed dunes and in base-rich grassland near the sea. Beach sand is generally rich in nutrients from comminuted shells (Fig. 295) and strand-line debris, and the marram dunes are home to several common weeds, including ragwort (*Senecio jacobaea*), groundsel (*S. vulgaris*), creeping thistle (*Cirsium arvense*) and curled dock (*Rumex crispus*)[SD6].

Particularly on the east coast of both Britain and Ireland, marram is sometimes replaced as the main dune builder by lyme-grass (Fig. 296a), a handsome grass with flat leaves a centimetre or two wide, contrasting with the tightly rolled leaves of marram. Lyme-grass is a less effective sand binder than marram, and much less tolerant of burial by accumulating sand, so sand couch remains much more frequent than in the marram community; various weedy species occur occasionally, including perennial sow-thistle (*Sonchus arvensis*)[SD5].

The dune-ridge habitat: water availability and water stress

Water for the vegetation of dune ridges must come almost entirely from rain. Although marram can put down roots to 2 m, the groundwater-table is often some metres below that (the highest ridges on Braunton Burrows are 27 m above the water-table), and capillary rise in the sand above the water-table is seldom more than 50 cm. The sand of the dune ridges is sharply drained and contains less than 0.1% organic matter. Some water may come from dew on clear summer nights, but this is unlikely to exceed 0.4 mm a night. It has been suggested that water may move upwards as vapour with the reversed temperature gradient in the sand at night, but from the physics of the situation this is unlikely to be significant. In dry spells in summer the soil more than 50 cm above the water-table has a water content of less than 5% by weight; sand near the surface may contain only 1–2%, and is effectively air-dry. Under these conditions, marram is dependent on its deep root system to tap the little available water, and its ability to minimise water loss by rolling its leaves and closing its stomata. This it can do because its slender tall-growing leaves efficiently shed heat to the air, while the small plants in grazed fixed-dune turf must keep transpiring if they are not to get intolerably hot.

Dune stabilisation: fixed dunes

Once the surface of the sand between the marram clumps becomes reasonably stable, it can be colonised by smaller plants. The resulting semi-fixed dunes present a picture of bewildering variability in which the common factors are marram, red fescue (*Festuca rubra*) and forms of smooth meadow-grass (*Poa pratensis* sensu lato, probably mostly spreading meadow-grass, *P. humilis*); only slightly less constant than these is cat's-ear (*Hypochaeris radicata*)[SD7]. A great diversity of other species may occur, and semi-fixed dunes can look extraordinarily different from one another; probably no two authors would agree on all the details of a workable classification (Fig. 297). Some of this variation may be due to differences in the nutrient status of the sand, some to its water-holding capacity or the proximity of the water-table, some to grazing and some to pure chance. The nitrogen-fixing legumes and lichens play a particularly crucial role in building up the nutrient capital of the soil. The commonest are bird's-foot trefoil (*Lotus corniculatus*), common restharrow (Fig. 294d), hare's-foot clover (*Trifolium arvense*, Fig. 324c) and white clover (*Trifolium repens*). Lichens of the genus *Peltigera* have cyanobacteria as their photosynthesising component, and these too can fix nitrogen.

FIG 297. Braunton Burrows, Devon. Semi-fixed dune with *Syntrichia ruralis* ssp. *ruraliformis* covering most of the ground between the remaining marram shoots. Scattered restharrow (*Ononis repens*), wild thyme (*Thymus polytrichus*), biting stonecrop (*Sedum acre*) and sparse red fescue. A leaf rosette of ragwort and a flower spike and several young rosettes of viper's-bugloss (*Echium vulgare*) can be seen.

North of the Scottish border, cat's-ear, dandelions, ragwort and sand sedge (*Carex arenaria*), mouse-ear hawkweed (*Pilosella officinarum*), lady's bedstraw (*Galium verum*) and mosses (especially *Hypnum cupressiforme* sensu lato, *Brachythecium albicans* and *Syntrichia* (*Tortula*) *ruralis* ssp. *ruraliformis*) are prominent on many semi-fixed dunes. In the rest of Britain and Ireland, a very widespread and distinctive community is dominated by extensive carpets of *Syntrichia ruralis* ssp. *ruraliformis*, with common restharrow, dune pansy (Fig. 294c), bird's-foot trefoil, lady's bedstraw, wild thyme (*Thymus polytrichus*), lesser hawkbit (*Leontodon saxatilis*), mouse-ear hawkweed and mosses such as *Hypnum lacunosum* and *Homalothecium lutescens*. This seems better regarded as a (transitional) semi-fixed dune, rather than a dune grassland. Often in the same dune system transitions can be found that are more direct, containing little but marram with an understorey of red fescue and *Poa humilis*. There is a great deal of point-to-point variation in this general kind of vegetation, in which a nice range of small annuals find niches in early summer, including thyme-leaved sandwort (*Arenaria serpyllifolia*), sand cat's-tail (*Phleum arenarium*) and dune fescue (*Vulpia fasciculata*)[SD19].

Most dune systems have more or less extensive expanses of fixed-dune grassland on the still-calcareous sand behind the mobile dunes (Fig. 298). These grasslands are variable, but as a whole they are a remarkably species-rich habitat. They are generally dominated by red fescue, forming a close turf which is usually grazed. Lady's bedstraw, ribwort plantain (*Plantago lanceolata*), white clover,

FIG 298. Braunton Burrows, Devon. Undulating fixed-dune grassland, with residual tufts of marram, damp hollows and some scrub growth in the middle distance. May 1967.

bird's-foot trefoil and *Poa humilis* are constant or nearly so, and such widespread grassland plants as common mouse-ear (*Cerastium fontanum*), ragwort, daisy (*Bellis perennis*), meadow buttercup (*Ranunculus acris*), yarrow (*Achillea millefolium*) and the moss *Rhytidiadelphus squarrosus* are common[SD8]. A conspicuous and characteristic plant where it occurs is burnet rose (*Rosa pimpinellifolia*), low bushes of neat foliage borne on bristly-spiny stems, with white flowers followed by globular maroon rosehips. In the more northerly parts of our area ragwort, eyebright (*Euphrasia officinalis* agg.), Yorkshire-fog (*Holcus lanatus*), fairy flax (*Linum catharticum*), harebell (*Campanula rotundifolia*), self-heal (*Prunella vulgaris*), red clover (*Trifolium pratense*), glaucous sedge (*Carex flacca*), autumn gentian (*Gentianella amarella*), lesser meadow-rue (*Thalictrum minus*) and the mosses *Rhytidiadelphus triquetrus* and *Pseudoscleropodium purum* are somewhat more prominent than elsewhere. Much of the machair of the Hebrides and northwestern Ireland is grassland broadly of this kind. In fixed dune (machair) on the shell-rich sands of the Burren and the west Galway coast squinancywort (*Asperula cynanchica*) and the calcicole moss *Homalothecium lutescens* can play a prominent role.

The dune water-table: dune slacks

Rain falling on a sand-dune system percolates into the sand to the water-table, which, if the dune system is essentially flat, will be gently domed (recall the 'hydrological dome' of the ombrogenous bogs of Chapter 15). A series of profiles across Braunton Burrows in North Devon is shown in Figure 290. The water-table is at about 3.5 m above mean sea level in the foredunes and in the boundary drain dividing the dunes from Braunton Marsh landwards, and about 10 m above sea level in the centre of the dunes – varying by around a metre in the course of the year. What is also clear from the profiles is that the mobile dune ridges stand high (up to *c.* 20 m) above the water-table. In simple terms, we can visualise the dune system as composed of mobile dunes marching downwind over a domed surface of damp sand; the areas of damp sand are *dune slacks*. In reality, it is not quite as simple as this. The water-table fluctuates, erosion is localised depending on the topography of the dunes at the time, and weather is capricious, so the level of the slacks is not strictly uniform. Parts of the slacks are under water for several months every winter, parts for a week or two at times of high flooding and parts ('dry slacks') are obviously related to the level of the water-table but never flood at all. Thus the dune-slack habitat is very variable, and correspondingly species-rich.

Dune slacks have in common the availability of water. If the sand on the foreshore contains a good proportion of shelly material, the water in the slacks will be calcareous. But because of the porous substratum the water-table fluctuates through the year in a rather different manner to that in many rich-fens (Chapter 14), and more akin to what happens in flood-meadows and the Irish turloughs. Probably the most universally distributed dune-slack plants are creeping bent and spreading meadow-grass. But the most conspicuous really common dune-slack plant is undoubtedly creeping willow (Fig. 299a). Other very widespread species include marsh pennywort (Fig. 299b), sand sedge, glaucous sedge, silverweed (*Potentilla anserina*), bog pimpernel, water mint (*Mentha aquatica*), self-heal (*Prunella vulgaris*) and the moss *Calliergonella cuspidata*.

Dune-slack communities are as bewilderingly diverse as semi-fixed dunes, but they can perhaps be broadly divided into five groups. Rather open slacks with only a low and patchy growth of creeping willow are probably in general at an early stage in development with growth limited by meagre nutrient supply. They have a patchy grass cover, largely of creeping bent, but often support extensive carpets of bryophytes. Creeping willow, knotted pearlwort (*Sagina nodosa*) and the moss *Bryum pseudotriquetrum* are constant; sand sedge, jointed rush (*Juncus articulatus*) and the thalloid liverwort *Aneura pinguis* are also constant or nearly so[SD13]. This is the habitat of a number of rare bryophytes, including the

FIG 299. Dune-slack colonists. (a) Young plants of creeping bent (*Agrostis stolonifera*), creeping willow (*Salix repens*) and sand sedge (*Carex arenaria*) newly established from seed on the damp sand at the foot of an eroding mobile dune. (b) A later stage in colonisation: bog pimpernel (*Anagallis tenella*), creeping bent, creeping willow and marsh pennywort (*Hydrocotyle vulgaris*) have formed a closed turf.

FIG 300. Dune-slack plants: (a) the liverwort *Petalophyllum ralfsii* (rosettes *c.* 1 cm across, November); (b) brookweed (*Samolus valerandi*); (c) round-leaved wintergreen (*Pyrola rotundifolia*); (d) early marsh-orchid (*Dactylorhiza incarnata* ssp. *coccinea*); (e) fen orchid (*Liparis loeselii*); (f) marsh helleborine (*Epipactis palustris*). All at Braunton Burrows, north Devon.

mosses *Amblyodon dealbatus, Bryum marattii* and *B. calophyllum,* and the thalloid
liverworts *Moerckia hibernica* and *Petalophyllum ralfsii* (Fig. 300a), from Devon to
the machair of northwest Ireland and the Hebrides.

In the *Salix repens–Campylium stellatum* community[SD14] the vegetation cover
is much more nearly closed, with a bushy carpet of creeping willow, ankle-deep
or rather more, usually with an extensive understorey of mosses. This is a very
species-rich community. Creeping willow and the moss *Campylium stellatum* are
constant; other constant or near-constant species are marsh pennywort, glaucous
sedge, creeping bent, water mint, marsh helleborine (Fig. 300f), variegated horsetail
(*Equisetum variegatum*) and the moss *Calliergonella cuspidata.* Sand sedge, autumn
hawkbit (*Leontodon autumnalis*), lesser spearwort (*Ranunculus flammula*), dewberry
(*Rubus caesius*), bird's-foot trefoil and jointed rush are frequent. This is the dune
slack that comes nearest to the rich-fens, with which it shares such species as
carnation sedge (*Carex panicea*), water mint, lesser spearwort, bog pimpernel, grass-
of-Parnassus (*Parnassia palustris,* Fig. 183b), few-flowered spike-rush (*Eleocharis
quinqueflora*), marsh bedstraw (*Galium palustre*), the moss *Campylium stellatum* and
the thalloid liverwort *Aneura pinguis.* The occurrence of silverweed, creeping
buttercup (*Ranunculus repens*), common sedge (*Carex nigra*), water germander and the
mosses *Drepanocladus sendtneri* and *Pseudocalliergon lycopodioides* in this community
at Braunton Burrows point to an affinity with some of the Burren turloughs in Co.
Clare. This community is home to a number of other rarities, including impressive
clumps of sharp rush (*Juncus acutus*), round-leaved wintergreen (Fig. 300c) and,
more rarely, coralroot (*Corallorhiza trifida*) and fen orchid (Fig. 300e).

Much dune-slack vegetation dominated by creeping willow is less rich than
this. The *Salix repens–Calliergonella cuspidata* dune-slack community[SD15] inclines
more in the direction of a fen meadow than a rich-fen. Marsh helleborine is
much less common, and the rich-fen moss *Campylium stellatum* and the calcicole
liverworts that occur in the last community are scarce or absent; sharp rush
and round-leaved wintergreen still occur but are less frequent. Conversely,
silverweed, creeping buttercup, common sedge, greater bird's-foot trefoil (*Lotus
pedunculatus*) and hemp agrimony (*Eupatorium cannabinum*) are rather more
prominent here.

Older and drier slacks are often occupied by a community dominated by a
bushy growth of creeping willow knee-high or more with abundant Yorkshire-
fog[SD16]. Apart from the two dominants, the most frequent species are red
fescue, bird's-foot trefoil and glaucous sedge; sand sedge, common restharrow,
spreading meadow-grass and self-heal are also frequent, with a long list of
associated species of low constancy and cover. This community has much more
the character of a grassy creeping-willow scrub than the last two.

FIG 301. Braunton Burrows, Devon. Old creeping willow (*Salix repens*) slack, still partly flooded in early May 1969. Grey sallow (*Salix cinerea*) bushes in the foreground.

Some slacks and dune hollows are occupied by a community of grasses and sedges, especially creeping bent and common sedge, with abundant silverweed, but little or no creeping willow[SD17]. Also frequent are marsh pennywort, lesser spearwort, red fescue, cuckooflower (*Cardamine pratensis*), common spike-rush (*Eleocharis palustris*), creeping buttercup, Yorkshire-fog, spreading meadow-grass, marsh bedstraw and the moss *Calliergonella cuspidata*. Vegetation of this kind can probably be found on most dune systems, but it becomes progressively more important northwards, and damp hollows in the machair of the Hebrides are very much of this character, intergrading with the surrounding fixed-dune grassland.

Old dune slacks become colonised by sallow (Fig. 301) and locally, as at Braunton Burrows and Bull Island in Dublin Bay, by alder, forming damp scrubby vegetation and potentially wet woodland.

VARIATIONS ON A THEME: SOME OTHER DUNE HABITATS

The importance of leguminous plants – clovers, vetches, common restharrow, etc. – in fixing nitrogen has already been mentioned in this book. Some other species have symbiotic nitrogen-fixing microorganisms associated with their roots, including alder, but most notably in a sand-dune context the thorny deciduous suckering shrub sea buckthorn (*Hippophae rhamnoides*). Sea buckthorn is regarded as native on the east coast of England, and widely introduced elsewhere (Pearson & Rogers 1962). On the east coast it usually grows as patches of scrub forming a mosaic with fixed-dune graslands[SD18]. The lower and more open patches retain a rather impoverished fixed-dune flora dominated by red fescue, but this is shaded out under denser and more mature *Hippophae* and replaced by a sparse weedy vegetation including stinging nettle, false oat-grass (*Arrhenatherum elatius*), cleavers (*Galium aparine*), bittersweet (*Solanum dulcamara*), spear thistle (*Cirsium vulgare*) and elder. Sea buckthorn can become established early in the dune succession (Fig. 302) and has been widely planted to stabilise dunes. At some sites it can be an invasive pest, and it is generally regarded with disfavour (and exterminated) by those responsible for managing species-rich west-coast dunes where it is not native.

Sand sedge is very common in coastal dunes, especially round the edges of slacks, where the seeds germinate in the damp sand and the rhizomes run up the sides of neighbouring bare dunes marking their progress with conspicuous straight lines of green shoots (Fig. 303). Less conspicuously, sand sedge is often a widespread minor component of dune-slack communities, but very seldom dominates more than small patches of dune vegetation. However, in East Anglia inland dunes are

FIG 302. Sea buckthorn colonising foredunes. Murlough dunes, near Newcastle, Co. Down, June 1991.

FIG 303. Sand sedge has long creeping rhizomes, which throw up leafy shoots at intervals. Here it is growing up the bare sand face of a mobile dune from the moist sand of a new dune slack where the seeds germinate.

developed locally on the 'cover-sands', notably in the Breckland between Mildenhall, Bury St Edmunds and Thetford. Sand sedge is dominant on these dunes (some of them 50 km from the sea), with a sparse associated flora in which only sheep's fescue (*Festuca ovina*), ragwort, common mouse-ear, sheep's sorrel (*Rumex acetosella*) and a few other common grassland species are frequent[SD10].

Machair

The beach sand on many west coasts contains a high proportion of calcium carbonate from broken sea-shells, locally supplemented by fragments of the calcareous red seaweeds *Corallina* and *Lithothamnion*. Indeed, in some places the bleached remains of *Lithothamnion* make up most of the beach ('coral strands': Fig. 304). The constant high winds on exposed western coasts militate against the development of high marram dunes, so that broad expanses of fixed-dune pasture develop, with a flattish surface a bit above the water-table, traditionally called by the Gaelic word *machair* (literally a plain or field). The machair provides the only cultivable ground in many far-western places. The persistent winds carry a constant top-dressing of calcareous sand across the machair, often for

FIG 304. *Lithothamnion* shingle from a 'coral strand' near Carraroe, Co. Galway. The coin (an Irish shilling) is 23 mm in diameter.

FIG 305. Machair. (a) An expanse of machair on Benbecula, Western Isles. The grassland in the foreground is well-drained, with abundant bird's-foot trefoil and ribwort plantain in flower, but there are often damper areas in machair as well, which add to its floristic diversity (© Lorne Gill/ Scottish Natural Heritage). (b) The back of the Fanore dunes, Co. Clare, the southernmost machair in Ireland, July 1971.

several kilometres back from the shore. This counterbalances the constant leaching by rain, so that the fixed dunes and machair remain calcareous even in an oceanic climate wet enough to leach most terrestrial soils to heath or blanket bog. Machair is best developed on the Hebridean islands (Fig. 305a), but also occurs widely on the western Scottish mainland and on the western and northern coasts of Ireland (Dargie 1998, Gaynor 2006). For good aerial views of machair in the Western Isles see Friend (2012).

Leaching, acidification and 'grey dunes'

However, some dunes, especially on the east coast of Britain, are built up from sand containing little shelly material. Table 12 illustrates the contrast between Scottish west-coast and east-coast dunes. Not only is the calcium carbonate ($CaCO_3$) content of the beach sand on the east coast much lower, but the fixed dunes become decalcified and acid, and ultimately bear dry *Calluna–Empetrum* 'dune heaths', which have no parallel on the west coast. The contrast between west- and east-coast dunes is nothing like as dramatic in southern Britain or Ireland as in Scotland. Most of the drainage of the Highlands is eastward, and rivers such as the Tay, Dee and Spey bring down an immense burden of sediment from largely non-calcareous mountain country; the major rivers of England and Ireland drain predominantly from limestone and soft-rock catchments. Nevertheless the effects of progressive leaching are apparent on English east-coast, and some west-coast, dunes. Sir Edward Salisbury (1952) showed that the $CaCO_3$ content of the dune soil at Southport, Lancashire, declines from over 6% in the foredunes to a little over 1% in dunes a century old, and to a fifth or a tenth of 1% in the course of two or three centuries, with a concomitant fall in pH from 8.2 to 5.5. At Blakeney Point in Norfolk, $CaCO_3$ fell from an initial 0.42% in young dunes (with pH 8.2) to very low values and pH 6.3 in dunes probably about 235 years old. The mildly acid conditions on these semi-fixed and fixed dunes are reflected in the prominence of mosses and especially lichens, mostly of the genus *Cladonia*[SD11]. This has led to them being called 'grey dunes', and because many dunes in southeast England (and hence accessible from London or Cambridge!) are of this nature this is sometimes used as general

TABLE 12. Average characteristics of soils from some Scottish dunes. Condensed from Gimingham (1964). Calcium carbonate ($CaCO_3$) as per cent dry weight of soil. E, east coast; W, west coast.

Stage of dune development	Tentsmuir, Fife (E)	St Cyrus, Kincardineshire (E)		Forvie, Aberdeenshire (E)	Tiree, Inner Hebrides (W)		Luskentyre, Harris, Western Isles (W)	
	pH	$CaCO_3$	pH	pH	$CaCO_3$	pH	$CaCO_3$	pH
Foreshore	7.5	2.7	7.5	6.8	–	7.9	63.3	7.2–7.6
Mobile dunes	6.8	1.6	7.8	6.7	66.0	–	55.0	through-
Fixed dunes	6.3	1.6	7.9	5.1	56.8	7.7	58.2	out
Dune pasture or young machair	–	1.0	6.4	4.8	57.3	7.5	58.2	
Mature machair	–	–	–	–	47.0	7.3	–	

term for fixed dunes – but as Salisbury remarks, 'the appropriateness of the epithet "grey dune" is apt to vary from one locality to another.'

The well-known dunes of the South Haven Peninsula south of the entrance to Poole Harbour are a special case. They are largely made up of sand from the Bracklesham Beds, decalcified sands that underlie the surrounding heathland, either eroded *in situ* or brought by tidal currents from the neighbouring coastline. They have grown up to seaward of a narrow heathy promontory in the course of the last 400 years. The present freshwater lagoon of the 'Little Sea' originated not as a dune slack, but as a tidal inlet cut off by the growth of a new dune ridge in front of it during the eighteenth and nineteenth centuries. 'Greenland Lake', the wet slack-like depression down the middle of Dawlish Warren in Devon, has a similar but more recent origin, and there are other examples. The South Haven Peninsula dunes show remarkably rapid transitions from typical foreshore vegetation and marram dunes to the typical *Ulex minor– Agrostis curtisii* heath[H3] of the heathland behind them.

COASTAL SHINGLE

Shingle occupies perhaps a third of our coastline. The greater part takes the form of *fringing beaches*, essentially steep foreshores washed by waves at every tide. Typically, there is a storm ridge high on the beach consisting of coarse shingle or pebbles washed up by the highest storm waves of the past season – or the past several seasons. The crest, above this, is the highest shingle thrown up by truly monumental storms. To seaward, the shingle is reshaped by the waves and currents of every tide, and often there is net movement of shingle along the shore, very obvious if groynes have been built. If there is an angle in the coast, as at the mouth of an estuary, the line of the fringing beach is often continued as a *shingle spit*. At the end of the spit, the shingle commonly turns abruptly landward as a *hook*; the spit may continue growth along the original line, leaving the back decorated with a succession of hooks. If the source of the shingle was in deep water and shingle has been driven shorewards by rising sea level, it may form an offshore *shingle bar*, which may later join up with the coast at one or both ends. If there is an abundant and continuing supply of shingle, successive ridges may build up to seaward of one another forming an *apposition beach*, as at Dungeness. These more elaborate shingle forms are concentrated where the average tidal range is no more than 3–4 m; greater tidal ranges than this generally favour wide sandy beaches and mud-flats.

Stability is the most important factor determining what will grow on shingle. The vagaries of tide, wind direction and weather can result in a single tide removing several metres of shingle; conversely shingle can at times build up rapidly over a short succession of tides. Consequently many fringing shingle beaches on exposed shores are too unstable for any vegetation to get established. At the level where the beach is disturbed only by winter storms and stable throughout the growing season, an open cover of annual plants can become established, akin to the strandline community on sandy beaches. It consists largely of oraches (*Atriplex prostrata, A. glabriuscula, A. laciniata*), and such species as prickly saltwort, sea sandwort and Ray's knotgrass (*Polygonum oxyspermum*)

LEFT: **FIG 306.** Yellow horned-poppy (*Glaucium flavum*), Slapton, Devon, August 1983.

BELOW: **FIG 307.** Sea-kale (*Crambe maritima*) in fruit on shingle at Pagham, Sussex, August 1983.

with occasional ruderals such as cleavers. Sea rocket, so conspicuous on sandy beaches, is less common on shingle. If the shingle is generally stable for several years in succession, perennials can colonise. These are the conditions for some of the most characteristic shingle plants, including the distinctive seaside form of curled dock (*Rumex crispus* ssp. *littoreus*), yellow horned-poppy (Fig. 306), sea-kale (Fig. 307), sea beet (*Beta vulgaris* ssp. *maritima*), sea mayweed (*Tripleurospermum maritimum*), sea campion (*Silene uniflora*) and, much more locally, sea pea (Fig. 308). These are often accompanied by a sprinkling of common weeds, including ragwort, perennial sow-thistle, bittersweet, creeping thistle, sticky groundsel (*Senecio viscosus*), herb-robert (*Geranium robertianum*) and silverweed[SD1]. These distinctive plants do not necessarily grow together, and there are many well-vegetated shingle beaches on which the plant cover is dominated by curled dock, sea beet, sea mayweed and sea campion, and of these curled dock is the most ubiquitous. With increasing stability grasses begin take on a role. Three species in particular are important. Sea couch (*Elytrigia atherica*) is common where shingle abuts on saltmarsh or on sea-walls. Red fescue is perhaps most common on sandy shingle, sometimes in company with bird's-foot trefoil, wild thyme and biting stonecrop (*Sedum acre*). False oat-grass can be seen as beginning a transition to more 'normal' grassland turf.

Sea pea is only really common on the Suffolk coast, yellow horned-poppy and sea-kale are rare north of the Clyde and Forth and in the west and north of Ireland, and sea beet is rare in most of Scotland. Here the vegetation of shingle is dominated by sea mayweed, curled dock, Babington's orache and Ray's knotgrass with cleavers, chickweed (*Stellaria media*), couch-grass (*Elytrigia repens*) and silverweed among a group of weedy associates. This general kind of vegetation[SD3] is the habitat of oysterplant (*Mertensia maritima*), a northern species that has become increasingly rare with us over the past century.

FIG 308. Sea pea (*Lathyrus japonicus*), Chesil Beach, Abbotsbury, Dorset, August 1972.

Clean, pure shingle is a virtually impossible habitat for plants to colonise. It has virtually no water-holding capacity after rain, and offers virtually no plant nutrients. At least some fine material amongst the shingle is essential for plant growth. On the face of it shingle is a dry habitat, and much has been written about the sources of water available to shingle plants. Shingle beaches have a freshwater-table beneath them, just as sand-dunes do, but unless this is near the surface even deep-rooted plants cannot reach it. It is well established that neither ordinary dew nor condensation of water vapour amongst the pebbles can contribute significantly to the water-budget of the vegetation. The conclusion has to be that shingle plants draw their water from the 'soil' in the normal way. Rain falls over the entire surface of the shingle and drains into it; the surface pebbles are an effective mulch, minimising evaporation. Typical shingle vegetation covers only a small fraction of the surface, so there is potentially a large supply to meet a relatively small demand.

Some nutrients come dissolved in rain (Chapter 15). The shingle traps wind-borne dust and sand. The churning of the waves at every tide acts as a pebble-grinder, so fine particles are constantly being produced, which weather, so producing mineral nutrients. Seabirds contribute their quota, most obviously around nesting colonies. Seaweed and other debris cast upon the shore contribute organic matter, nitrogen and phosphorus, and of course once plants becomes established their roots are an ongoing source of organic matter. The net result of all these processes is that the soil tends progressively to become more 'normal', so that what began as open (and very characteristic) shingle vegetation evolves gradually into a grassy turf, initially dominated by red fescue but later invaded by false oat-grass, cock's-foot (*Dactylis glomerata*) and other grassland plants[MG1a[d,e]].

These grasslands on stable shingle usually have a distinctly seaside character; the splendid dark-green rosettes of sea radish (*Raphanus raphanistrum* ssp. *maritimus*) are often a striking feature of them. The closed grassland loses water over its whole surface so, unlike the sparse pioneer shingle plants, it is often severely droughted in late summer. Sometimes bryophytes (e.g. *Dicranum scoparium, Brachythecium albicans*) and lichens (*Xanthoria parietina, Cladonia* spp.) are important colonisers of stable shingle, leading to a thin, acid species-poor *Festuca–Agrostis* grassland with sheep's sorrel and annuals such as whitlowgrass (*Erophila verna*), early hair-grass (*Aira praecox*), stork's-bill (*Erodium cicutarium*), dove's-foot crane's-bill (*Geranium molle*), bird's-foot (*Ornithopus perpusillus*) and various annual clovers (*Trifolium* spp.). Succession on stable shingle can progress to scrub of gorse, bramble and blackthorn, and at some sites to woodland, but that takes us back to earlier chapters.

FIG 309. Foredunes with marram and cottonweed (*Otanthus maritimus*), Lady's Island Lake, Co. Wexford, July 1970.

SOME CLASSIC SHINGLE LOCALITIES

Most of the foregoing has been aimed at giving a general picture relevant to shingle beaches all round our coasts. However, so much attention has been concentrated on a few classic sites that we cannot close without saying something about them. More detail is given by Tansley (1939), Steers (1946) and Packham & Willis (1997). All of the sites are in Britain; Ireland seems to have less shingle development than its sister island, though there is extensive cobble-type shingle in Galway Bay and, as noted by Praeger (1934), 'The SE coast of Ireland is characterised by great stretches of sand and gravel … From Bray Head in Wicklow down as far as Waterford stretch after stretch of shingle and sand extend … sometimes closing the mouths of inlets still or once marine, as at Lady's Island Lake, Tacumshin Lake and Bannow Bar, all in Wexford.' That stretch of coast in Wexford is well known as the only remaining Irish or British locality for cottonweed (Fig. 309). The south coast from Wexford to Cork is probably the richest in Ireland for shingle plants.

Blakeney Point and Scolt Head Island, Norfolk

Blakeney Point is a complex shingle spit, about 6 km long, running in a west-northwest direction from the coast near Cley-next-the-Sea on the north Norfolk coast. The shingle provides the foundation on which extensive dunes have built

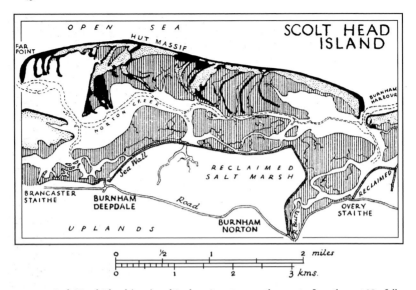

FIG 310. Scolt Head Island (1925), a shingle-spit system on the coast of north-west Norfolk. The shore current runs from east to west and the growing tip is at the western end. The hook of the shingle spit (black) marked as 'Far point' is the most recent; the others preserve stages in the history of the spit. This coast changes in detail from year to year, but the main outlines remained much the same in 2010 as they were in 1925. Sand-dunes (stippled) overlie the broader expanses of shingle; saltmarshes (vertical hatching) have grown up on the tidal mud in the shelter of the shingle banks. From Tansley (1939), slightly amended; see also the more detailed maps of Blakeney Point (1946) and Scolt Head Island (1933) in Steers (1946).

up. There are saltmarshes between the hooks and broad expanses of saltmarsh along the coast facing the spit from Blakeney village westwards. Scolt Head Island (Figs 310, 311), some 20 km to the west, is essentially an offshore shingle bar that has grown by spit development westwards, leaving a succession of long curved hooks in its wake. There are some dunes, and extensive saltmarshes between the hooks, and fronting the mainland. Both sites combine good shingle and saltmarsh vegetation with dunes. Floristically they are notable for matted sea-lavender (*Limonium bellidifolium*) and the common occurrence of shrubby sea-blite (*Suaeda vera*) at the contact of saltmarsh and shingle, and for grey hair-grass (*Corynephorus canescens*) on the dunes.

Orford Ness, Suffolk

This is a magnificent shingle spit, some 17 km long, separating the River Alde from the sea between Aldeburgh and its mouth at Shingle Street (Fig. 312). From

FIG 311. The eastern end of Scolt Head Island, at the entrance to Burnham Harbour and Overy Staithe, showing vividly the distribution of shingle, sand-dunes and saltmarsh. Compare Fig. 310; since the 1920s and 1930s the shingle ridge has moved back making the coast straighter and encroaching on the dunes, and the saltmarsh has extended and matured (© Adrian Warren & Dae Sasitorn/www.lastrefuge.co.uk).

Aldeburgh the shingle ridge starts narrow and straight a few degrees west of south. After a few kilometres a succession of old hooks can be discerned on the landward (river) side, but the seaward face remains almost straight to a point east of Orford village, where it swings southwest. This coincides with a broad section in which a close succession of hooks pass into apposition ridges before the spit settles into its new direction for another 5 km, finally following the curve of the coast to the mouth of the river. This stretch of the Suffolk coast has

FIG 312. Orford Ness, a long and complex shingle spit extending for some 17 km south from Aldeburgh along the Suffolk coast, and enclosing the mouths of the Rivers Alde and Ore. Shingle Street is at the bottom of the picture (© Adrian Warren & Dae Sasitorn/ www.lastrefuge.co.uk).

outstandingly rich shingle vegetation, and is notable as the headquarters in our area for sea pea, elsewhere thinly scattered round the British and Irish coasts.

Dungeness, Kent

The largest shingle spit in Europe, this is the classic example of an apposition beach, formed by successive deposition of ridges of shingle. It is suggested that it may have passed through a stage similar to Orford Ness, as an eastward-growing spit along the lower course of the River Rother, which then emerged to the sea east of New Romney (Fuller 1989). A breach in the shingle ridge near

Rye changed accretion and erosion patterns, leading to the present-day cuspate foreland. Dungeness is not particularly notable for shingle vegetation as such (many sites on the Sussex coast are as good), but is fascinating for its diversity of habitat and richness in examples of succession on old shingle (Scott 1965), including scrub of gorse, brambles, blackthorn and elder, and the famous holly wood.

Chesil Beach, Dorset

This is another site that would be famous even if nothing grew there. It starts as a fringing beach at Burton Bradstock, and leaves the low indented coastline just south of Abbotsbury (Figs 313, 314) to sweep in a smooth gentle curve to the Isle of Portland. It is a single ridge around 7–8 m high; the shingle grades in average size from less than 1 cm at Abbotsbury to 2–3 cm near Portland. Surprisingly, flints from the chalk make up a large proportion of the shingle, as they do at Slapton on the Devon coast (Fig. 315). The lagoon to landward of Chesil Beach, the Fleet, is open to the sea at its southern end. Shingle vegetation is largely confined to the back of the beach; the most prominent species are curled dock, sea beet and sea campion, but yellow horned-poppy, sea-kale and sea pea are there as well. Shrubby sea-blite forms an intermittent zone at the base of the

FIG 313. Chesil Beach, Abbotsbury, Dorset, September 1981. Shingle vegetation dominated by curled dock (*Rumex crispus* ssp. *littoreus*), sea campion (*Silene uniflora*) and sea beet (*Beta vulgaris* ssp. *maritima*).

FIG 314. Chesil Beach, near Abbotsbury, Dorset, August 1972. Shrubby sea-blite (*Suaeda vera*) growing along the shoreline of the Fleet, at the back of the shingle ridge.

FIG 315. The shingle bar at Slapton Ley, Devon, from the south, May 1973. There is open shingle vegetation with yellow horned-poppy (*Glaucium flavum*) etc. between the crest of the shingle and the road; the backslope between the road and the freshwater Ley was formerly grazed by rabbits but is now rank grassland and scrub.

shingle along the shore of the Fleet; high water in the Fleet is marked by a strandline of dead *Zostera* leaves, from the beds of *Z. angustifolia* growing in the quiet saline water. The brackish marsh at the Abbotsbury end of the Fleet (near the swannery) has marsh-mallow (*Althaea officinalis*) and divided sedge (*Carex divisa*). At the Portland end wild thyme carpets expanses of stabilised shingle.

CHAPTER 20

Sea-cliffs and Rocky Shores

AS ISLANDERS WE ARE VERY aware of our sea-cliffs. The white cliffs of Dover have long been symbolic of England to the returning traveller. Anyone who has been to the west of Ireland (or who saw the film *Ryan's Daughter*) will have carried away memories of spectacular cliff scenery (Fig. 316). The sea-cliffs of Wales and southwest England are no less spectacular, and we are constantly reminded on television of the magnificent sea-bird cliffs of the Scottish islands. Sea-cliffs are where land and sea come closest together. Their vegetation is distinctive, not particularly rich in species but sometimes kaleidoscopic in its variety within a small space.

THE MARITIME CLIFF HABITAT

Salt spray, plant growth and vegetation

Sea-cliffs are close to the breaking waves, and many of our west-coast cliffs overlook dramatically rocky, wave-beaten shores. Wave-wash reaches higher than high water of spring tides and wave-splash higher than that. Both reach highest in exposed situations, and in stormy weather. Beyond the reach of wave-splash, fine sea-spray is carried to higher levels and farther inland. Where low cliffs face open ocean, as on the coast of the Burren or locally on the west Cornish coast there is what could be described as a 'perched saltmarsh' behind the cliff edge. In the Burren, this is dominated by red fescue (*Festuca rubra*) with halophytes including thrift (*Armeria maritima*), distant sedge (*Carex distans*), sea-milkwort (*Glaux maritima*) and sea plantain (*Plantago maritima*) along with such common grassland plants as creeping bent (*Agrostis stolonifera*), buck's-

FIG 316. The Cliffs of Moher,
Co. Clare, July 1971. Yorkshire-fog
(*Holcus lanatus*) in the foreground;
sea campion (*Silene uniflora*), bird's-
foot trefoil (*Lotus corniculatus*) and
thrift (*Armeria maritima*) to the left.
Roseroot (*Sedum rosea*) can just be
glimpsed to the right of the *Silene*.

horn plantain (*Plantago coronopus*) and white clover (*Trifolium repens*). These
'saltmarshes' may owe as much to impermeable soil and waterlogging as to
salinity. Often, vegetation comparable with this is fragmentary, and more or less
halophile species are scattered above high tide mark, frequently where there is
a little fresh-water seepage. Apart from obvious halophytes, brookweed (*Samolus
valerandi*), slender club-rush (*Isolepis cernua*), long-bracted sedge (*Carex extensa*)
and the rarer dotted sedge (*Carex punctata*) often occur in situations like this.

Most of the cliff is affected by only the finer spray, and sea-salt deposition
declines rapidly with height above sea level, and distance inland. Table 13 shows
the deposition of sodium at sites across the Lizard peninsula of Cornwall over
two years at the end of the 1960s, with some comparative figures across southern
England for 1986–88. As would be expected, deposition is very much higher on
the west coast than the east; the Lizard measurement site, 500 m from the west
coast, received almost as much sodium as Coverack, only 30 m from the sea on
the east side. Much higher deposition was recorded close to the west-coast cliffs.

TABLE 13. Annual deposition of sodium at some sites on the Lizard peninsula estimated from measurements, January 1968–April 1970 (Malloch 1972), with average wet deposition in 1986–88 at some sites across southern England for comparison (UKAGAR 1990). The Lizard transect started at the edge of the cliff 67 m above sea level at Pigeon Ogo, north of Kynance. (The distances given for sites east of the Lizard are rough indications of distance to the nearest open ocean coast to the southwest.)

Site	Distance from west coast	Average annual deposition of sodium (kg/ha)
Pigeon Ogo, site 2	22 m	2442
Sample site 4	100 m	420
Lizard (between town and sea)	500 m	197
Goonhilly	5 km	51
Coverack	11.5 km [30 m from east coast]	204
Goonhilly, 1986–88	5 km	46.9
Yarner Wood, Devon, 1986–88	(40 km)	31.3
Bowood House, Wilts, 1986–88	(c. 100 km)	10.4
Woburn, Beds, 1986–88	(c. 200 km)	9.3
Stoke Ferry, Cambs, 1986–88	(c. 300 km)	7.2

At both Lizard and Coverack windspeed had a marked effect. Deposition of sodium averaged around 15 kg per hectare per day in periods with around 20% windy days (windspeed > 14 m/s), and only a third of that in periods with < 10% windy days.

In terms of vegetation, red fescue–thrift (*Festuca rubra–Armeria*) maritime grassland often received an average of more than 10 kg/ha of sodium per day. Red fescue–cock's-foot–Yorkshire-fog maritime grassland farther from the sea averaged c. 1.5–5 kg/ha per day, and cliff-top heather–spring squill (*Calluna–Scilla verna*) heath c. 1 kg/ha per day. However, daily deposition is very variable, and what matters most in the long run for plant growth is the amount of salt in the soil. Table 14 summarises some representative soil-analysis data for cliff grasslands.

Wind exposure
The prevailing winds over Britain and Ireland are southwesterly. All coasts are windier than inland, and exposed sites on the west coast are very windy indeed. This has striking effects on the vegetation, with short wind-clipped turf on exposed west-facing cliff edges, and blackthorn scrub in gullies or around outcrops looking as if it has been sculpted with a hedge-trimmer. Even several

TABLE 14. Average characteristics of the soils of some sea-cliff plant communities, summarised from Rodwell (2000). Water content and loss on ignition (a rough measure of organic matter) are expressed as per cent air-dry soil; ions are expressed in μequiv/g air-dry soil. The ratio of sodium to organic matter (last column) is a more realistic measure of effective soil salinity than the sodium figure alone. (Cations determined in M ammonium acetate, and phosphate (H_3PO_4) in acid 0.03 M ammonium fluoride extract.)

Vegetation type	Surface pH	Water content	Loss on ignition	Na	K	Mg	Ca	P	Na/LOI
(a) rock-crevice communities									
Crithmum maritimum–Spergularia rupicola (MC1)	7.4	46	15	83	12	64	66	2.0	6.8
(b) grassland communities									
Armeria maritima–Festuca rubra (MC8)	6.3	89	25	82	12	57	46	1.9	3.6
Festuca rubra–Holcus lanatus (MC9)	5.6	96	28	56	11	46	34	1.4	2.0
Festuca rubra–Plantago spp. (MC10)	5.5	119	32	74	13	51	27	1.4	2.5
Festuca rubra–Hyacinthoides non-scripta (MC12)	5.5	123	32	60	15	65	20	2.9	1.9
Festuca rubra–Daucus carota ssp. gummifer (MC11)	7.2	29	16	24	10	24	135	1.1	1.9
(c) Sedum–therophyte communities									
Armeria maritima–Cerastium diffusum (MC5)	5.8	53	26	40	12	41	55	5	1.7

kilometres inland trees lean away from the wind. Wind probably affects plant growth more by increasing the rate of water loss and cooling the growing-points than by mechanical damage. Low-growing and prostrate forms of many species are common on exposed sea-cliffs. Sometimes these are just a plastic response of the plant to the harsh environment, but a number have been shown by cultivation experiments to be genetically fixed ecotypes or subspecies, some distinctive enough to be given formal names, such as the prostrate broom

FIG 317. The west coast of the Lizard, Cornwall, looking northwest from near Kynance; thrift (*Armeria maritima*), sea beet (*Beta vulgaris* ssp. *maritima*), dyer's greenweed (*Genista tinctoria* ssp. *littoralis*) and wild thyme (*Thymus polytrichus*) in the foreground. June 1995.

(*Cytisus scoparius* ssp. *maritimus*), which occurs on exposed cliff-tops in Devon, Cornwall, Wales, the Channel Islands (Fig. 323) and two sites in Ireland, and the corresponding form of dyer's greenweed (*Genista tinctoria* ssp. *littoralis*) in wind-clipped turf on Devon and Cornish cliffs (Fig. 317). Many plants of such habitats are always prostrate; examples are hairy greenweed (*Genista pilosa*; Cornwall, Pembrokeshire plus a few inland sites), fringed rupturewort (*Herniaria ciliolata*; Lizard, Alderney) and such common species as wild thyme and buck's-horn plantain. The intense exposure to wind relieves them of the competition of taller-growing plants, and they are able to take advantage of the few centimetres of more congenial microclimate close to the ground. The effect of wind is often obvious at a larger scale too. In general, on the south coast of Alderney and Guernsey (Fig. 318), gorse is dominant on the crests and exposed faces of the headlands, while bracken is dominant on the less exposed slopes. Blackthorn occurs in areas less exposed again within the bracken, but these are far from being sheltered in any absolute sense, and the blackthorn is often severely wind-cut. The picture is similar on the south Devon cliffs.

FIG 318. Common gorse (*Ulex europaeus*) alternating with wind-trimmed blackthorn (*Prunus spinosa*) scrub, leafless at this time of year, on the south coast of Guernsey, April 2010.

Aspect: sun and shade

Aspect – the compass direction a cliff faces – is no less important on sea-cliffs than inland. South-facing cliffs are warmer and drier, and it is here that southern species are usually found. Toadflax-leaved St John's-wort (*Hypericum linariifolium*) and purple gromwell (*Lithospermum purpureocaeruleum*) have their maritime occurrences with us on the south coast, the moss *Grimmia laevigata* grows on sunny south-facing rocks on the south Devon cliffs, and there are countless other examples. Bluebells, common on cliff-slopes in the west, are densest and most luxuriant on northerly aspects. Bryophytes too are generally most luxuriant on north-facing slopes. Sea spleenwort (*Asplenium marinum*) is at its best with a degree of shade, and the same is true of the salt-tolerant moss *Schistidium maritimum* towards the southern end of its range.

ZONATION ON SEA-CLIFFS AND ROCKY SHORES

Life between the tidemarks and the splash zone: the province of seaweeds and lichens

On rocky shores the only plants that can get a foothold between the tidemarks are seaweeds and lichens. At the lowest levels on the shore, around low water of spring tides, kelps dominate almost everywhere – big brown seaweeds (mostly species of *Laminaria*) forming submarine forests of flexible fronds that can be a metre or more in length. The kelps often shelter a rich flora of delicate feathery red seaweeds. The shore between low water and high water of neap tides is uncovered and covered again at every tide and takes the full force of the waves. Where there is some degree of shelter this zone in dominated by another group of brown algae, the wracks, species of *Fucus* and their relatives. Serrated wrack (*Fucus serratus*) commonly dominates the lower part of the intertidal zone,

FIG 319. (a) Rocky-shore zonation in a sheltered inlet on the Donegal coast, September 1965. Wave-splash is minimal here. The middle shore is occupied by fucoid seaweeds (see text). The black *Verrucaria* zone above the seaweeds is narrow, soon giving way to predominantly grey lichens, and grassy and heathy vegetation. (b) A more exposed shore on Burhou, Channel Islands, July 1969. The seaweeds are much less prolific, the black *Verrucaria* zone is wider, there is a broad zone of predominantly yellow lichens below the grey-lichen zone, and the whole (small) island is affected by salt-spray deposition (see Fig. 326).

replaced by bladder-wrack (*F. vesiculosus*) higher on the shore. In sheltered places knotted wrack (*Ascophyllum nodosum*) covers the rocks (Fig. 319a). This zonation is topped off by a zone of *Fucus spiralis*, and a narrow fringe of *Pelvetia canaliculata* wetted only by spring tides. On wave-exposed shores, the fucoids become stunted and sparser, letting in smaller leathery red seaweeds such as *Chondrus crispus* (carrageen) and *Gigartina stellata*, until on very exposed shores most of the intertidal zone is left to barnacles, a few small tough red seaweeds like *Laurencia pinnatifida* and *Corallina officinalis*, and a few lichens (Fig. 319b). Of these, *Lichina pygmaea* looks like a tiny black seaweed a few millimetres high; *Verrucaria maura* grows in smooth black tar-like patches that form a dark zone visible from a distance around high tide mark – giving a graphic demonstration of how much higher wave-splash reaches on exposed headlands than on sheltered shores.

Once above the level regularly wetted by spring tides, terrestrial lichens begin to colonise the top of the black *Verrucaria* zone – first the dull yellow *Caloplaca marina*, then several bright orange-yellow species, including the common foliose *Xanthoria parietina* and the neat crustose thalli of *Caloplaca flavescens* (Fletcher 1973a, 1973b). These pass upwards into a zone of predominantly greyish lichens, some crustose, some fruticose (shrub-like). The commonest of these, *Ramalina siliquosa*, often forms a shaggy covering on rocks facing the sea. We are now in the region where flowering plants – thrift, plantains, rock samphire – and even a few mosses can become established.

Rock-crevice vegetation

The Earl of Gloucester and his son Edmund in Shakespeare's *King Lear* stood above a cliff near Dover:

> *How fearful and dizzy 'tis to cast one's eyes so low! The crows and choughs that wing the mid-way air, show scarce as gross as beetles. Half-way down, hangs one that gathers samphire – dreadful trade!*

Rock samphire (Fig. 320a) is one of the most characteristic plants of the common rock-crevice community of the south and west coasts of England and Wales, and the southern coast of Ireland. It occurs over a wide range of levels, from cliffs and rocks on the shore just above the zone of wave-splash to high on the cliffs; by no means all the places it grows are 'dreadful' in the old literal sense of 'dread-inspiring'. Rock samphire is often accompanied by thrift, red fescue and rock sea-spurrey (*Spergularia rupicola*). The only other plants that are at all common, and that only locally, are sea plantain, sea aster (*Aster tripolium*), golden-samphire (Fig. 321d), common scurvygrass (Fig. 321b) and rock sea-lavender (*Limonium binervosum* sensu lato)[MC1]. It is surprising that several of these cliff plants seem to be more frequent on the south and east coasts of Ireland than in the west. A few other species occur

FIG 320. Maritime rock-crevice plants. (a) Rock samphire (*Crithmum maritimum*), common round much of the coast of England, Wales and Ireland. Prawle Point, Devon, August 1987. (b) Scottish lovage (*Ligusticum scoticum*), which replaces *Crithmum* on the west coast of Scotland north of Galloway. Scourie, Sutherland, June 1981.

occasionally, including sea mayweed (*Tripleurospermum maritimum*), sea campion (*Silene uniflora*), sea fern-grass (*Catapodium marinum*), curved hard-grass (*Parapholis incurva*), sea spleenwort and, on acid rocks from Cornwall and Devon northwards, *Schistidium maritimum*, one of our few halophytic mosses.

Rock samphire is conspicuously absent from the east coast of Britain north of Suffolk, and almost absent from the west coast of Scotland north of Galloway and Ayrshire. It occurs all round the coast of Ireland (except Mayo?), but is commonest in the southern half of the country. Around the whole Scottish coast and on the extreme north coast of Ireland its place is largely taken by Scottish lovage (Fig. 320b), which occupies much the same range of habitats. Lovage characterises the common rock-crevice community of the north and west Scottish coast; its constant (or near-constant) companions are thrift (Fig. 321a), red fescue and the moss *Schistidium maritimum*[MC2]. Frequent associates are sea plantain, sea campion, sea mayweed, common scurvygrass and roseroot (Figs 316, 322b). Rock sea-spurrey reaches only the southwest corner of the Scottish coast and the southern Hebrides, so it is not widespread in this community, and this area of Scotland and the adjacent northern Irish coast form a zone of transition between this community and the last.

FIG 321. (a) Thrift (*Armeria maritima*) in flower and Danish scurvygrass (*Cochlearia danica*) in fruit on the edge of a Purbeck limestone cliff, Winspit, Dorset, May 1970. (b) Common scurvygrass (*Cochlearia officinalis*), Lamorna, Cornwall, April 1980. (c) Nottingham catchfly (*Silene nutans*), Branscombe, Devon, June 1966. (d) Golden-samphire (*Inula crithmoides*), Annestown, Co. Waterford, August 1974.

Cliff-tops and ledges

Some conspicuous plants have their main habitat on the ledges and edges of sea-cliffs. Wild cabbage (*Brassica oleracea*) is particularly a plant of chalk and limestone cliffs, but it grows on other rock types in Devon, Cornwall and elsewhere. It is usually accompanied by red fescue, cock's-foot (*Dactylis glomerata*), sea beet (*Beta vulgaris* ssp. *maritima*) and sea carrot (*Daucus carota* ssp. *gummifer*); thrift and cleavers (*Galium aparine*), groundsel (*Senecio vulgaris*) and the annual grass *Bromus hordeaceus* ssp. *ferronii* are other frequent associates[MC4]. In some places, the community intergrades with chalk or limestone grassland, lacking sea beet and thrift, but including such species as common restharrow (*Ononis repens*), knapweeds (*Centaurea nigra* and *C. scabiosa*), Nottingham catchfly (Fig. 321c), kidney vetch (*Anthyllis vulneraria*), ragwort (*Senecio jacobaea*) and wood sage (*Teucrium scorodonia*). Several of the commonest plants of this community are calcicole and southwestern in their distribution in Britain, and are sparse or absent in Ireland. There, their place is filled by the ubiquitous maritime plants, sea beet, sea campion, thrift and red fescue.

In northern Scotland and the Scottish islands cliff ledges may be colonised by plants drawn from the rock-crevice community on the one hand, and cliff grasslands on the other, with constant red fescue, thrift, roseroot, common sorrel (*Rumex acetosa*), frequent Yorkshire-fog (*Holcus lanatus*), ribwort and sea plantains (*Plantago lanceolata* and *P. maritima*) and occasional sea campion, creeping bent, lovage, wild angelica (*Angelica sylvestris*), red campion (*Silene dioica*), sea mayweed and primrose (*Primula vulgaris*)[MC3].

Maritime grasslands

Most sea-cliffs have some grassland in their vegetation, and 'grassland' implies soil, not just bare rock. The bulk of the herbage in these grasslands is made up of a very limited range of species: red fescue, Yorkshire-fog, cock's-foot and creeping bent amongst the grasses, and the broad-leaved herbs thrift, sea carrot, buck's-horn plantain, ribwort plantain, sea plantain, common sorrel, bird's-foot trefoil (*Lotus corniculatus*), white clover and common scurvygrass. These are the main ingredients. They are combined in various quantities and with various minor ingredients to make up what at first sight is a bewildering range of intergrading grasslands in which some noda (Chapter 4) can usefully be recognised but within which no sharp boundaries can be drawn. The zone nearest the sea is occupied by a species-poor but often luxuriant turf of red fescue and thrift. Both species are constant, but they vary greatly in relative proportions. The only other species that are at all common are creeping bent and sea plantain, but a notable rarity that occurs occasionally in this community is wild asparagus (*Asparagus officinalis* ssp. *prostratus*)[MC8]. With rather less maritime influence and with little or no grazing a much more varied *Festuca rubra–Holcus lanatus* grassland develops[MC9]. This is dominated by any one of red fescue, Yorkshire-fog, sea plantain and cock's-foot, or some combination of these.

The most frequent associated species are bird's-foot trefoil, white clover, lady's bedstraw (*Galium verum*), yarrow (*Achillea millefolium*), common sorrel, cat's-ear (*Hypochaeris radicata*), creeping bent and spring squill (western, but curiously, only on the east coast of Ireland). This grassland is exceedingly variable in moisture, aspect, shelter, degree of saline influence, nutrient status and no doubt past history, but defies formal subdivision. Some areas lean towards inland *Festuca–Agrostis* pasture (Chapter 16) or agricultural grasslands (Chapter 9), some have a heathy component, while some, with primrose, lesser celandine (*Ranunculus ficaria*), common dog-violet (*Viola riviniana*) and false brome (*Brachypodium sylvaticum*), show leanings towards scrub. Prostrate broom grows in very exposed sites, probably most often in what is essentially this community but close to the transition with the *Calluna–Scilla verna* heath discussed below.

This variable grassland, ankle-deep or somewhat more, grades into a much shorter grazed and wind-clipped maritime grassland in which the dominant red fescue is conspicuously accompanied by three species of plantain (buck's-horn, ribwort and sea plantain). The only other species to occur with any regularity are creeping bent, thrift, eyebrights (*Euphrasia* spp.), common mouse-ear (*Cerastium fontanum*) and spring squill, and less commonly carnation sedge (*Carex panicea*), bird's-foot trefoil, autumn hawkbit (*Leontodon autumnalis*), wild thyme and white clover[MC10]. Many other species occur more sparsely, or in particular places. Praeger (1934) vividly describes how on Inishturk, 11 km off the west Galway coast, 'The *Plantago* formation, here often composed of *P. maritima* and *P. coronopus* without any other ingredient, occupies a large area at the W of the island, as close and smooth as if shaved with a razor.' This *Festuca rubra–Plantago* community and the *Festuca rubra–Holcus lanatus* grassland both occur from Land's End to Shetland and the far west of Ireland, and they are to an extent interchangeable, but the *Festuca–Holcus* grassland predominates in southwest England and Wales (and, interestingly, on north-facing coasts in northern Scotland), while the *Festuca–Plantago* grassland predominates in northwest Scotland, northwest Ireland and the Hebrides.

Locally on deep, moist soils, generally on northerly aspects, a community occurs in which bluebell, red fescue, Yorkshire-fog and common sorrel are constant or nearly so, and dominate the ground – sometimes to the virtual exclusion of other plants, sometimes with a contingent of common maritime species and sometimes with a flora that recalls open woodland or a wood-margin but without shrubs or trees[MC12]. Species that occur in this facies include lesser celandine, cock's-foot, hogweed (*Heracleum sphondylium*), bracken, brambles, primrose, common dog-violet, ivy and false brome. Vegetation of this kind is scattered in suitable spots from south Devon and Cornwall to Skye.

The maritime grasslands that we have considered so far cover most of the possibilities on the hard acid rocks that make up most of our coastal cliffs. Transitions between them (and their soils) are gradual, as is clear from Table 14. Chalk and limestone are different (Fig. 322). Calcium is the predominant ion in the soil, and the dominance of saline influence affects a narrower zone of vegetation on the cliff-top. Some maritime influence remains, but it is heavily diluted by the more calcareous character of the vegetation. Locally, a grassland is found on limestone cliffs dominated by red fescue, with constant cock's-foot and sea carrot and frequent thrift, ribwort plantain, bird's-foot trefoil and salad burnet (*Sanguisorba minor*)[MC11]. Buck's-horn plantain, common restharrow, greater knapweed, lady's bedstraw, kidney vetch, tor grass (*Brachypodium pinnatum*) and tall fescue (*Festuca arundinacea*) also commonly occur. The maritime element in this hardly amounts to more than the constancy of sea carrot and the frequent presence of thrift and buck's-horn plantain, but it is there. The affinity with inland calcareous grasslands is very plain, and apparent too in the soil data of Table 14. The red fescue–sea carrot community occurs

FIG 322. Contrasting sea-cliffs. (a) Carboniferous limestone cliffs near St Govan's Head, Pembrokeshire, with golden-samphire (*Inula crithmoides*), August 1977. (b) Roseroot (*Sedum rosea*) on a Lewisian gneiss sea-cliff at Scourie, Sutherland, June 1981.

on the Purbeck and Portland limestone cliffs, on the Carboniferous limestone of Gower and Pembrokeshire in South Wales, and sporadically elsewhere, probably where chalk or blown sand have created suitably calcareous soils. The community (unless redefined) cannot occur outside the range of the sea carrot. A poorly characterised grassland of similar maritime flavour along the west-coast Burren cliffs had constant red fescue, near-constant creeping bent, sea plantain and bird's-foot trefoil, occasional thrift, and much in common with the neighbouring calcareous grasslands on limestone or boulder-clay.

Cliff-top heaths

On hard acid rocks heath often covers the cliff-tops and spurs and broken rocky ridges facing the sea. These heaths are variable amongst themselves, and intergrade with the more acid maritime grasslands: prostrate broom favours grassland–heath transitions on exposed cliffs (Fig. 323). Maritime heaths generally have many more constant or near-constant species than most inland heaths. Heather (*Calluna vulgaris*) and bell heather (*Erica cinerea*) are generally

FIG 323. Prostrate broom (*Cytisus scoparius* ssp. *maritimus*) on the southwest cliffs of Alderney, with heathers and maritime grassland, in (a) April and (b) July. Notice the parasitic greater broomrape (*Orobanche rapum-genistae*), just starting into growth in April (bottom LH corner), but tall dry fruiting spikes by July amongst the dry flowering heads of thrift and Yorkshire-fog.

dominant. Other species of high constancy are sheep's fescue (*Festuca ovina*), sea plantain, spring squill, bird's-foot trefoil, wild thyme, tormentil (*Potentilla erecta*), Yorkshire-fog, ribwort plantain and cat's-ear; red fescue and common dog-violet are frequent. Like most cliff communities, this *Calluna vulgaris–Scilla verna* heath[H7] is variable. The variant showing the strongest maritime influence has constant thrift and English stonecrop (*Sedum anglicum*), and frequent cock's-foot, kidney vetch and sheep's-bit (*Jasione montana*). What might be seen as the 'central' form of the community is characterised by near-constant common dog-violet and glaucous sedge (*Carex flacca*), with frequent common milkwort (*Polygala vulgaris*), spring-sedge (*Carex caryophyllea*), devil's-bit scabious (*Succisa pratensis*) and heath-grass (*Danthonia decumbens*). A damp-heath version, with cross-leaved heath (*Erica tetralix*) largely replacing bell heather, near-constant devil's-bit scabious, heath-grass and common bent, and frequent common sedge (*Carex nigra*), sweet vernal-grass (*Anthoxanthum odoratum*), red fescue and eyebright species, is common from Anglesey northwards and is the predominant form of the community in northwest Scotland and the Western Isles. A version with crowberry (*Empetrum nigrum*) largely replacing bell heather occurs in the north of Scotland. Species-poor stands with little beyond the basic list of near-constants can be found throughout the community's range.

Open communities of (mostly) dry places: stonecrops, clovers and others

Thin soils on cliff edges, around rock outcrops or on sun-exposed stony slopes often bear low open vegetation, moist and green in winter and spring, largely brown, dry and shrivelled in late summer, in which small annuals are prominent. These communities are very diverse in species composition and ecological relationships.

Braun-Blanquet and Tüxen (1952) described a *Plantago coronopus–Cerastium diffusum* community from a rocky shore on Achill, low cliffs in the Burren and dunes at Rossbeigh (Kerry), and conjectured that it is common round the Irish coast. This appears to be a near match for part of the *Armeria maritima–Cerastium diffusum* community[MC5a] in Britain. Thrift, buck's-horn plantain (*Plantago coronopus*), red fescue, sea mouse-ear (*Cerastium diffusum*, Fig. 324b) and sea fern-grass are constant in the British community; sea fern-grass seems to have a slightly different but related niche in western Ireland, but the difference may be more apparent than real.

Braun-Blanquet and Tüxen also described from Ireland an '*Aira praecox–Sedum anglicum* association', of shallow siliceous soils, not specifically bound to the coast. They saw this as related to a complex of communities distributed along the North Atlantic seaboard from mid Portugal north to Ireland and

OPPOSITE: **FIG 324.** Open communities of sea-cliffs. (a) Habitat, Outer Froward Point, near Dartmouth, Devon, July 1989. (b) Sea mouse-ear (*Cerastium diffusum*), Durness, Sutherland, June 1981. (c) Hare's-foot clover (*Trifolium arvense*), English stonecrop (*Sedum anglicum*) and the annual grass *Bromus ferronii*, Alderney, July 1967. (d) Open community by cliff-top path near Le Gouffre, Guernsey, April 2010. Sand crocus (*Romulea columnae*) is several weeks past flowering; a developing seed-capsule can be seen near the centre of the picture but the curly leaves are already withered and pale brown. The broader leaves of autumn squill (*Scilla autumnalis*) are still green, though yellowing. Patches of reddish mossy stonecrop (*Crassula tillaea*) are prominent. (e) Spotted rock-rose (*Tuberaria guttata*) with English stonecrop and scattered small grasses, Alderney, April 1957. (f) Spotted rock-rose in bare patch in grazed and windcut heather, with early hair-grass (*Aira praecox*). Inishbofin, Co. Galway, August 1958.

Britain – and probably southwest Scandinavia. In their 11 species-lists, early hair-grass (*Aira praecox*), English stonecrop (*Sedum anglicum*) and the moss *Polytrichum juniperinum* were constant, and silver hair-grass (*Aira caryophyllea*), squirreltail fescue (*Vulpia bromoides*), the moss *Hypnum lacunosum* and red fescue almost so. They found their community 'particularly common in SW-Ireland, but also in the NW of the island'. It is the habitat of spotted rock-rose (*Tuberaria guttata*) in Ireland. Essentially the same community is common on coastal cliffs in the Channel Islands, where it is the habitat of sheep's-bit, sand crocus (Fig. 324d), spotted rock-rose (Fig. 324e), four-leaved allseed (*Polycarpon tetraphyllum*), hairy bird's-foot trefoil (*Lotus subbiflorus*), hare's-foot clover (Fig. 324c) and a number of rarer clovers and other annuals. It is also common on sea-cliffs and cliff-slopes in southwest England and Wales, but (as in Ireland) with a more restricted flora. This community spans a wide range of altitude (and salt deposition) on sea-cliffs. At low levels, where marine influence is greatest, most of the species of the *Plantago coronopus–Cerastium diffusum* community can occur[cf. MC5c]. At high levels on the cliffs the two communities have almost nothing in common.

There is a parallel community to this on limestone, with biting stonecrop (*Sedum acre*) replacing *S. anglicum*, and small annuals including thyme-leaved sandwort (*Arenaria serpyllifolia*) and often rue-leaved saxifrage (*Saxifraga tridactylites*). This too can appear associated with the *Armeria maritima–Cerastium diffusum* community[MC5d], or in association with other communities of limestone rocks and walls[OV39, OV42].

Although most of the annual clovers (and a number of other rare annuals) can appear occasionally in the *Aira praecox–Sedum anglicum* community, this is not their preferred habitat, which is something more extended and grassier – but still dry in summer. Many of these species are southern rather than southwestern plants in their European ranges. This shows up too in their distribution patterns within Britain; most have their centre of gravity in southeast England (good examples are *Trifolium glomeratum, T. suffocatum* and *T. subterraneum*), or in the southern half of the country (*Ornithopus perpusillus, Moenchia erecta*). These species seem to need a hot sunny summer, which probably accounts for their rarity in Ireland.

Orange bird's-foot (*Ornithopus pinnatus*) grows with us only in short grassy turf in the Channel Islands and Isles of Scilly. Small hare's-ear (*Bupleurum baldense*) and small restharrow (*Ononis reclinata*, Fig. 141b) grow in a few places, mostly on the south coast of Britain and in the Channel Islands. The most celebrated group of plants of this kind are the Lizard clovers. Three species, twin-headed clover (*Trifolium bocconei*), long-headed clover (*T. incarnatum* ssp. *molinerii*) and upright clover (*T. strictum*) are confined to the Lizard district of Cornwall or nearly so, where they grow along with all the other annual clovers of our flora. The Rev. C. A. Johns (author of the popular Victorian *Flowers of the Field*) visited the Lizard in 1848, and observed 'I actually covered with my hat growing specimens of all together *Lotus hispidus, Trifolium bocconi, T. molinerii,* and *T. strictum*.' The twentieth-century photograph (Fig. 325) shows that the Rev. Johns's hat need not have been large!

The influence of seabirds

Seabirds and sea-cliffs go together, and the birds inevitably have an impact on the vegetation. The impact is most intense in and around the colonies of the gregariously nesting species, but the birds have a more diffuse influence too in manuring the cliff vegetation. The birds must be responsible for much of the nitrogen and phosphate in sea-cliff soils.

In the densest breeding colonies (gannets, cormorants) virtually nothing grows. Gulls and terns like a little more space, and there is generally room for some greenery between the nests. Species that nest in burrows (puffins, Manx shearwaters, petrels) need space between the burrows and vegetation is important to the stability of the ground, but it suffers a lot of trampling, manuring and traffic. Outside the nesting season, bird colonies are often obvious from their conspicuously luxuriant vegetation, and bird islands are often conspicuously green.

Compared with the general run of cliff grasslands, the maritime and ruderal plants are more conspicuous at bird-influenced sites. On the sunnier and more southern coasts of both Britain and Ireland, this shows itself in the abundance of oraches (*Atriplex* spp.), sea beet, sea mayweed and rock sea-spurrey (Fig. 326). The ubiquitous red fescue remains common, along with thrift, curled dock (*Rumex crispus* ssp. *littoreus*), cock's-foot, knotgrass (*Polygonum aviculare*), sea

LEFT: **FIG 325.** Lizard clovers. Twin-headed clover (*Trifolium bocconei*), upright clover (*T. strictum*), knotted clover (*T. striatum*), rough clover (*T. scabrum*), slender trefoil (*T. micranthum*), lesser trefoil (*T. dubium*) and common restharrow (*Ononis repens*). Caerthillian Cove, Lizard, Corwall, June 1968.

BELOW: **FIG 326.** Burhou, an uninhabited but seabird-frequented small island near Alderney, July 1969; dominant rock sea-spurrey (*Spergularia rupicola*) in flower, a patch of wall pennywort (*Umbilicus rupestris*) in the foreground and bracken (*Pteridium aquilinum*) in the middle distance. Little Burhou, just to the west, is covered with sea campion (*Silene uniflora*).

FIG 327. Tree-mallow (*Lavatera arborea*) on Devonian limestone. Berry Head, Devon, May 1983.

stork's-bill (*Erodium maritimum*) and common scurvygrass. A conspicuous and beautiful plant frequent in this community is tree-mallow (Fig. 327)[MC6]. Tree-mallow sometimes grows on exposed sea-stacks. One may wonder how such an apparently soft and vulnerable plant survives in a habitat where even red fescue is wind-cut. Much is possible for a quick-growing annual or biennial, which can take maximum advantage of the warm summer months, and does not have to ride out the full force of the chilly winter storms – perhaps an example of the relative métiers of r-selected and k-selected species.

Sea beet, rock sea-spurrey and tree-mallow are all missing from much of Scotland and rather thinly distributed in northwest Ireland. Northwards, bird-influenced sites tend to be increasingly dominated by chickweed and grasses – creeping bent, Yorkshire-fog and red fescue – with thrift, common sorrel, sea plantain, common scurvygrass, curled dock, sea campion and spear-leaved orache (*Atriplex prostrata*)[MC7]. However, sites vary greatly with local conditions, and no doubt with the chance consequences of which plant happened to get there first.

Cliff scrub and woodland
The sheltered parts of sea-cliffs are open to colonisation by woody vegetation. Bracken is an occasional ingredient of the red fescue–bluebell grassland of

sheltered cliff slopes, and in southwest England it is often prominent on cliff slopes that are not too exposed. Some of this bracken community at higher levels and on acid rocks is of the bracken–heath bedstraw type: essentially an upland *Festuca ovina–Agrostis capillaris* pasture with a bracken canopy[U20]. But some is more nearly related to scrub, with a thin understorey of brambles[W25], foxgloves, false brome and occasional low stunted bushes of blackthorn in a matrix of common cliff-grassland species. Bloody crane's-bill (*Geranium sanguineum*) grows in vegetation of this general kind both on the Lizard and on hornblende–schist cliffs on the south Devon coast. In more sheltered places blackthorn can grow up to form a continuous scrub[W22]. On poor, acid soils, the most exposed ground is occupied by wind-cut heather, and where there is deeper soil and some moderation of the exposure patches of common gorse can develop[W23]. The pattern of the main vegetation types on the south coast of Guernsey (Fig. 318) is repeated with variations in many places on the coast of southwest England, Wales and Ireland. Further development to cliff woodland returns us to Chapter 8.

CULTIVATED PLANTS, INTRODUCTIONS AND RUDERALS

Some our most-used vegetables originated on the coast. Sea beet and wild carrot are coastal plants, and wild cabbage on chalk and limestone cliffs is instantly recognisable by anyone who has ever grown or cooked one of its numerous cultivated varieties. Add to that basic list some more esoteric vegetables – asparagus, sea-kale, radish, celery, marsh samphire, fennel – and some that were eaten more in the past than now – rock samphire, scurvygrass, alexanders – and the list becomes quite impressive. Probably the succulence of many halophytes made them a natural choice as food plants. It is easy to see why our ancestors cultivated beet, with its fleshy roots, sweet in autumn (the winter store of sugar is used up by spring). It is less easy to see why man chose to cultivate carrot, with its much tougher woody roots, unless it was for their flavour. Wild celery (*Apium graveolens*) grows in the brackish upper fringe of the saltmarsh zonation, and its characteristic smell is one of the safest ways to recognise it (though a Dutch family camping in England had their holiday rudely cut short in hospital when they put hemlock water-dropwort, which does not occur in the Netherlands, in their stew). Fennel (*Foeniculum vulgare*) and alexanders (*Smyrnium olusatrum*) are generally acknowledged to be introductions, but in crop plants and weeds there is a fine line between 'native' and 'introduced'. Useful plants were traded by prehistoric societies, and weeds went with them. Fennel has been in our gardens (and kitchens) since at least the eighteenth century; it has come into more general use as a vegetable in recent decades. Alexanders was formerly used as a potherb, but has found a congenial niche in our maritime ruderal flora; sometimes it is a major part of the vegetation of seabird islands. Fennel too is often encountered near the sea, where it joins the ruderal flora in which black mustard (*Brassica nigra*) is often prominent on cliff-tops by the sea. Silver ragwort (*Senecio cineraria*) is commonly naturalised on cliffs and cliff slopes in southern England; hoary stock (*Matthiola incana*) less often so. Both are west Mediterranean plants, introduced to English gardens in Tudor times and now naturalised with us. A more recent immigrant is the South African hottentot fig (*Carpobrotus edulis*), which sprawls over cliff slopes in the Channel Islands and southwest England. It has been in our gardens for several centuries, but was first recorded wild in 1886. Everyone notices its gaudy flowers, and many conservationists view its spread with apprehension. Are their fears justified? It contributes to the seaside ambience, as do slender thistle (Fig. 328), a generally accepted native, tamarisk (*Tamarix gallica*) from the west Mediterranean, and the Duke of Argyll's tea plant (*Lycium* spp.) from China.

FIG 328. Slender thistle (*Carduus tenuiflorus*), a very characteristic seaside plant in Britain, less so in Ireland. Branscombe, Devon, June 1973.

CHAPTER 21

Conclusion

W E HAVE COME A LONG WAY since the opening pages of this book, and since the days of Sir Arthur Tansley. One can only admire what he and the other pioneers of plant ecology achieved, but we now know much more about the historical context of the British and Irish landscape – in terms of geology, and social and vegetation history. This has given us a keener appreciation of the part played by geological events and by the past influence of man in the present-day landscape, and it has made it easier to see the present as just an episode in an unfolding story, and to see past, present and future as an indissoluble whole.

HOW WIDELY DOES THE 'CULTURAL LANDSCAPE' EXTEND?

Inaccessible cliffs, uninhabited islands, the highest parts of our mountains, and some ombrotrophic bogs, dunes and saltmarshes may be the only places in our islands to have truly 'natural' vegetation, and even these are not wholly exempt from the past or present influence of man. The rest of our landscape (and that of our Continental neighbours) has been shaped to a greater or less extent by man and his domesticated animals. It is in that sense 'semi-natural', the product of the interaction of man and nature. In my student days the prairies of North America and the steppes of the Hungarian plain and the Ukraine were widely regarded as natural grasslands, as a climatic climax. Now, deprived of their traditional grazing herds (and the traditional way of life that went with them), both regions have shown that they are capable of supporting at least an open cover of trees. A similar effect can be seen in Britain and Ireland with the decline in casual rough grazing.

Our best guesses about the potential natural vegetation of our two islands have been sketched in Chapters 2 and 5, and the picture that emerges is very different from the landscape we are used to. It is too easy to be complacent about our long-settled countryside, and to take it for granted as the natural order of things. We Europeans should learn from the American experience. In my student days, there were many papers in American scientific journals on 'old-field succession', and we thought little of it, except perhaps, 'well, that's America'. In *A Walk in the Woods*, Bill Bryson (1998) remarks, 'In 1850, New England was 70 percent open farmland and 30 percent woods. Today the proportions are exactly reversed', which neatly sums up the cumulative effect of the abandonment of labour-intensive farming in difficult country for easier ways of earning a living. Much the same could easily happen here – and to an extent it has already happened. In our more densely populated and well-regulated countryside, and checked by farm subsidies, it may creep up on us more slowly. The pressures and trend are there, nevertheless. How many of us have noticed open views we used to know now blocked by trees?

Nature conservation

Nature conservation means different things to different people. Some (understandably) are jealous of *all* sites of *all* rare species in their locality (parish, townland, county, country), and cite 'protecting biodiversity' as their justification. But biodiversity is a product of a number of things, amongst them habitat diversity. Our present diverse countryside is largely the product of labour-intensive traditional farming. To preserve this diversity demands an input of labour commensurate with that which created it; we do not expect our houses or gardens to weather the years without maintenance, and maintenance is expensive.

Given that we are now caught up willy-nilly in a move from a fine-grained largely hand-made countryside to a coarser-grained largely machine-made countryside, it is at least arguable that we should be worrying much less about outlying localities of rare species, and much more about maintaining good populations of local species in their core areas. Visiting Continental botanists get much more excited about our bluebells, common gorse, bell heather, foxgloves and our abundance of holly and ivy, than about the rarities we ourselves most cherish. Global biodiversity benefits more if we preserve and enhance habitats of the plants that are most characteristic with us on the Atlantic fringe of Europe. Most rare species are common somewhere. It is nice to be able to see such plants in British or Irish localities, but far more important for global biodiversity that they should continue to flourish where they are common.

All wild populations fluctuate for various reasons, some more than others. Many chance factors are responsible, including the weather from year to year. This means that there is always a chance that a population will go extinct after a run of 'bad' years, and small populations are particularly vulnerable to extinction. If they are isolated too, the site is unlikely to be recolonised. This emphasises the importance of protecting large sites and, wherever possible, groups of sites – to minimise the likelihood of extinction, and to maximise the probability of recolonisation, especially in those species prone to wide population fluctuations. We should regard our countryside much less as an old master painting bequeathed to us to care for and preserve, and much more as a theatre for an ongoing play in which things happen. That is particularly so for plant communities that are inherently successional, like hydroseres and sand-dunes. Generally, it is a good precept for long-term nature conservation to look after the habitat, and let the species look after themselves. Over-concentration on species too easily degenerates into gardening – not that I have anything against gardening, but it is a different activity with a different philosophy!

A more fundamental challenge is to develop patterns of land use that create rich and (bio)diverse environments, and are at the same time economically viable. Traditional woods remained in existence for millennia because they were a renewable resource that met many economic needs in society, including energy and building material. Hedgerows likewise served multiple functions. The traditional woods that are still regularly coppiced are mostly either nature reserves or woods used for shooting. The income from shooting pays for the gamekeeper and the coppicing. We pay lip-service to the principle of sustainability. Maybe we could, at small economic cost, but to our substantial benefit in other ways, take lessons from our predecessors.

Introduced species, and what is 'native'?

Arguably, a 'native' species is one that reached our islands before the invention of writing. The authors of the *New Atlas of the British and Irish Flora* recognise a category of 'archaeophytes', such as common poppy (*Papaver rhoeas*), which presumably came in with our prehistoric ancestors, and are printed in the *Atlas* in red like other introduced species. People's responses to introduced plants are curiously capricious. Sycamore seems to come in for more 'racial prejudice' than most plants. It is a common native tree only a short way across the Channel, and it is probably only a historical accident that it did not make its own way to Britain before its introduction was recorded in the sixteenth century. Objectively, sycamore poses fewer problems than another introduction from the Continent, the common rhododendron, which can form dense thickets excluding almost all other species in acid oakwoods, but most complaints about rhododendron come from conservation managers to whom its invasiveness really is a problem, and few simply because it is an 'alien'. Canadian waterweed (*Elodea canadensis*) has become a settled and well-behaved member of our flora since the days when it was dubbed 'Babingtonia pestifera', after the Professor of Botany from whose Botanic Garden at Cambridge it was thought to have escaped. We should now miss it. Plantlife has recently been campaigning against Indian balsam (*Impatiens glandulifera*). That in my judgement is misconceived. This beautiful plant clothes many kilometres of canal and riversides in the industrial north of England where little else grows, and poses no threat to native vegetation in the many places elsewhere that I know it. Most non-native species do little harm to our established flora (Maskell *et al.* 2006).

It is hard to predict invasiveness – until a plant makes a nuisance of itself, and then it is usually too late. Who would have thought that our own purple-loosestrife (*Lythrum salicaria*) would be a problem invasive species in North America? Of the newcomers that do get established we should concentrate on the ones that pose real problems, and not dissipate limited resources on species simply because we perceive them to be 'alien'. Japanese knotweed (*Fallopia japonica*) is an expensive problem species, but can be eradicated if enough money and persistent effort are thrown at it. Bird-sown cotoneasters, especially *C. microphyllus* and *C. horzontalis*, locally cause conservation problems by overrunning limestone cliffs where better things grow. But they are a problem that can be contained, and are such common and useful garden plants that exterminating them is unthinkable.

CLIMATIC CHANGE

As we saw in Chapter 2, climatic change has always been with us. What has sparked current concerns is the combined evidence of rising temperatures, and of rising CO_2 concentrations in the atmosphere. The climate has on average become warmer by nearly 1 °C over the past century, and increase in atmospheric CO_2 concentration is well attested. The evidence is becoming increasingly conclusive that at least a major part of the rise in atmospheric CO_2 since the industrial revolution, and the rise in temperature over the past half-century, is due to human activity (Woodward *et al.* 2010).

Climatic change, however caused, will present us with new conservation challenges. Is the apparent decrease in dwarf willow (*Salix herbacea*) noted in the *New Atlas of the British and Irish Flora* a symptom of global warming? There will inevitably be retractions and losses of northern species. The BSBI Local Change project (Braithwaite *et al.* 2006) recorded an overall significant decrease from 1987 to 2004 in montane plants generally, but a significant *increase* in *Salix herbacea*. Is this increase real, or a consequence of better recording of this modest plant in the second survey? This highlights the methodological problems of designing a survey to detect and measure change.

The plants of many habitats have shown little overall change, but calcareous grasslands have recorded more decreases than most, particularly among the classic chalk and limestone grassland species such as dwarf thistle (*Cirsium acaule*), horseshoe vetch (*Hippocrepis comosa*), small scabious (*Scabiosa columbaria*) and squinancywort (*Asperula cynanchica*) and, rather surprisingly, tall species such as greater knapweed (*Centaurea scabiosa*) and wild parsnip (*Pastinaca sativa*). On the other hand, bee orchid (*Ophrys apifera*), pyramidal orchid (*Anacamptis pyramidalis*) and bloody crane's-bill (*Geranium sanguineum*) have shown significant increases. Another habitat that has shown a significant overall decrease is dwarf shrub heath. Overall, broad-leaved woodlands have shown a marginally significant decrease, particularly in the species regarded as 'ancient woodland' indicators, but there have been increases in such species as stinking iris (*Iris foetidissima*), wood-sedge (*Carex sylvatica*), tutsan (*Hypericum androsaemum*), wood forget-me-not (*Myosotis sylvatica*), pendulous sedge (*Carex pendula*), ramsons (*Allium ursinum*) and hart's-tongue (*Phyllitis scolopendrium*).

Southern species would be expected to move northward. Some of these may at first not be welcomed. Species that have recorded big recent increases include some long-established 'native' species such as the sedges *Carex divulsa* and *C. muricata*, wild teasel (*Dipsacus fullonum*), common ramping-fumitory (*Fumaria muralis*), black nightshade (*Solanum nigrum*), weld (*Reseda luteola*), slender trefoil (*Trifolium micranthum*), keeled-fruited cornsalad (*Valerianella carinata*), mistletoe (*Viscum album*) and the fescues *Vulpia bromoides* and *V. myuros*; all of these are southern plants. The list also includes saltmarsh plants that have expanded inland along main roads salted in winter, such as Danish scurvygrass (*Cochlearia danica*), reflexed saltmarsh-grass (*Puccinellia distans*) and lesser sea-spurrey (*Spergularia marina*). Garden escapes figure prominently. Red valerian (*Centranthus ruber*), purple toadflax (*Linaria purpurea*), cherry laurel (*Prunus laurocerasus*) and silver ragwort (*Senecio cineraria*) come from southern Europe and are so familiar they almost qualify as natives, but *Buddleja davidii*, *Cotoneaster horizontalis* (both from China), orange balsam (*Impatiens capensis*), Canadian fleabane (*Conyza canadensis*) and Mexican fleabane (*Erigeron karvinskianus*) (all from North America) have made themselves almost equally at home with us. Evergreen oak (*Quercus ilex*), a widespread climax dominant in the Mediterranean area, has expanded its native distribution in France in recent decades. It has long been planted in Britain, but is now widely naturalised in the south of England, and locally invasive and displacing native vegetation. Is this 'a natural extension of its range', or yet another 'invasive exotic species'?

It is surprisingly difficult to disentangle the effects of climatic change from all the other factors that may influence plant distribution and vegetation – farming and land-use changes, eutrophication and so on. Changes in agricultural practice, particularly intensification of land use, and eutrophication tend to emerge most clearly as drivers of change (Smart *et al.* 2003, 2005), but climate already has had some effect. We should remember that temperatures have been at least as high in the past as anything envisaged in the next few decades, and our Bronze Age predecessors survived. That should take our fingers from the panic button, but not make us complacent.

We have the good fortune to live in two biodiverse and beautiful islands. To keep them so, we face challenges enough to absorb all our ingenuity, intelligence and effort through the years that lie ahead. 'No man is an island', and no island can ignore the wider world. The whole planet has ongoing problems to solve, economic and (more fundamentally) ecological, including that most fundamental problem of all, growing human population.

Further Reading and References

FURTHER READING

Braun-Blanquet & Tüxen (1952): a classic and a remarkable achievement, this paper (in German) was the fruit of two distinguished Continental phytosociologists' participation in the International Phytogeographical Excursion in Ireland in 1949.

Hill, Preston & Roy (2004): a tabular summary of the status, size, life-history, geography and habitats of British and Irish vascular plants.

McVean & Ratcliffe (1962): a pioneering book, not entirely superseded by Rodwell's volumes.

Oberdorfer (1983): a fascinating book for those with some German; gives a neat ecological and phytosociological characterisation of the habitat of each species.

Praeger (1934): a guidebook to the notable plants of Ireland; dated, but still a mine of information.

Preston, Pearman & Dines (2002).

Preston & Hill (1997) and Hill & Preston (1998) tabulate the geographical relationships of British and Irish vascular plants and bryophytes.

Rackham (1986): widely relevant; includes Ireland.

Ratcliffe (1977): vol. 2 contains short descriptions of many classic and interesting sites.

Rodwell (1991–2000): the standard modern account of British vegetation ('NVC'), with comprehensive references up to the 1990s; valuable features are the essays introducing each group of communities, and the index of species in each volume listing the communities in which each occurs.

Tansley (1939): dated, but a magnificent achievement in its day, and still a valuable source.

REFERENCES

Abraham, F. & Rose, F. (2000). Large-leaved limes on the South Downs. *British Wildlife*, Dec. 2000, 86–90.

Adam, P. (1978). Geographical variation in British salt-marsh vegetation. *Journal of Ecology*, **66**, 339–66.

Adam, P. (1981). The vegetation of British salt-marshes. *New Phytologist*, **88**, 143–96.

Albertson, N. (1960). Das grosse südliche Alvar der Insel Öland. *Svensk Botanisk Tidskrift*, **44**, 270–331.

Armstrong, W., Wright, E. J., Lythe, S. & Gaynard, T. J. (1985). Plant zonation and the effects of the spring-neap tidal cycle on soil aeration in a Humber salt marsh. *Journal of Ecology*, **73**, 323–39.

Atherton, I., Bosanquet, S. & Lawley, M. (eds) (2010). *Mosses and Liverworts of Britain and Ireland: a Field Guide*. British Bryological Society.

Averis, A. M., Averis, A. B. G., Birks, H. J. B. *et al.* (2004). *An Illustrated Guide to British Upland Plant Communities*. Joint Nature Conservation Committee, Peterborough.

Avery, B. W. (1980). *Soil Classification for England and Wales [Higher Categories]*. Soil Survey Technical Monograph No. 14. Harpenden.

Balme, O. E. (1953). Edaphic and vegetational zoning on the Carboniferous Limestone of the Derbyshire Dales. *Journal of Ecology*, **41**, 331–44.

Barnes, G. & Williamson, T. (2006). *Hedgerow History: Ecology, History and Landscape Character*. Windgather Press, Macclesfield.

Bates, J. W. (1992). Influence of chemical and physical factors on *Quercus* and *Fraxinus* epiphytes at Loch Sunart, western Scotland: a multivariate analysis. *Journal of Ecology*, **80**, 163–79.

Bates, J. W. & Preston, C. D. (2011) Can the effects of climate change on British bryophytes be distinguished from those resulting from other environmental changes? In: *Bryophyte Ecology and Climate Change* (ed. Z. Tuba, N. G. Slack & L. R. Stark). Cambridge, Cambridge University Press, pp. 371–407.

Beddows, A. R. (1967). Biological Flora of the British Isles. *Lolium perenne* L. *Journal of Ecology*, **55**, 567–87.

Bellamy, D. (1986). *Bellamy's Ireland: the Wild Boglands*. Country House, Dublin.

Belyea, L. R. & Clymo, R. S. (2001). Feedback control of the rate of peat formation. *Proceedings of the Royal Society of London B*, **268**, 1315–21.

Bennett, K. D. (1983). Devensian Late-glacial and Flandrian vegetational history at Hockham Mere, Norfolk, England. I. Pollen percentages and concentrations. *New Phytologist*, **85**, 457–87.

Beresford, G. (1979). Three deserted medieval settlements on Dartmoor: a report on the late E. Marie Minter's excavations. *Medieval Archaeology*, **23**, 98–158.

Beresford, M. W. & St Joseph, J. K. S. (1979). *Medieval England: an Aerial Survey*. 2nd edn. Cambridge University Press, Cambridge.

Bevis, J. F. & Jeffery, H. J. (1911). *British Plants: Their Biology and Ecology*. Alston Rivers, London.

Birks, H. J. B. (1973). *The Past and Present Vegetation of the Isle of Skye: a Palaeoecological Study*. Cambridge University Press, Cambridge.

Blackstock, T. H., Duigan, C. A., Stevens, D. P. & Yeo, M. M. (1993). Vegetation zonation and invertebrate fauna in Pant-y-llyn, an unusual seasonal lake in South Wales, U.K. *Aquatic Conservation*, **3**, 253–68.

Boatman, D. J. (1983). The Silver Flowe National Nature Reserve, Galloway, Scotland. *Journal of Biogeography*, **10**, 163–274.

Boatman, D. J., Goode, D. A. & Hulme, P. D. (1981). The Silver Flowe. III. Pattern development on Long Loch B and Craigeazle mires. *Journal of Ecology*, **69**, 897–918.

Bradshaw, M. E. (1962). The distribution and status of five species of the *Alchemilla vulgaris* L. aggregate in Upper Teesdale. *Journal of Ecology*, **50**, 681–706.

Braithwaite, M. E., Ellis, R. W. & Preston, C. D. (2006). *Change in the British Flora*. Botanical Society of the British Isles, London.

Braun-Blanquet, J. (1927, 1954, 1964). *Pflanzensoziologie*. Springer, Berlin, Wien.

Braun-Blanquet, J. (1932). *Plant Sociology* (transl. G. D. Fuller & H. S. Conard). McGraw Hill, New York.

Braun-Blanquet, J. & Tüxen, R. (1952). Irische Pflanzengesellschaften. In *Die Pflanzenwelt Irlands* (ed. W. Lüdi). Hans Huber, Bern. (*Veröffentlichungen des geobotanischen Institutes Rübel in Zürich*, **25**, 224–415.)

Brontë, P. (1824a). *The Phenomenon: or, An Account in Verse of the Extraordinary Disruption of a Bog, which took place in the Moors of Haworth on the 12th [sic] Day of September, 1824*. T. Inkersley, Bradford.

Brontë, P. (1824b). *A Sermon preached in the Church of Haworth on Sunday, the 12th Day of September, 1824, in reference to an Earthquake, and extraordinary Eruption of Mud and Water, that had taken place ten days before, in the Moors of that Chapelry.* T. Inkersley, Bradford.

Brooks, M., Cherrington, J. & Proctor, M. (1988). The distribution of plant species and vegetation on the saltmarsh in the Otter Estuary at Budleigh Salterton. *Transactions of the Devonshire Association,* **120**, 155–76.

Brun-Hool, J. & Wilmanns, O. (1982). Plant communities of human settlements in Ireland. 2. Gardens, parks and roads. *Journal of Life Sciences of the Royal Dublin Society,* **3**, 91–103.

Bryson, B. (1998). *A Walk in the Woods: Rediscovering America on the Appalachian Trail.* Broadway Books (Random House), New York.

Chapman, V. J. (1976). *Coastal Vegetation.* Pergamon, Oxford.

Christy, M. & Worth, R. H. (1922). The ancient dwarfed woods of Dartmoor. *Transactions of the Devonshire Association,* **54**, 291–342.

Clapham, A. R. (ed.) (1978). *Upper Teesdale: the Area and its Natural History.* Collins, London.

Clausen, J. (1954). Partial apomixis as an equilibrium system in evolution. *Atti del IX Congresso Internazionale di Genetica. I. Caryologia,* **6** (suppl.), 469–79.

Clements, F. E. (1916). *Plant Succession: an Analysis of the Structure of Vegetation.* Carnegie Institution, Washington.

Clements, F. E. (1936). Nature and structure of the climax. *Journal of Ecology,* **24**, 252–84.

Clymo, R. S. (1984). The limits to peat growth. *Philosophical Transactions of the Royal Society of London B,* **303**, 605–54.

Connor, E. F. & McCoy, E. D. (1979). The statistics and biology of the species–area relationship. *American Naturalist,* **113**, 791–833.

Conolly, A. P. & Dahl, E. (1970). Maximum summer temperature in relation to the modern and Quaternary distributions of certain arctic-montane species in the British Isles. In: *Studies in the Vegetational History of the British Isles* (ed. D. A. Walker & R. G. West). Cambridge University Press, Cambridge, pp. 159–223.

Coté, D., Dubois, J. M. M. & Gwyn, Q. H. J. (2006). Les lacs karstiques de l'Île d'Anticosti: analyse

hydrogéomorphologique. Bulletin de Recherche 181. Université de Sherbrooke, Quebec, Canada.

Couwenberg, J. & Joosten, H. (2005). Self organisation in raised bog patterning: the origin of microtope zonation and mesotope diversity. *Journal of Ecology,* **93**, 1238–48.

Crocker, R. L. & Major, J. (1955). Soil development in relation to vegetation and surface age at Glacier Bay, Alaska. *Journal of Ecology,* **43**, 427–48.

Cross, J. R. (2006). The potential natural vegetation of Ireland. *Biology and Environment: Proceedings of the Royal Irish Academy,* **106B**, 65–116.

Cross, J. R. & Kelly, D. L. (2003). Wetland woods. In: *Wetlands of Ireland* (ed. M. L. Otte). University College Dublin Press, Dublin, pp. 160–72.

Curtis, J. T. (1959). *The Vegetation of Wisconsin: an Ordination of Plant Communities.* University of Wisconsin, Madison.

Curtis J. T. & McIntosh, R. P. (1951). An upland forest continuum in the prairie-forest border region of Wisconsin. *Ecology* **32**, 476–96.

Curtis, T. F. G. & Sheehy Skeffington, M. J. (1998). The salt marshes of Ireland: an inventory and account of their geographical variation. *Biology and Environment: Proceedings of the Royal Irish Academy,* **98B**, 87–104.

Dargie, T. C. D. (1998). *Sand Dune Vegetation Survey of Scotland. Western Isles. Vol. 3: NVC Maps.* Scottish Natural Heritage, Battleby.

Davy, A. J. (1980). Biological Flora of the British Isles. *Deschampsia cespitosa* (L.) Beauv. *Journal of Ecology,* **68**, 1075–96.

Dobson, F. S. (2011). *Lichens: an Illustrated Guide to the British and Irish Species.* 6th edn. Richmond Publishing, Slough.

Dodd, M. E., Silvertown, J., McConway, K., Potts, J. & Crawley, M. (1994). Application of the British National Vegetation Classification to the communities of the Park Grass Experiment through time. *Folia Geobotanica et Phytotaxonomica,* **29**, 321–34.

Dony, J. G. (1963). The expectation of plant records from prescribed areas. *Watsonia,* **5**, 377–85.

Doogue, D. & Kelly, D. L. (2006). Woody plant assemblages in the hedges of eastern Ireland: products of history or of ecology? *Biology and Environment: Proceedings of the Royal Irish Academy,* **106B**, 237–50.

Doyle, G. J. (1982). Minuartio-Thlaspietum alpestris (*Violetea calaminariae*) in Ireland. *Journal of Life Sciences, Royal Dublin Society,* **3**, 143–46.

Doyle, G. J. & O Críodáin, C. (2003). Peatlands: fens and bogs. In: *Wetlands of Ireland* (ed. M. L. Otte). University College Dublin Press, Dublin, pp. 79–108.

Dunnett, J. P., Willis, A. J., Hunt, R. & Grime, J. P. (1998). A 38-year study of relations between weather and vegetation dynamics in road verges near Bibury, Gloucestershire. *Journal of Ecology,* **86**, 610–23.

Du Rietz, G. E. (1948). Huvudenheter och huvudgränser i svensk myrvegetation. *Svensk Botanisk Tidskrift,* **43**, 274–309.

Eckwall, E. (1960). *The Concise Oxford Dictionary of English Place-Names.* 4th edn. Oxford University Press, Oxford.

Ellenberg, H. (1988). *Vegetation Ecology of Central Europe.* 4th edn. Cambridge University Press, Cambridge.

Etherington, J. R. (1981). Limestone heaths in southwest Britain: their soils and the maintenance of their calcicole-calcifuge mixtures. *Journal of Ecology,* **69**, 277–94.

Feeser, I. & O'Connell, M. (2009). Fresh insight into long-term changes in flora, vegetation, land use and soil erosion in the karstic environment of the Burren, western Ireland. *Journal of Ecology,* **97**, 1085–1100.

Fitter, A. H. & Jennings, R. D. (1975). The effects of sheep grazing on the growth and survival of seedling junipers (*Juniperus communis* L.). *Journal of Applied Ecology,* **12**, 637–42.

FitzPatrick, E. A. (1980). *Soils: Their Formation, Classification and Distribution.* Longman, London.

Fletcher, A. (1973a). The ecology of marine (littoral) lichens on some rocky shores of Anglesey. *Lichenologist,* **5**, 368–400.

Fletcher, A. (1973b). The ecology of maritime (supralittoral) lichens on some rocky shores of Anglesey. *Lichenologist,* **5**, 401–22.

Fox, H. S. A. (1972). Field systems of East and South Devon: Part I: East Devon. *Transactions of the Devonshire Association,* **104**, 81–135.

Friend, P. (2008). *Southern England.* New Naturalist 108. Collins, London.

Friend, P. (2012). *Scotland.* New Naturalist 119. Collins, London.

Fuller, R. M. (1989). Orfordness and Dungeness: a comparative study. *Botanical Journal of the Linnean Society,* **101**, 91–101.

Gaynor, K. (2006). The vegetation of Irish machair. *Biology and Environment: Proceedings of the Royal Irish Academy,* **106B**, 311–21.

George, M. (1992). *The Land Use, Ecology and Conservation of Broadland.* Packard Publishing, Chichester.

Gil, L., Fuentes-Utrilla, P., Soto, Á., Cervera, M. T. & Collada, C. (2004). English elm is a 2,000-year-old Roman clone. *Nature,* **431**, 1053.

Gimingham, C. H. (1964). Maritime and sub-maritime communities. In: *The Vegetation of Scotland* (ed. J. H. Burnett). Oliver & Boyd, Edinburgh, pp. 67–142.

Gimingham, C. H. (1972). *Ecology of Heathlands.* Chapman & Hall, London.

Gleason, H. A. (1917). The structure and development of the plant association. *Bulletin of the Torrey Botanical Club,* **43**, 463–81.

Gleason, H. A. (1939). The individualistic concept of the plant association. *American Midland Naturalist,* **21**, 92–110.

Godwin, H. (1975). *History of the British Flora.* 2nd edn. Cambridge University Press, Cambridge.

Godwin, H. (1978). *Fenland: its Ancient Past and Uncertain Future.* Cambridge University Press, Cambridge.

Godwin, H. (1981). *The Archives of the Peat Bogs.* Cambridge University Press, Cambridge.

Godwin, H. & Turner, J. S. (1933). Soil acidity in relation to vegetational succession in Calthorpe Broad, Norfolk. *Journal of Ecology,* **21**, 235–62.

Gonner, E. C. K. (1912). *Common Land and Inclosure*. Macmillan, London.

Goode, D. A. & Lindsay, R. A. (1979). The peatland vegetation of Lewis. *Proceedings of the Royal Society of Edinburgh B*, **77**, 279–93.

Goodman, P. J., Braybrooks, E. M. & Lambert, J. M. (1959). Investigations into 'die-back' in *Spartina townsendii* agg. I. The present status of *Spartina townsendii* in Britain. *Journal of Ecology*, **47**, 651–77.

Goodwillie, R. (2003). Vegetation of turloughs. In: *Wetlands of Ireland* (ed. M. L. Otte). University College Dublin Press, Dublin, pp. 135–44.

Goudie, A. S. & Brunsden, D. (1994). *The Environment of the British Isles: an Atlas*. Clarendon Press, Oxford.

Green, B. H. & Pearson, M. C. (1977). The ecology of Wybunbury Moss, Cheshire. II. Post-glacial history and the formation of the Cheshire mere and mire landscape. *Journal of Ecology*, **65**, 793–814.

Grime, J. P. (1979). *Plant Strategies and Vegetation Processes*. John Wiley, Chichester. (2nd edition, 2001.)

Grime, J. P., Hodgson, J. G. & Hunt, R. (2007). *Comparative Plant Ecology: a Functional Approach to Common British Species*. 2nd edn. Unwin Hyman, London.

Hall, J. E., Kirby, K. J. & Whitbread, A. M. (2004). *National Vegetation Classification: Field Guide to Woodland*. JNCC, Peterborough.

Haslam, S. M. (1978). *River Plants*. Cambridge University Press, Cambridge.

Hayman, G., Vincent, K., Hasler, S. *et al.* (2000). *Acid Deposition Monitoring in the UK, 1986–1998*. Abingdon: AEA Technology.

Heery, S. (1991). The plant communities of the grazed and mown grasslands of the River Shannon callows. *Proceedings of the Royal Irish Academy*, **91B**, 1–19.

Heery, S. (2003). Callows and floodplains. In: *Wetlands of Ireland* (ed. M. L. Otte). University College Dublin Press, Dublin, pp. 109–23.

Hill, M. O. & Preston, C. D. (1998). The geographical relationships of British and Irish bryophytes. *Journal of Bryology*, **20**, 127–226.

Hill, M. O., Preston, C. D. & Roy, D. B. (2004). *PLANTATT: Attributes of British and Irish Plants: Status, Size, Life-History, Geography and Habitats*. Abbots Ripton, Huntingdon, Biological Records Centre.

Hopkins, B. (1955). The species–area relations of plant communities. *Journal of Ecology*, **43**, 409–26.

Hoskins, W. G. (1955). *The Making of the English Landscape*. Hodder & Stoughton, London. Reprinted by Penguin Books, Harmondsworth, 1970.

Hudson, W. H. (1919). *The Book of a Naturalist*. George H. Doran & Co., New York.

Hughes, R. E. (1958). Sheep population and environment in Snowdonia (North Wales). *Journal of Ecology*, **46**, 169–89.

Hughes, R. E., Dale, J., Mountford, M. D. & Williams, I. E. (1975). Studies in sheep populations and environment in the mountains of Northwest Wales. II. Contemporary distribution of sheep populations and environment. *Journal of Applied Ecology*, **12**, 165–78

Hulme, M. & Barrow, E. (eds) (1997). *Climates of the British Isles: Past, Present and Future*. Routledge, London.

Hunter, R. F. (1962). Hill sheep and their pasture: a study of sheep-grazing in southeast Scotland. *Journal of Ecology*, **50**, 651–80.

Huntley, B. (1979). The past and present vegetation of the Caenlochan National Nature Reserve. I. Present vegetation. *New Phytologist*, **83**, 215–83.

Hutchinson, J. N. (1980). The record of peat wastage in the East Anglian fenlands at Holme Post, 1848–1978 AD. *Journal of Ecology*, **68**, 229–49.

Ingram, H. A. P. (1983). Hydrology. In: *Ecosystems of the World, 4A. Mires: Swamp, Bog, Fen and Moor; General Studies* (ed. A. J. P. Gore). Elsevier, Amsterdam, pp. 67–158.

Jessen, K. (1952). An outline of the history of the Irish vegetation. In: *Die Pflanzenwelt Irlands* (ed. W. Lüdi). Hans Huber, Bern. (*Veröffentlichungen des geobotanischen Institutes Rübel in Zürich*, **25**, 79–84.)

Johns, C. A. (1848). *A Week at the Lizard*. SPCK, London.

Kelly, D. L. (1981). The native forest vegetation of Killarney, southwest Ireland: an ecological account. *Journal of Ecology*, **69**, 437–72.

Kelly, D. L. & Iremonger, S. F. (1997). Irish wetland woods: the plant communities and their ecology. *Biology and Environment: Proceedings of the Royal Irish Academy*, **97B**, 1–32.

Kelly, D. L. & Kirby, E. N. (1982). Irish native woodlands over limestone. *Journal of Life Sciences, Royal Dublin Society*, **3**, 181–98.

Lambert, J. M. & Jennings, J. N. (1951). Alluvial stratigraphy and vegetational succession in the region of the Bure Valley broads. II. Detailed vegetational-stratigraphical relationships. *Journal of Ecology*, **39**, 120–48.

Lambert, J. M., Jennings, J. N., Smith, C. T., Green, C. & Hutchinson, J. N. (1960). *The Making of the Broads*. R.G.S. Research Series 3. John Murray, London.

Lewis, J. R. (1964). *The Ecology of Rocky Shores*. English Universities Press, London.

Lindsay, R. A., Charman, D. J., Everingham, F., et al. (1988). *The Flow Country: the Peatlands of Caithness and Sutherland*. Nature Conservancy Council, Peterborough.

Lousley, J. E. (1969). *Wild Flowers of Chalk and Limestone*. Collins, London.

Lund, J. W. G. (1961). The algae of the Malham Tarn district. *Field Studies*, **1**(3): 85–119.

Mabey, R. (1973). *The Unofficial Countryside*. Reprinted (2010) by Little Toller Books, Wimborne Minster, Dorset.

Macan, T. T. (1970). *Ecological Studies of the English Lakes*. Longmans, Harlow.

McIntosh, R. P. (1967). The continuum concept of vegetation. *Botanical Review*, **33**, 130–87.

McVean, D. N. & Ratcliffe, D. A. (1962). *Plant Communities of the Scottish Highlands*. HMSO, London.

Malloch, A. J. C. (1972). Salt-spray deposition on the maritime cliffs of the Lizard peninsula. *Journal of Ecology*, **60**, 103–12.

Mandelbrot, B. (1967). How long is the coast of Britain? Statistical self-similarity and fractional dimension. *Science*, **156**, 636–8.

Maskell, L. C., Firbank, L. G., Thompson, K., Bullock, J. M. & Smart, S. M. (2006).

Interactions between non-native plant species and the floristic composition of common habitats. *Journal of Ecology*, **94**, 1052–60.

Michelmore, A. P. G. & Proctor, M. C. F. (1994). The hedges of Farley Farm, Chudleigh. *Transactions of the Devonshire Association*, **126**, 59–84.

Mitchell, F. (1986). *The Shell Guide to Reading the Irish Landscape*. Country House, Dublin.

Mitchell, F. J. G. (2005). How open were European primeval forests? Hypothesis testing using palaeoecological data. *Journal of Ecology*, **93**, 168–77.

Moore, J. J. (1960). A re-survey of the district lying south of Dublin (1905–1956). *Proceedings of the Royal Irish Academy*, **61B**, 1–36.

Moore, N. W. (1962). The heaths of Dorset and their conservation. *Journal of Ecology*, **50**, 369–91.

Moore, N. W., Hooper, M. D. & Davis B. N. K. (1967). Hedges. I. Introduction and reconnaissance studies. *Journal of Applied Ecology*, **4**, 201–20.

Moore, P. D. (1977). Stratigraphy and pollen analysis of Claish Moss, northwest Scotland: significance of the origin of surface pools and forest history. *Journal of Ecology*, **65**, 375–97.

Moran, J., Kelly, S., Sheehy Skeffington, M. & Gormally, M. (2008). The use of GIS techniques to quantify the hydrological regime of a karst wetland in Ireland. *Applied Vegetation Science*, **11**, 25–36.

Mountford, E. P., Backmeroff, C. E. & Peterken, G. F. (2001). Long-term patterns of growth, mortality, regeneration and natural disturbance in Wistman's Wood, a high-altitude oakwood on Dartmoor. *Transactions of the Devonshire Association*, **133**, 227–62.

Oberdorfer, E. (1949, 1973). *Pflanzensoziologische Exkursionsflora*. Eugen Ulmer, Stuttgart.

O'Sullivan, A. M. (1982). The lowland grassland of Ireland. *Journal of Life Sciences, Royal Dublin Society*, **3**, 131–42.

Osvald, H. (1923). *Die Vegetation des Hochmoores Komosse*. Svenska Växtsociologiska Sällskapets Handlingar 1, 1–436. Uppsala.

Packham, J. R. & Willis, A. J. (1997). *Ecology of Dunes, Salt Marsh and Shingle.* Chapman & Hall, London.

Paton, J. A. (1999). *The Liverwort Flora of the British Isles.* Harley Books, Colchester.

Pears, N. V. (1967). Present tree-lines of the Cairngorm mountains, Scotland. *Journal of Ecology,* **55**, 815–30.

Pearsall, W. H. (1950). *Mountain and Moorlands.* New Naturalist 11. Collins, London.

Pearson, M. C. & Rogers, J. A. (1962). Biological Flora of the British Isles. *Hippophae rhamnoides* L. *Journal of Ecology,* **50**, 501–13.

Perring, F. (1959). Topographical gradients in chalk grassland. *Journal of Ecology,* **47**, 447–81.

Perring, F. H. & Walters, S. M. (1962). *Atlas of the British Flora.* Nelson, London.

Peterken, G. F. (1996). *Natural Woodland: Ecology and Conservation in Northern Temperate Regions.* Cambridge University Press, Cambridge.

Peterken, G. F. (2008). *Wye Valley.* New Naturalist 105. Collins, London.

Peterken, G. F. & Mountford, E. P. (1998). Long-term change in an unmanaged population of wych elm subjected to Dutch elm disease. *Jounal of Ecology,* **86**, 205–18.

Pigott, C. D. (1956). The vegetation of Upper Teesdale in the North Pennines. *Journal of Ecology,* **44**, 545–86.

Pigott, C. D. (1958). Biological Flora of the British Isles: *Polemonium caeruleum* L. *Journal of Ecology,* **46**, 507–25.

Pigott, C. D. & Huntley, J. P. (1981). Factors controlling the distribution of *Tilia cordata* at the northern limits of its geographical range. 3. Nature and causes of seed sterility. *New Phytologist,* **87**, 817–39.

Pigott, C. D. & Wilson, J. F. (1978). The vegetation of North Fen at Esthwaite in 1967–9. *Proceedings of the Royal Society of London A,* **200**, 331–51.

Pigott, M. E. & Pigott, C. D. (1959). Stratigraphy and pollen analysis of Malham Tarn and Tarn Moss. *Field Studies,* **1**, 84–101.

Pollard, E., Hooper, M. D. & Moore, N. W. (1974). *Hedges.* Collins, London.

Poore, M. E. D. (1955). The use of phytosociological methods in ecological investigations. I. The Braun-Blanquet system. *Journal of Ecology,* **43**, 226–44.

Poore, M. E. D. & McVean, D. N. (1957). A new approach to Scottish mountain vegetation. *Journal of Ecology,* **45**, 401–39.

Poore, M. E. D. & Walker, D. (1959). Wybunbury Moss, Cheshire. *Memoirs & Proceedings of the Manchester Literary and Philosophical Society,* **101**, 1–24.

Praeger, R. L. (1897). Bog bursts. with special reference to the recent disaster in Co. Kerry. *The Irish Naturalist* **6**, 141–62.

Praeger, R. L. (1934). *The Botanist in Ireland.* Hodges, Figgis, & Co., Dublin. (Republished 1974 by EP Publishing, Wakefield.)

Preston, C. D. & Hill, M. O. (1997). The geographical relationships of British and Irish plants. *Botanical Journal of the Linnean Society,* **124**, 1–120.

Preston, C. D., Pearman, D. A. & Dines, T. D. (eds) (2002). *New Atlas of the British and Irish Flora.* Oxford University Press, Oxford.

Preston, F. W. (1948). The commonness, and rarity, of species. *Ecology,* **29**, 254–83.

Preston, F. W. (1962). The canonical distribution of commonness and rarity. *Ecology,* **43**, 185–215; 410–32.

Proctor, M., Yeo, P. & Lack, A. (1996). *The Natural History of Pollination.* New Naturalist 83. Collins, London.

Proctor, M. C. F. (1980). Vegetation and environment in the Exe Estuary. In: *Essays on the Exe Estuary* (ed. G. T. Boalch). The Devonshire Association, Exeter and Devonshire Press, Torquay, pp. 117–34.

Proctor, M. C. F. (2003). Malham Tarn Moss: the surface-water chemistry of an ombrotrophic bog. *Field Studies,* **10**, 553–78.

Proctor, M. C. F. (2006). Temporal variation in the surface-water chemistry of a blanket bog on Dartmoor, southwest England: analysis of 5 years' data. *European Journal of Soil Science,* **57**, 167–78.

Proctor, M. C. F. (2008). Water analyses from some Irish Bogs and fens, with thoughts on 'the *Schoenas* problem'. *Biology and*

Environment: Proceedings of the Royal Irish Academy, **108B**, 81–95.

Proctor, M. C. F. (2010). Environmental and vegetational relationships of lake, fens and turloughs in the Burren. *Biology and Environment: Proceedings of the Royal Irish Academy*, **110B**, 17–34.

Proctor, M. C. F., Spooner, G. M. & Spooner, M. F. (1980). Changes in Wistman's Wood, Dartmoor: photographic and other evidence. *Transactions of the Devonshire Association*, **112**, 43–79.

Rackham, O. (1986). *The History of the Countryside*. J. M. Dent, London.

Rackham, O. (2003). *Ancient Woodland: its History, Vegetation and Uses in England*. New edition. Castlepoint Press, Dalbeattie, Kirkcudbrightshire. (First published by Edward Arnold, London, 1980).

Ranwell, D. S. (1972). *Ecology of Salt Marshes and Sand Dunes*. Chapman & Hall, London.

Ratcliffe, D. A. (ed.) (1977). *A Nature Conservation Review*. 2 vols. Cambridge University Press, Cambridge.

Ratcliffe, D. A. & Walker, D. (1958). The Silver Flowe, Galloway, Scotland. *Journal of Ecology*, **46**, 407–45.

Raunkiaer, C. (1907). *Planterigets Livsformer og deres Betydning for Geografien*. Gyldendalske Boghandel – Nordisk Forlag, København and Kristiania.

Raunkiaer, C. (1934) *The Life Forms of Plants and Statistical Plant Geography*. Introduction by A. G. Tansley. Oxford University Press, Oxford.

Raven, J. A., Handley, L. L., Macfarlane, J. J. *et al.* (1988). The role of CO_2 uptake by roots and CAM in acquisition of inorganic C by plants of the isoetid life-form: a review, with new data on *Eriocaulon decangulare* L. *New Phytologist*, **128**, 125–48.

Raven, J. E. & Walters, S. M. (1956). *Mountain Flowers*. New Naturalist 33. Collins, London.

Reynolds, C. S. (1984). *The Ecology of Freshwater Phytoplankton*. Cambridge University Press, Cambridge.

Rich, T., Houston, L., Robertson, A. & Proctor, M. C. F. (2010). *Whitebeams, Rowans and Service Trees of Britain and Ireland. A Monograph of British and Irish Sorbus L.*

B.S.B.I Handbook No.14. Botanical Society of the British Isles, London.

Richards, P. W. (1938). The bryophyte communities of a Killarney oakwood. *Annales Bryologici*, **11**, 108–30.

Rishbeth, J. (1948). The flora of Cambridge walls. *Journal of Ecology*, **36**, 136–48.

Roberts, R. A. (1959). Ecology of human occupation and land-use in Snowdonia. *Journal of Ecology*, **47**, 317–23.

Robinson, T. (1986). *Stones of Aran: Pilgrimage*. Faber & Faber, London.

Robinson, T. (2006). *Connemara: Listening to the Wind*. Penguin Ireland, Dublin (Penguin Books, London, 2007).

Robinson, T. (2008). *Connemara: the Last Pool of Darkness*. Penguin Ireland, Dublin (Penguin Books, London, 2009).

Rodwell, J. S. (ed.) (1991a). *British Plant Communities. 1. Woodlands and Scrub*. Cambridge University Press, Cambridge.

Rodwell, J. S. (ed.) (1991b). *British Plant Communities. 2. Mires and Heaths*. Cambridge University Press, Cambridge.

Rodwell, J. S. (ed.) (1992). *British Plant Communities. 3. Grasslands and Montane Communities*. Cambridge University Press, Cambridge.

Rodwell, J. S. (ed.) (1995). *British Plant Communities. 4. Aquatic Communities, Swamps and Tall-Herb Fens*. Cambridge University Press, Cambridge.

Rodwell, J. S. (ed.) (2000). *British Plant Communities. 5. Maritime Communities and Vegetation of Open Habitats*. Cambridge University Press, Cambridge.

Rodwell, J. S., Morgan, V., Jefferson, R. G. & Moss, D. (2007). *The European Context of British Lowland Grasslands*. JNCC Report 394. Joint Nature Conservation Committee, Peterborough.

Rose, F. (1976). Lichenological indicators of age and ecological continuity in woodlands. In: *Lichenology: Progress and Problems* (ed. D. H. Brown, D. L. Hawksworth & R. H. Bailey). Academic Press, London, pp. 279–307.

Rose, F. (1992). Temperate forest management: its effects on bryophyte and lichen floras and habitats. In: *Bryophytes and Lichens in a Changing Environment* (ed. J. W.

Bates & A. M. Farmer). Oxford Scientific Publications, Oxford, pp. 211–33.

Rübel, E. A. (1912). The International Phytogeographical Excursion in the British Isles. V. The Killarney woods (Co. Kerry, Ireland). *New Phytologist*, **11**, 54–7.

Rydin, H. & Jeglum, J. (2006). *The Biology of Peatlands*. Oxford University Press, Oxford.

Salisbury, E. J. (1916). The oak–hornbeam woods of Hertfordshire. *Journal of Ecology*, **4**, 83–117.

Salisbury, E. J. (1952). *Downs and Dunes: their Plants Life and its Environment*. Bell, London.

Salisbury, E. J. (1964) *Weeds and Aliens*, 2nd edn. New Naturalist 43. Collins, London.

Scott, G. A. M. (1965). The shingle succession at Dungeness. *Journal of Ecology*, **53**, 21–31.

Sculthorpe, C. D. (1967). *The Biology of Aquatic Vascular Plants*. Edward Arnold, London.

Sheehy Skeffington, M. & Curtis, T. G. F. (2000). The Atlantic element in Irish salt marshes. In: *Biodiversity: the Irish Dimension* (ed. B. S. Rushton). Royal Irish Academy, Dublin, pp. 179–96.

Sheehy Skeffington, M., Moran, J., O'Connor, A. *et al.* (2006). Turloughs: Ireland's unique wetland habitat. *Biological Conservation* **133**, 265–90.

Silvertown, J., Poulton, P., Johnston, E., Edwards, G., Heard, M. & Bliss, P. M. (2006). The Park Grass Experiment 1856–2006: its contribution to ecology. *Journal of Ecology*, **94**, 801–14.

Sinker, C. A. (1962). The North Shropshire meres and mosses: a background for ecologists. *Field Studies*, **1**, 101–38.

Smart, S. M., Clarke, R. T., van de Poll. H. M. *et al.* (2003). National-scale vegetation change across Britain: an analysis of sample-based surveillance data from the Countryside Surveys of 1990 and 1998. *Journal of Environmental Management*, **67**, 239–54.

Smart, S. M., Bunce, R. G. H., Marrs, R. *et al.* (2005). Large-scale changes in the abundance of common higher plant species across Britain between 1978, 1990 and 1998 as a consequence of human activity: test of hypothesised changes in trait representation. *Biological Conservation*, **124**, 355–71.

Smith, A. J. E. (2004). *The Moss Flora of Britain and Ireland*, 2nd edn. Cambridge University Press, Cambridge.

Soil Survey of England and Wales (1983). *Soil Map of England and Wales: Scale 1:250,000*. Lawes Agricultural Trust (Soil Survey of England and Wales), Harpenden.

Stace, C. (1997). *New Flora of the British Isles*, 2nd edn. Cambridge University Press, Cambridge.

Stace, C. (1999). *Field Flora of the British Isles*. Cambridge University Press, Cambridge.

Standen, V., Tallis, J. H. & Meade, R. (eds) (1998). *Patterned Mires and Mire Pools: Origin and Development; Flora and Fauna*. British Ecological Society, London.

Steers, J. A. (1946). *The Coastline of England and Wales*. Cambridge University Press, Cambridge.

Stevens, C. J., Duprè, C., Dorland, E. *et al.* (2010). Nitrogen deposition threatens species richness of grasslands across Europe. *Environmental Pollution*, **158**, 2940–5.

Stevens, C. J., Duprè, C., Gaudnik, C. *et al.* (2011). Changes in species composition of European acid grasslands observed along a gradient of nitrogen deposition. *Journal of Vegetation Science*, **22**, 207–15.

Svenning, J.-C., Normand, S. & Kigayama, M. (2008). Glacial refugia of temperate trees in Europe: insights from species distribution modelling. *Journal of Ecology*, **96**, 1117–27.

Tallis, J. H. (1991). Forest and moorland in the south Pennine uplands in the mid-Flandrian period. III. The spread of moorland: local, regional and national. *Journal of Ecology*, **79**, 401–15.

Tallis, J. H. (1995). Blanket mires in the upland landscape. In: *Restoration of Temperate Wetlands* (ed. B. D. Wheeler, S. C. Shaw, W. J. Fojt & R. A. Robertson). Wiley, Chichester, pp. 495–508.

Tallis, J. H. (2001). Bog bursts. *Biologist*, **48**(5), 218–23.

Tansley, A. G. (1939). *The British Islands and their Vegetation*. Cambridge University Press, Cambridge.

Tansley, A. G. (1968). *Britain's Green Mantle.* 2nd edn, revised by M. C. F. Proctor. George Allen & Unwin, London.

Taylor, J. A. (1967). Growing season as affected by land aspect and soil texture. In: *Weather and Agriculture.* Pergamon Press, Oxford, pp. 15–36.

Thompson, K., Petchley, O. L., Askew, A. P. et al. (2010). Little evidence for limiting similarity in a long-term study of a roadside plant community. *Journal of Ecology,* **98**, 480–7.

Townsend, C. R., Begon, M. & Harper, J. L. (2008). *Essentials of Ecology,* 3rd edn. Wiley-Blackwell, Oxford.

UKAGAR (1990). *Acid Deposition in the United Kingdom.* Third report of the United Kingdom Advisory Group on Acid Rain. Warren Spring Laboratory, Stevenage.

Van Vlymen, C. D. (1980). The water balance, physico-chemical environment and phytoplankton studies of Slapton Ley, Devon. Unpublished PhD thesis, University of Exeter.

Vincent, K. J., Campbell, G. W., Stedman, J. R. et al. (1995). *Acid Deposition in the United Kingdom: Wet Deposition 1994.* AEA Technology, Culham.

Walker, D. (1970). Direction and rate in some British Post-glacial hydroseres. In: *Studies in the Vegetational History of the British Isles* (ed. D. A. Walker & R. G. West). Cambridge University Press, Cambridge, pp. 117–39.

Watt, A. S. & Jones, E. W. (1948). The ecology of the Cairngorms. Part 1: the environment and the altitudinal zonation of the vegetation. *Journal of Ecology,* **36**, 283–304.

Webb, D. A. (1954). Is the classification of vegetation either possible or desirable? *Botanisk Tidsskrift,* **51**, 362–70.

Weber, C. A. (1902). *Über die Vegetation und Entstehung des Hochmoors von Augstumal im Memeldelta mit vergleichenden Ausblicken auf andere Hochmoore der Erde. Eine formationsbiologisch-historische Studie.* Verlag Paul Parey, Berlin.

Westhoff, V. & Den Held, A. J. (1975). *Plantengemeenschappen in Nederland.* W. J. Thieme, Zutphen.

Wheeler, B. D. & Proctor, M. C. F. (2000). Ecological gradients, subdivisions and terminology of northwest European mires. *Journal of Ecology,* **88**, 187–203.

Wheeler K. G. R. (2007). *A Natural History of Nettles.* Trafford Publishing Co. (www.trafford.com).

White, J. (1985). The Geragh woodland, Co. Cork. *Irish Naturalists' Journal* **21**, 391–6.

White, J. & Doyle, G. J. (1982). The vegetation of Ireland: a catalogue raisonné. *Journal of Life Sciences, Royal Dublin Society,* **3**, 289–368.

Whittaker, R. H. (1956). Vegetation of the Great Smoky Montains. *Ecological Monographs,* **26**, 1–80.

Whittaker, R. H. (1960). Vegetation of the Siskiyou mountains, Oregon and California. *Ecological Monographs,* **30**, 279–338.

Wiehe, P. O. (1935). A quantitative study of the influence of tide upon populations of *Salicornia europaea. Journal of Ecology,* **23**, 323–33.

Willis, A. J., Folkes, B. F. & Yemm, E. W. (1959). Braunton Burrows: the dune system and its vegetation. I. *Journal of Ecology,* **47**, 1–24.

Wilmanns, O. (1978). *Ökologische Pflanzensoziologie.* 2nd edn. Quelle & Meyer, Heidelberg.

Wilmanns, O. & Brun-Hool, J. (1982). Irish *Mantel* and *Saum* vegetation. *Journal of Life Sciences, Royal Dublin Society,* **3**, 165–74.

Wilson, R. C. L., Drury, S. A. & Chapman, J. L. (eds) (1976). *The Great Ice Age.* Routledge: London.

Woodcock, N. H. & Strachan, R. A. (eds) (2000). *Geological History of Britain and Ireland.* Blackwell, Oxford.

Woodhead, T. W. (1906). The ecology of woodland plants in the neighbourhood of Huddersfield. *Journal of the Linnean Society (Botany),* **37**, 333–406.

Woodward, F. I., Quaife, T. & Lomas, M. I. (2010). Changing climate and the Irish landscape. *Biology and Environment: Proceedings of the Royal Irish Academy,* **110B**, 1–16.

Indexes

SPECIES INDEX

GENERAL INDEX